PRINCIPLES OF

THIRD EDITION

Vincent K. Omachonu, Ph.D.
Joel E. Ross, Ph.D.

CRC PRESS

Boca Raton London New York Washington, D.C.

Library of Congress Cataloging-in-Publication Data

Omachonu, Vincent K.
Principles of total quality / Vincent K. Omachonu, Joel E. Ross.--3rd ed.
p. cm.
Rev. ed. of: Principles of total quality / J.A. Swift, Joel E. Ross, Vincent K. Omachonu.
Includes bibliographical references and index.
ISBN 0-57444-326-7 (alk. paper)
1.Total quality management. 2. Quality control. 3. Quality control--standards. I. Ross, Joel E. II. Swift, J.A. Principles of total quality. III. Title.

HD62.15.O43 2004
658.4′.013--dc22 2004041857

Direct all inquiries to CRC Press LLC, 2000 N.W. Corporate Blvd., Boca Raton, Florida 33431.

Visit the CRC Press Web site at www.crcpress.com

International Standard Book Number 0-57444-326-7
Library of Congress Card Number 2004041857
Printed in the United States of America 2 3 4 5 6 7 8 9 0
Printed on acid-free paper

TABLE OF CONTENTS

SECTION I MANAGEMENT OF TOTAL QUALITY

1 Total Quality Management and the Revival of Quality in the U.S. 3

The Concept of TQM 5
Antecedents of Modern Quality Management 6
The Quality Pioneers 7
Accelerating Use of TQM 12
Quality and Business Performance 13
Service Quality vs. Product Quality 15
Exercises 17
Illustrative Case 18
  Questions 18
Endnotes 18
References 21

2 Leadership 23

Attitude and Involvement of Top Management 26
Communication 27
  How Employees Receive Information 27
Culture 30
  Embedding a Culture of Quality 32
Management Systems 33
Control 35
Exercises 37
Illustrative Cases 37
  Questions 38
Endnotes 38
References 40

3 Information and Analysis 41
Organizational Implications 41
Information Technology 42
Decision Making 43
Information Systems in Japan 43
Strategic Information Systems 44
Environmental Analysis 45
Shortcomings of Accounting Systems 45
Organizational Linkages 47
White-Collar Measures 49
Advanced Processes/Systems 49
Information and the Customer 51
Information Needs 51
The Information Systems Specialists 52
The Chief Information Officer 52
Systems Design 53
Exercises 54
Illustrative Case 54
Questions 55
Endnotes 55
References 57

4 Strategic Quality Planning 59
Strategy and the Strategic Planning Process 60
Strategic Quality Management 62
Mission 63
Environment 65
Product/Market Scope 65
Differentiation 67
Definition of Quality 68
Which Approach(es)? 69
Market Segmentation (Niche) Quality 70
Objectives 70
Supporting Policies 72
Testing for Consistency of Policies 73
Control 73
Service Quality 74
Summary 75
Exercises 75
Illustrative Cases 76
Questions 76
Endnotes 77
References 79

5 Human Resource Development and Quality Management ...81
Involvement: A Central Idea of Human Resource Utilization 83
Organizing for Involvement 84

Training and Development 86
Selection 87
Performance Appraisal 88
Compensation Systems 89
    Individual or Team Compensation? 90
    Summary 91
Total Quality Oriented Human Resource Management 91
Exercises 92
Illustrative Cases 93
    Question 93
Endnotes 94
References 95

6 Management of Process Quality 97
A Brief History of Quality Control 98
Product Inspection vs. Process Control 100
Moving from Inspection to Process Control 101
Statistical Quality Control 102
Basic Approach to Statistical Quality Control 103
    The Deming Cycle 103
Manufacturing to Specification vs. Manufacturing to Reduce Variations 104
Process Control in Service Industries 105
    Customer Defections: The Measure of Service Process Quality 105
Process Control for Internal Services 106
Quality Function Deployment 107
Just-in-Time (JIT) 110
Just-in-Time or Just-in-Case 112
    Benefits of JIT 113
The Human Side of Process Control 114
Exercises 115
Illustrative Cases 115
    Question 116
Endnotes 116
References 119

7 Customer Focus and Satisfaction 121
Process vs. Customer 123
Internal Customer Conflict 124
Defining Quality 125
A Quality Focus 125
    Break Points 126
    A Central Theme 127
The Driver of Customer Satisfaction 127
Getting Employee Input 128
Measurement of Customer Satisfaction 129
The Role of Marketing and Sales 130
The Sales Process 131

Service Quality and Customer Retention 131
Customer Retention and Profitability 132
Buyer–Supplier Relationships 133
Exercises 135
Illustrative Cases 136
    Questions 136
Endnotes 137
References 139

8 Benchmarking 141
The Evolution of Benchmarking 141
    Xerox 142
    Ford 143
    Motorola 143
The Essence of Benchmarking 144
Benchmarking and the Bottom Line 144
The Benefits of Benchmarking 145
    Cultural Change 145
    Performance Improvement 145
    Human Resources 145
Strategic Benchmarking 146
Operational Benchmarking 148
The Benchmarking Process 149
    Determine the Functions/Processes to Benchmark 149
    Select Key Performance Variables 149
Identify the Best-in-Class 151
Measure Your Own Performance 154
Actions to Close the Gap 154
Pitfalls of Benchmarking 154
Exercises 155
Illustrative Case 156
    Questions 156
Endnotes 157
References 158

9 Organizing for Total Quality Management 159
Organizing for TQM: The Systems Approach 160
Organizing for Quality Implementation 165
The People Dimension: Making the Transition from a Traditional to a TQM Organization 168
    The Inverted Organizational Chart 169
    Internal Quality 169
Roles in Organizational Transition to TQM 171
Small Groups and Employee Involvement 172
Teams for TQM 174
    Quality Circles 174
    Cross-Functional Teams 175

Exercises 177
Illustrative Cases 177
Question 177
Endnotes 178
References 181

10 Quality and Productivity 183
The Leverage of Productivity and Quality 185
Management Systems vs. Technology 185
Productivity in the United States 187
Reasons for Slow Growth 187
Measuring Productivity 189
Basic Measures of Productivity: Ratio of Output to Input 189
Total Productivity Measurement Model (TPM) 190
White-Collar Productivity 192
Measuring the Service Activity 192
Improving Productivity (and Quality) 193
Five Ways to Improve Productivity (and Quality) 195
Examples of Increasing Productivity While Improving Quality 196
Capital Equipment vs. Management Systems 197
Activity Analysis 198
Exercises 200
Illustrative Case 200
Question 201
Endnotes 201
References 203

11 The Cost of Quality 205
Cost of Quality Defined 205
The Cost of Quality 206
Three Views of Quality Costs 207
Quality Costs 208
Measuring Quality Costs 211
The Use of Quality Cost Information 213
Accounting Systems and Quality Management 214
Activity-Based Costing 214
The Multiproduct Problem 216
Strategic Planning and Activity-Based Costing 217
Summary 218
Exercises 218
Endnotes 218
References 221

SECTION II PROCESSES AND QUALITY TOOLS

12 The Concept of a Process 225
What is a Process? 225

Examples of Processes 226
Types of Processes 228
The Total Process 228
The Feedback Loop 229
Exercises 229
References 230

13 Understanding Data 231
Introduction 231
Data and Information 231
The Concept of a Dashboard 232
Significance of Data 232
Deciding What to Measure 233
Questions to Ask Prior to Data Collection 233
Data Collection Methods 233
Types of Data 234
Data Reliability 235
Stratification 235
How to Present/Describe Data 237
Visual Description: Tabular Displays 237
Visual Description: Graphical Displays 237
Numerical Description 238
Sampling 241
Basic Definitions 241
Types of Sampling 242
Sampling Error 246
Summary 247
Exercises 248
Endnotes 248
References 248

14 The Seven Basic Quality Control Tools 251
Background 251
Check Sheets 252
Defect-Location Check Sheets 253
Tally Check Sheet 253
Defect-Cause Check Sheet 254
Flowcharts 256
Graphs 256
Line Graph 256
Bar Graphs 257
Circle Graph 259
Histograms 260
Pareto Charts 260
Cause-and-Effect Diagrams 262
Dispersion Analysis Cause-and-Effect Diagram 265
Production Process Classification Cause-and-Effect Diagram 265

Cause Enumeration Cause-and-Effect Diagram 269
Scatter Diagrams 269
Control Charts 269
Exercises 271
Endnotes 273
References 273

15 Control Charts for Variables 275
Background 275
Uses of Control Charts 275
Variables Control Charts 276
Applications of Variables Control Charts 277
Preparing to Use Variables Control Charts 277
Problem Definition 278
Choice of Quality Characteristic 278
Size and Number of Samples 278
Sampling Frequency 279
Rational Subgroups 279
Choice of Control Limits 279
Collecting the Samples 279
Examples of Variables Control Charts 280
Example 1: X and R Charts 280
Example 2: X and S Charts 282
Example 3: X and MR Charts 284
Interpreting Control Charts 287
Summary 290
Exercises 290
Questions 290
Problems 290
References 292
Appendix A 294

16 Control Charts for Attributes 295
Control Chart for Fraction Non-Conforming (p Chart) 296
Example 1 297
Example 2 300
Control Chart for Number Non-Conforming (np Chart) 301
Example 3 302
Control Chart for Non-Conformities 304
c Chart 304
Example 4 305
u Chart 306
Example 5 307
Summary 309
Exercises 309
Questions 309
Problems 310

References 312

17 When to Use the Different Control Charts 313

Introduction 313
Example 1 313
    Solution 315
Example 2 315
    Solution 315
Example 3 315
    Solution 315
Summary 315
Exercises 316

18 Quality Improvement Stories 319

What Is a Quality Improvement Story? 319
Step 1: Identify the Problem Area 321
Step 2: Observe and Identify Causes of the Problem 321
Step 3: Analyze, Identify, and Verify Root Cause(s) of the Problem 321
Step 4: Plan and Implement Preventive Action 322
    Purpose 322
Step 5: Check Effectiveness of Action Taken 323
Step 6: Standardize Process Improvement 323
Step 7: Determine Future Action 324
Other Considerations 324
    Time Frame 324
    Quality Improvement Story Requirements 324
QI Story Line 326
Data Collection 328
    What Data Will the Team Need? 328
Exercises 329
References 329

19 Quality Function Deployment 331

History 331
What Is Quality Function Deployment? 331
    Voice of the Customer 335
        A Case Study 335
    Customer Requirements 336
    Technical Requirement 337
    Strength of Relationships 337
    The Vertical Entries 338
        1. Importance Rating 338
        2. Surveying Company 338
        3. Chief Competitor 338
        4. Plan 339
        5. Rate of Improvement 339
        6. Absolute Weight 339
        7. Demanded Weight 339

Competitive Technical Assessment 340
1. Total 340
2. Percentage (%) 340
3. Company Now 340
4. Chief Competitor 341
5. Plan 341
Benefits 341
Conclusion 342
Exercises 342
Endnotes 342
References 343

SECTION III CRITERIA FOR QUALITY PROGRAMS

20 ISO 9000 347
ISO around the World 347
ISO 9000 in the U.S. 349
ISO 9000 351
Components of ISO 9000 Standard Series 352
Historical Perspectives 352
Management Responsibility 353
Functional Standards 355
Benefits of ISO 9000 Certification 355
Getting Certified: The Third-Party Audit 356
Documentation 356
Post-Certification 357
Choosing an Accredited Registration Service 358
ISO 9000 and Services 358
The Cost of Certification 359
ISO 9000 vs. the Baldrige Award 359
Implementing the System 360
Final Comments 360
Exercises 361
Endnotes 361
References 363

21 The Baldrige Award 365
Endnotes 370

22 QS-9000 371
Historical Perspective 371
Basic QS-9000 371
Structure of QS-9000 372
Document Control and Registration 372
Summary 373
Exercises 373
References 373

23 ISO 14000 375
Components of ISO 14000 375
ISO 14001 377
Registration 378
Benefits 378
Exercises 379
References 379

24 ISO 9000: A Practical Step-by-Step Approach 381
Preparing for ISO 9000 Registration 381
Pre-Audit Conducted 384
Documenting Control Procedures and Work Activities 385
Implementing a Policies and Procedures Training Program 386
The Final Assessment 386
GSP's Recommendations for Achieving ISO 9000 387
Exercises 388

SECTION IV SPECIAL TOPICS IN QUALITY

25 Process Capability 393
Introduction 393
Attribute Process Capability 394
Variable Process Capability 395
Tolerance and Specifications 396
Capability Indexes 399
Shortcomings of C_p 400
Comments Concerning C_p and C_{pk} 401
Process Capability Based on Range (R) Values 402
Process Capability Based on Standard Deviation (s) values 402
Colony Fasteners 403
Design and Introduction of Products and Services 405
How Products and Delivery Processes are Designed 405
Translation of Customer Requirements 405
Requirements Translated into Processes 407
How All Requirements Are Addressed Early in the Design Cycle 408
How Processes Are Reviewed Prior to Launch 408
How Processes Are Evaluated and Improved 409
Process Management: Product and Service Production and Delivery 410
How the Company Maintains Performance Production 410
Key Processes 410
Measurement Plan 413
How Processes Are Improved 414
Process Analysis and Research 414
Benchmarking 415
Use of Alternative Technology 415
Information from Customers 415

Process Management: Support Services 416
How Support Service Processes Are Designed 416
How Key Requirements Are Determined 416
How Requirements Are Translated into Processes 417
Requirement Addressed Early in the Design 418
How the Company Maintains Performance 419
Key Processes and Requirements 419
The Measurement Plan 421
How Processes Are Improved 421
Process Analysis and Research 421
Benchmarking 421
Use of Alternative Technology 422
Information from Customers 422
Management of Supplier Performance 423
Summary of Company Requirements 424
Principal Requirements for Key Suppliers 424
How the Company Determines Performance 425
How Performance Is Fed Back 425
How the Company Improves Supplier Performance 426
Improve Suppliers' Abilities 426
Improve Procurement 426
Minimize Costs of Inspection 427
Exercises 427
Endnotes 429
References 429

26 Introduction to Reliability 431
Introduction 431
Reliability 432
Why the Emphasis on Reliability? 433
Product Life Cycle Curve 433
Causes of Product Unreliability 434
Measures of Reliability 435
Failure Rate, Mean Life, and Availability 435
System Reliability 438
Implications for Design 440
Exercises 440
Questions 440
Problems 441
Endnotes 442
References 442

27 Introduction to Six Sigma 443
The History of Six Sigma 445
The Six Sigma Concept 446
Define Phase 446
Measure Phase 446

Analyze Phase 446
Improve Phase 446
Control Phase 447
Six Sigma Features 447
Customer-Focused 447
Benefits of Six Sigma 447
Exercises 448
Endnotes 448
References 449

28 Healthcare Service Excellence 451
Service Is the Competitive Edge 451
Implications for a Hospital 454
Why Customer Service in Healthcare? 454
What's In It for the Employee? 454
What's In It for the Organization? 455
What's In It for the Community? 455
The Healthcare Customer 456
Who Is a Customer? 456
Understanding the Healthcare Customer 458
How to Use Customer Profiles 458
Patient Profiling 459
Service Design 460
The "Find It, Fix It" Approach to Medicine 460
Some Suggestions on How to Handle the Disease and the Person 461
Quality of Conformance 462
Two Components of Quality 463
Theories of Service Quality 464
How a Patient's Expectations Are Formed 467
The Art of Caring 469
The Quality of Behavior 469
Behavior 469
Attitudes 471
Reports from Healthcare Surveys 471
Moving toward a Total Service Quality Management Process 471
Critical Factors in TSQM 472
Marketing Research 472
The Dimensions of Quality in Healthcare 473
Summary 474
Exercises 475
Endnotes 475
References 476

Index...479

PREFACE

At a time when the United States faces tremendous pressure from global competition, and the voices of external customers are growing louder and stronger, the quest for quality has never been more urgent. The very survival of organizations is acutely in jeopardy. The notion of quality has gone from being a socially provocative one to being a deliberate strategy for long-term viability. In the third edition of *Principles of Total Quality,* we have captured the essence of this strategy for both the service and manufacturing sectors.

WHAT'S NEW ABOUT THIS EDITION?

Three New Chapters

- With the increasing amount of litigation in the area of product performance, the concept of reliability has commanded the attention of lawmakers, manufacturers, and consumers. This new edition of *Principles of Total Quality* discusses the concept of reliability in Chapter 27.
- Many organizations, in their pursuit of perfection, have embraced the concept of Six Sigma. This edition introduces the concept of Six Sigma in Chapter 28, along with a historical perspective.
- The growing popularity of the service sector has brought much attention to the healthcare industry. In Chapter 29, we discuss service excellence in the healthcare industry. Chapter 29 presents practical applications of the concepts of service excellence in healthcare organizations.

Other Changes

Expanded Chapters

The following chapters have been considerably expanded to reflect modern emphases in the field of quality management:

Chapter 2: Leadership
Chapter 3: Information and Analysis
Chapter 6: Management of Process Quality
Chapter 7: Customer Focus and Satisfaction
Chapter 9: Organizing for Total Quality Management
Chapter 13: Understanding Data
Chapter 15: Control Charts for Variables
Chapter 19: Quality Function Deployment

ABOUT THE AUTHORS

Vincent Omachonu, Ph.D., is an Associate Professor of Industrial Engineering at the University of Miami. He received his Ph.D. in Industrial Engineering from the Polytechnic University of New York. He has two masters degrees — one in Operations Research from Columbia University in New York, and the other in Industrial Engineering from the University of Miami, Florida. His B.S. degree is also in Industrial Engineering, from the University of Miami. Dr. Omachonu is a Licensed Professional Engineer (PE) in the State of Florida.

Dr. Omachonu is a nationally recognized trainer and consultant in the areas of Quality Management, Service Excellence, Productivity Measurement and Improvement, Practical Management Techniques, Statistical Process Control, Business Process Re-engineering and Design, Organizational Development and Strategic Planning.

He is the author of two other books — *Health Care Performance Improvement* (1999) and *Total Quality and Productivity Management in Health Care Organizations* (1991), which won the IIE Book-of-the-Year Award. He has written several articles in technical and professional journals, proceedings, and books. Dr. Omachonu has conducted hundreds of workshops and seminars and has implemented quality management, service excellence, and performance improvement processes in several organizations in the United States and overseas.

Dr. Omachonu has been cited in *Who's Who in the World* and *Who's Who in America.* He has been featured on CNN Business (to discuss Health Care Quality Management). He is the recipient of several teaching awards from both the Engineering and Business Schools of the University of Miami.

Joel Ross, Ph.D., is Emeritus Professor of Management at Florida Atlantic University in Boca Raton, Florida. He graduated from Yale University and received his doctorate in business administration from George Washington University. He has been Chairman of Management and Director of the MBA Program. Prior to his academic career, Dr. Ross was a Commander in the U.S. Navy.

Dr. Ross is widely known as a platform speaker, seminar leader, consultant, and author. He has developed and conducted management developmental programs for over 100 companies and organizations in the areas of general management, strategy, productivity, and quality. He has been an invited teacher on management topics in Israel, South Africa, Venezuela, Panama, India, Ecuador, the Philippines, and Japan.

His articles have appeared in journals such as *Journal of Systems Management, Business Horizons, Long Range Planning, Industrial Management, Personnel, Management Accounting,* and *Academy of Management Review.* He is the author of thirteen books, including the landmark *Management Information Systems, People, Profits, and Productivity,* and *Total Quality Management: Text, Cases and Readings*, which has been adopted by over 250 colleges and universities.

I

MANAGEMENT OF TOTAL QUALITY

The concepts of quality and good management principles have been around for some time, but each has been treated separately and the two have sometimes been considered unrelated topics. Both concepts are integrated in Part I, where the idea is advanced that quality requires the continuing application of management principles.

In Chapter 1, the concept of total quality management (TQM) is introduced, the emergence of the movement is traced, and the pioneers who developed the principles and techniques are identified. In Chapter 2, the need for top management support and involvement is outlined, and how this should be reflected in the corporate culture and supporting management systems is described.

How information systems serve both strategic and operational needs and link organizational functions is described in Chapter 3. Elements of system design are addressed.

In Chapter 4, the process of strategy development is explained and the role of quality as the differentiating factor in strategy is explored. The idea of involvement and empowerment as the critical dimension of human resource management is presented in Chapter 5, and the need to make quality a central ingredient of these methods is examined.

The emergence of process control rather than final inspection as a means to continuous improvement is traced in Chapter 6. Quality function deployment and just-in-time are discussed. The measurement and improvement of customer satisfaction and standards for customer retention are covered in Chapter 7. In Chapter 8, the steps involved in benchmarking — comparing oneself to best-in-class organizations — are provided.

The systems approach to a TQM organization style and how to achieve cross-functional integration with teams are described in Chapter 9. Included in Chapter 10 are the basics of productivity management and how productivity is achieved through quality improvement.

The cost of quality is covered in Chapter 11, as well as how to measure the cost of not meeting customer requirements — the cost of doing things wrong. The use of quality cost information is also discussed.

1

TOTAL QUALITY MANAGEMENT AND THE REVIVAL OF QUALITY IN THE U.S.

The total quality concept as a business strategy began to grow in popularity in the United States in the late 1980s and early 1990s. However, individual elements of the concept — such as team building, problem-solving tools, statistical process control, design of experiments, customer service, and process documentation — have been used by some organizations for years. Total quality management (TQM) is the integration of all functions and processes within an organization in order to achieve continuous improvement of the quality of goods and services. The goal is customer satisfaction.

> Xerox, one of two 1989 winners of the Malcolm Baldrige National Quality Award, has come a long way since the 1970s. The company that invented the dry-paper copier saw its share of the North American market plunge from 93% to 40%. Their Japanese competitors were selling copiers for less than the cost of manufacture at Xerox. By using TQM (known as Leadership through Quality at Xerox), the company has gained market share in all key markets worldwide and builds five of the six highest quality copiers in the world. The company has since learned to apply quality beyond the manufacturing confines in all the functions of the organization.

> Motorola, another winner of the Malcolm Baldrige Award, has received more awards for excellence as a supplier than any other U.S. company and is widely acknowledged as a quality leader. This is quite an improvement over the early 1980s, when Chairman Bob Galvin was calling for a 3-year, 20% surcharge on all imported manufactured goods in an attempt to counteract the threat from the Orient. The company applies TQM to every aspect of its operations and six sigma to every significant business process.

Of all the management issues faced in the last decade, none has had the impact of or caused as much concern as quality in U.S. products and services. A report by the Conference Board indicates that senior executives in the United States agree that the banner of total quality is essential to ensure competitiveness in global markets. Quality expert J. M. Juran calls it a major phenomenon in this age.[1] This concern for quality is not misplaced.

The interest in quality is due, in part, to foreign competition and the trade deficit.[2] Analysts estimate that the vast majority of U.S. businesses will continue to face strong competition from the Pacific Rim and the European Union.[3] This comes in the face of a serious erosion of corporate America's ability to compete in global markets over the past 25 years.

The problem has not gone unnoticed by government officials, corporate executives, and the public at large. The concern of the president and Congress culminated in the enactment of the Malcolm Baldrige National Quality Improvement Act of 1987 (Public Law 100-107), which established an annual United States National Quality Award. The concern of business executives is reflected in their perceptions of quality. In a 1989 American Society for Quality Control (ASQC) survey, 54% of executives rated quality of service as extremely critical and 51% rated quality of product as extremely critical.[4] Seventy-four percent gave U.S.-made products less than eight on a ten-point scale for quality. Similarly, a panel of Fortune 500 executives agreed that U.S. products deserved no better than a grade of C+.

Public opinion regarding U.S.-made products is somewhat less than enthusiastic. In a 1988 ASQC survey of consumer perceptions, less than one-half gave U.S. products high marks for quality.[5] Employees also have misgivings about quality in general and, more specifically, about quality in the companies in which they work. They believe that there is a significant gap between what their companies say and what they do. More importantly, employees believe that their talents, abilities, and energies are not being fully utilized for quality improvement.[6]

Despite the pessimism reflected by these groups, progress is being made. In a 1991 survey of U.S. owners of Japanese-made cars, 32%

indicated that their next purchase will be a domestic model, and the reason given most often was the improved quality of cars built in the U.S.[7] Ford's "Quality Is Job One" campaign may have been a contributing factor. There is also evidence that quality has become a competitive marketing strategy in the small business community, as U.S. consumers are beginning to shun mass-produced, poorly made, disposable products.

Other promising developments include the increasing acceptance of TQM as a philosophy of management and a way of company life. It is essential that this trend continue if U.S. companies are to remain competitive in global markets. Customers are becoming more demanding and international competition more fierce. Companies that deliver quality will prosper in the next century.

THE CONCEPT OF TQM

Total quality management is based on a number of ideas. It means thinking about quality in terms of all functions of the enterprise, a start-to-finish process that integrates interrelated functions at all levels. It is a systems approach that considers every interaction between the various elements of the organization. Thus, the overall effectiveness of the system is higher than the sum of the individual outputs from the subsystems. The subsystems include all the *organizational functions* in the life cycle of a product, such as (1) design, (2) planning, (3) production, (4) distribution, and (5) field service. The *management* subsystems also require integration, including (1) strategy with a customer focus, (2) the tools of quality, and (3) employee involvement (the linking process that integrates the whole). A corollary is that any product, process, or service can be improved, and a successful organization is one that consciously seeks and exploits opportunities for improvement at all levels. The load-bearing structure is customer satisfaction. The watchword is *continuous improvement*.

Following an international conference in May 1990, the Conference Board summarized the key issues and terminology related to TQM:

- The ***cost of quality*** as the measure of non-quality (not meeting customer requirements) and a measure of how the quality process is progressing.
- A ***cultural change*** that appreciates the primary need to meet customer requirements, implements a management philosophy that acknowledges this emphasis, encourages employee involvement, and embraces the ethic of continuous improvement.
- ***Enabling mechanisms of change***, including training and education, communication, recognition, management behavior, teamwork, and customer satisfaction programs.

- ***Implementing TQM*** by defining the mission, identifying the output, identifying the customers, negotiating customer requirements, developing a "supplier specification" that details customer objectives, and determining the activities required to fulfill those objectives.
- ***Management behavior*** that includes acting as role models, using quality processes and tools, encouraging communication, sponsoring feedback activities, and fostering and providing a supporting environment.[8]

ANTECEDENTS OF MODERN QUALITY MANAGEMENT

Quality control as we know it probably had its beginnings in the factory system that developed following the Industrial Revolution. Production methods at that time were rudimentary at best. Products were made from non-standardized materials using non-standardized methods. The results were products of varying quality. The only real standards used were measures of dimension, weight, and, in some instances, purity. The most common form of quality control was inspection by the purchaser, under the common law rule of *caveat emptor*.[9]

Much later, around the turn of this century, Frederick Taylor developed his system of scientific management, which emphasized productivity at the expense of quality. Centralized inspection departments were organized to check for quality at the end of the production line. An extreme example of this approach was the Hawthorne Works at Western Electric Company, which at its peak in 1928 employed 40,000 people in the manufacturing plant, 5,200 of whom were in the inspection department. The control of quality focused on final inspection of the manufactured product, and a number of techniques were developed to enhance the inspection process. Most involved visual inspection or testing of the product following manufacture. Methods of statistical quality control and quality assurance were added later. Detecting manufacturing problems was the overriding focus. Top management moved away from the idea of managing to achieve quality and, furthermore, the work force had no stake in it. The concern was limited largely to the shop floor.

Traditional quality control measures were (and still are) designed as defense mechanisms to prevent failure or eliminate defects.[10] Accountants were taught (and are still taught) that expenditures for defect prevention were justified only if they were less than the cost of failure. Of course, cost of failure was rarely computed.[11] (Cost of quality is discussed further in Chapter 11.)

Following World War II, the quality of products produced in the United States declined as manufacturers tried to keep up with the demand for non-military goods that had not been produced during the war. It was during this period that a number of pioneers began to advance a methodology of quality control in manufacturing and to develop theories and practical techniques for improved quality. The most visible of these pioneers were W. Edwards Deming, Joseph M. Juran, Armand V. Feigenbaum, and Philip Crosby.[12] It was a great loss to the quality movement when Deming died in December 1993 at the age of 93.

THE QUALITY PIONEERS

Deming, the best known of the "early" pioneers, is credited with popularizing quality control in Japan in the early 1950s. Today he is regarded as a national hero in that country and is the father of the world-famous Deming Prize for Quality. He is best known for developing a system of statistical quality control, although his contribution goes substantially beyond those techniques.[13] His philosophy begins with top management but maintains that a company must adopt the 14 points of his system at all levels. He also believes that quality must be built into the product at all stages in order to achieve a high level of excellence. Although it cannot be said that Deming is responsible for quality improvement in Japan or the United States, he has played a substantial role in increasing the visibility of the process and advancing an awareness of the need to improve.

Deming defines quality as a predictable degree of uniformity and dependability at low cost and suited to the market. Deming teaches that 96% of variations have common causes and 4% have special causes. He views statistics as a management tool and relies on statistical process control as a means of managing variations in a process.

Deming developed what is known as the Deming chain reaction: As quality improves, costs will decrease and productivity will increase, resulting in more jobs, greater market share, and long-term survival. Although it is the worker who will ultimately produce quality products, Deming stresses worker pride and satisfaction rather than the establishment of quantifiable goals. His overall approach focuses on improvement of the process, in that the system, rather than the worker, is the cause of process variation.

Deming's *universal 14 points* for management are summarized as follows:

1. ***Create consistency of purpose with a plan*** — The objective is constancy of purpose for continuous improvement. An unwavering commitment to quality must be maintained by management. Quality, not short-term profit, should be at the heart of organization purpose. Profit will follow when quality becomes the objective and purpose.

2. ***Adopt the new philosophy of quality*** — The modern era demands ever-increasing quality as a means of survival and global competitiveness. Inferior material, poor workmanship, defective products, and poor service must be rejected. Reduction of defects is replaced by elimination of defects. The new culture of quality must reflect a commitment to quality and must be supported by all employees.
3. ***Cease dependence on mass inspection*** — Quality can't be inspected in; it must be built-in from the start. Defects discovered during inspection cannot be avoided — it is too late; efficiency and effectiveness have been lost, as has continuous process improvement. Continuous process improvement reduces costs incurred by correcting errors that should not have been made in the first place.
4. ***End the practice of choosing suppliers based on price*** — Least cost is not necessarily the best cost. Buying from a supplier based on low cost rather than a quality/cost basis defeats the need for a long-term relationship. Vendor quality can be evaluated with statistical tools.
5. ***Identify problems and work continuously to improve the system*** — Continuous improvement of the system requires seeking out methods for improvement. The search for quality improvement is never-ending and results from studying the process itself, not the defects detected during inspection.
6. ***Adopt modern methods of training on the job*** — Training involves teaching employees the best methods of achieving quality in their jobs and the use of tools such as statistical quality control.
7. ***Change the focus from production numbers (quantity) to quality*** — The focus on volume of production instead of quality leads to defects and rework that may result in inferior products at higher costs.
8. ***Drive out fear*** — Employees need to feel secure in order for quality to be achieved. Fear of asking questions, reporting problems, or making suggestions will prevent the desired climate of openness.
9. ***Break down barriers between departments*** — When employees perceive themselves as specialists in one function or department without too much regard for other areas, it tends to promote a climate of parochialism and sets up barriers between departments. Quality and productivity can be improved when departments have open communication and coordination based on the common organization goals.

10. ***Stop requesting improved productivity without providing methods to achieve it*** — Continuous improvement as a general goal should replace motivational or inspirational slogans, signs, exhortations, and workforce targets. The major causes of poor productivity and quality are the management systems, not the workforce. Employees are frustrated when exhorted to achieve results that management systems prevent them from achieving.
11. ***Eliminate work standards that prescribe numerical quotas*** — Focus on quotas, like a focus on production, may encourage and reward people for numerical targets, frequently at the expense of quality.
12. ***Remove barriers to pride of workmanship*** — A major barrier to pride of workmanship is a merit or appraisal system based on targets, quotas, or some list of personal traits that have little to do with incentives related to quality. Appraisal systems that attempt to coerce performance should be replaced by systems that attempt to overcome obstacles imposed by inadequate material, equipment, or training.
13. ***Institute vigorous education and retraining*** — Deming emphasizes training, not only in the methods of the specific job but in the tools and techniques of quality control, as well as instruction in teamwork and the philosophy of a quality culture.
14. ***Create a structure in top management that will emphasize the preceding 13 points every day*** — An organization that wants to establish a culture based on quality needs to emphasize the preceding 13 points on a daily basis. This usually requires a transformation in management style and structure. The entire organization must work together to enable a quality culture to succeed.

Juran, like Deming, was invited to Japan in 1954 by the Union of Japanese Scientists and Engineers (JUSE). His lectures introduced the managerial dimensions of planning, organizing, and controlling and focused on the responsibility of management to achieve quality and the need for setting goals.[14] Juran defines quality as *fitness for use* in terms of design, conformance, availability, safety, and field use. Thus, his concept more closely incorporates the point of view of the customer. He is prepared to measure everything and relies on systems and problem-solving techniques. Unlike Deming, he focuses on top-down management and technical methods rather than worker pride and satisfaction.

Juran's ten steps to quality improvement are as follows:

1. Build awareness of opportunities to improve.
2. Set goals for improvement.
3. Organize to reach goals.
4. Provide training.
5. Carry out projects to solve problems.
6. Report progress.
7. Give recognition.
8. Communicate results.
9. Keep score.
10. Maintain momentum by making annual improvement part of the regular systems and processes of the company.

Juran is the founder of the Juran Institute in Wilton, Connecticut. He promotes a concept known as Managing Business Process Quality, which is a technique for executing cross-functional quality improvement. Juran's contribution may, over the longer term, be greater than Deming's because Juran has the broader concept, while Deming's focus on statistical process control is more technically oriented.[15]

Armand Feigenbaum, like Deming and Juran, achieved visibility through his work with the Japanese. Unlike the latter two, he used a total quality control approach that may very well be the forerunner of today's TQM. He promoted a system for integrating efforts to develop, maintain, and improve quality by the various groups in an organization. To do otherwise, according to Feigenbaum, would be to inspect for and control quality after the fact rather than build it in at an earlier stage of the process.

Philip Crosby, author of the popular book *Quality Is Free,*[16] may have achieved the greatest commercial success by promoting his views and founding the Quality College in Winter Park, Florida. He argues that poor quality in the average firm costs about 20% of revenues, most of which could be avoided by adopting good quality practices. His "absolutes" of quality are as follows:

- Quality is ***defined*** as conformance to requirements, not "goodness."
- The ***system*** for achieving quality is prevention, not appraisal.
- The performance ***standard*** is zero defects, not "that's close enough."
- The ***measurement*** of quality is the price of non-conformance, not indexes.[17]

Crosby stresses motivation and planning and does not dwell on statistical process control and the several problem-solving techniques of Deming and Juran. He states that quality is free because the small costs of prevention will always be lower than the costs of detection, correction, and failure. Like Deming, Crosby has his own *14 points*:

1. ***Management commitment*** — Top management must become convinced of the need for quality and must clearly communicate this to the entire company by written policy, stating that each person is expected to perform according to the requirement or cause the requirement to be officially changed to what the company and the customers really need.
2. ***Quality improvement team*** — Form a team composed of department heads to oversee improvements in their departments and in the company as a whole.
3. ***Quality measurement*** — Establish measurements appropriate to every activity in order to identify areas in need of improvement.
4. ***Cost of quality*** — Estimate the costs of quality in order to identify areas where improvements would be profitable.
5. ***Quality awareness*** — Raise quality awareness among employees. They must understand the importance of product conformance and the costs of non-conformance.
6. ***Corrective action*** — Take corrective action as a result of steps 3 and 4.
7. ***Zero defects planning*** — Form a committee to plan a program appropriate to the company and its culture.
8. ***Supervisor training*** — All levels of management must be trained in how to implement their part of the quality improvement program.
9. ***Zero defects day*** — Schedule a day to signal to employees that the company has a new standard.
10. ***Goal setting*** — Individuals must establish improvement goals for themselves and their groups.
11. ***Error cause removal*** — Employees should be encouraged to inform management of any problems that prevent them from performing error-free work.
12. ***Recognition*** — Give public, non-financial appreciation to those who meet their quality goals or perform outstandingly.
13. ***Quality councils*** — Composed of quality professionals and team chairpersons, quality councils should meet regularly to share experiences, problems, and ideas.
14. ***Do it all over again*** — Repeat steps 1 to 13 in order to emphasize the never-ending process of quality improvement.

All of these pioneers believe that management and the system, rather than the workers, are the cause of poor quality. These and other trailblazers have largely absorbed and synthesized each other's ideas, but generally speaking they belong to two schools of thought: those who focus on technical processes and tools and those who focus on the managerial dimensions.[18] Deming provides manufacturers with methods to measure the variations in a production process in order to determine the causes of poor quality. Juran emphasizes setting specific annual goals and establishing teams to work on them. Crosby stresses a program of zero defects. Feigenbaum teaches total quality control aimed at managing by applying statistical and engineering methods throughout the company.

Despite the differences among these experts, a number of common themes arise, as follows:

1. Inspection is never the answer to quality improvement, nor is "policing."
2. Involvement of and leadership by top management are essential to the necessary culture of commitment to quality.
3. A program for quality requires organization-wide efforts and long-term commitment, accompanied by the necessary investment in training.
4. Quality is first and schedules are secondary.

Admiration for Deming's contribution is not confined to Japan. At the Yale University commencement in May 1991, Deming was awarded an honorary degree. The citation read as follows:

> W. Edwards Deming, '28 PhD, *consultant in statistical studies*. For the past four decades, you have been the champion of quality management. You have developed a theory of management, based on scientific and statistical principles in which people remain the least predictable and most important part. Your scholarly insights and your wisdom have revolutionized industry. Yale is proud to confer upon you the degree of Doctor of Laws.[19]

ACCELERATING USE OF TQM

The increased acceptance and use of TQM is the result of three major trends: (1) reaction to increasing domestic and global competition, (2) the pervasive need to integrate the several organizational functions for improvement of total output of the organization as well as the quality of output within each function, and (3) the acceptance of TQM in a variety of service industries.

Aside from existing competitive pressures from Japan and the Pacific Rim countries, U.S. firms are faced with the prospect of increasing competition from members of the European Union (EU). This concern is justified by the very nature of manufacturing strategy among European firms, where quality has replaced technology as the primary consideration.

Basic to the concept of TQM is the notion that quality is essential in all functions of the business, not just manufacturing. This is justified by reason of organization synergism: the need to provide quality output to internal as well as external customers and the facilitation of a quality culture and value system throughout the organization. Companies that commit to the concept of TQM apply quality improvement techniques in almost every area of product development, manufacturing, distribution, administration, and customer service.[20] Nowhere is the philosophy of "customer is king" more prevalent than in TQM. Customers are both external (including channels) and internal (including staff functions) to the business.

The paradigm of TQM applies to all enterprises, both manufacturing and service, and many companies in manufacturing, service, and information industries have reaped the benefits. Industries as diverse as telecommunications, public utilities, and healthcare have applied the principles of TQM.

Government agencies and departments have also joined the movement, although private sector efforts have been considerably more effective.[21] According to a 1992 General Accounting Office (GAO) special report, 68% of the federal organizations and installations surveyed had some kind of TQM effort underway. Productivity and quality improvement programs have been implemented in many of the federal programs, including the defense departments.[22,23] Oregon State University was among the first academic institutions to make a commitment to adopt the principles of TQM throughout the organization.[24]

The widespread adoption of one or more approaches or principles of TQM does not mean that results have met expectations. According to the GAO survey mentioned earlier, only 13% of government agency employees actively participate in the TQM efforts.[25] Today, the principles of TQM have been adopted by various regulatory agencies responsible for granting accreditation to colleges and universities. Colleges and universities are now required to demonstrate evidence of a TQM-based srategy toward improving thier academic programs.[26]

QUALITY AND BUSINESS PERFORMANCE

The relationship between quality, profitability, and market share has been studied in depth by the Strategic Planning Institute of Cambridge,

Massachusetts. The conclusion, based on performance data of about 3000 strategic business units, is unequivocal:

> One factor above all others — quality — drives market share. And when superior quality and large market share are both present, profitability is virtually guaranteed.[27]
>
> There is no doubt that relative perceived quality and profitability are strongly related. Whether the profit measure is return on sales or return on investment, businesses with a superior product/service offering clearly outperform those with inferior quality.[28]

Even producers of commodity or near-commodity products seek and find ways to distinguish their products through cycle time, availability, or other quality attributes.[29] In addition to profitability and market share, quality drives growth. The linkages between these correlates of quality are shown in Figure 1-1.

Quality can also reduce costs. This reduction, in turn, provides an additional competitive edge. Note that Figure 1-1 includes two types of quality: customer-driven quality and conformance or internal specification

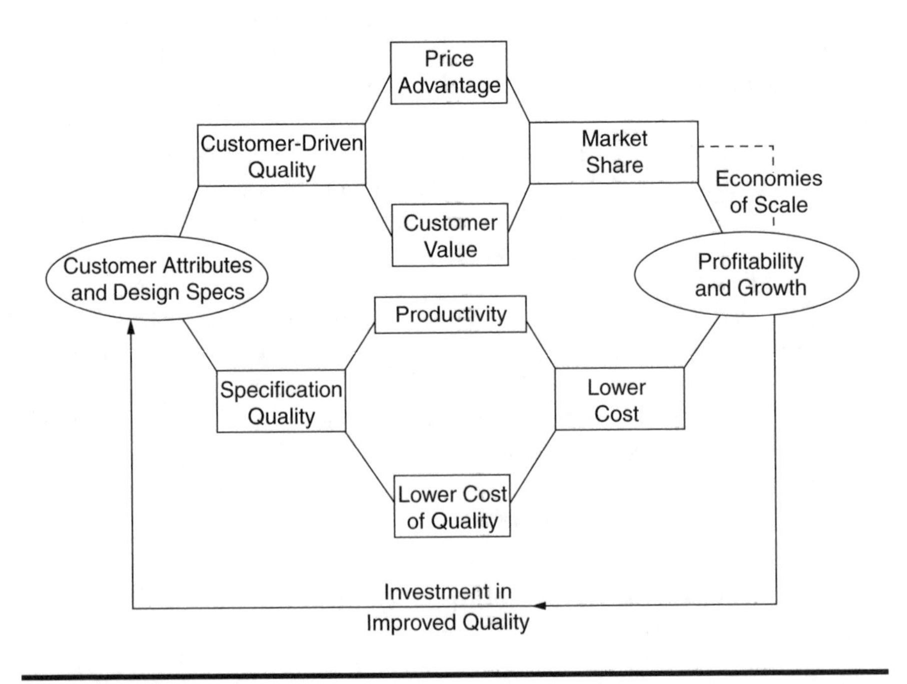

Figure 1-1 The Quality Circle

quality. The latter relates to appropriate product specifications and service standards that lead to cost reduction. As will be discussed in Chapter 11, internal or conformance quality is inversely related to costs, and thus the phrase coined by Crosby: "Quality Is Free."[30] As quality improves, so does cost, resulting in improved market share and hence profitability and growth. This, in turn, provides a means for further investment in such quality improvement areas as research and development. The cycle goes on. In summary, improving both internal (conformance) quality and external (customer-perceived) quality not only lowers the cost of poor quality or "non-quality" but also serves as a driver for growth, market share, and profitability.

The rewards of higher quality are positive, substantial, and pervasive.[31] Findings indicate that attaining quality superiority produces the following organizational benefits:

1. Greater customer loyalty
2. Market share improvements
3. Higher stock prices
4. Reduced service calls
5. Higher prices
6. Greater productivity

SERVICE QUALITY VS. PRODUCT QUALITY

In the United States and other highly industrialized countries, the economy has shifted away from manufacturing toward service industries. Nearly 80% of workers globally are employed in the service sector. Service accounts for 75% of the gross domestic product (GDP) for the United States.[32] If quality improvement can only be achieved through the actions of people, more than 90% of the potential for improvement may lie in service industries and service jobs in manufacturing firms. The concept of "white-collar quality" has become increasingly recognized as the service sector grows.[33]

Despite this rather obvious need for quality service, people directly employed in manufacturing functions tend to focus on production first and quality second. "Get out the production" and "meet the schedule" are common cries on many shop floors. A study conducted by David Garvin of the Harvard Business School revealed that U.S. supervisors believed that a deep concern for quality was lacking among workers and that quality as an objective in manufacturing was secondary to the primary goal of meeting production schedules. This conclusion was suggested by the experiences of over 100 companies. Supervisors almost invariably set targets related to productivity and cost reduction rather than quality improvement.[34]

This seeming manufacturing–service paradox is unusual in view of the several considerations suggesting that the emphasis on services should be substantially increased. The first of these considerations is the "bottom-line" factor. Studies have shown that companies rated highly by their customers in terms of service can charge close to 10% more than those rated poorly.[35] People will go out of their way and pay more for good service, which indicates the importance placed on service by customers. Conference Board reports concluded that the strongest complaints of customers were registered not for products but rather for services. Recognizing this, executives rate quality of service as a more critical issue than quality of product.[36] Tom Peters, co-author of *In Search of Excellence,* scolds U.S. manufacturers for allowing quality to deteriorate into a mindless effort to copy the Japanese and suggests that the best approach is to learn from America's leading service companies.[37]

> Taking a cue from Domino's Pizza, their Michigan-based neighbor, Doctor's Hospital in Detroit, is promising to see its emergency room patients in 20 minutes or the care will be free. During the first three weeks of the offer, no patients have been treated free of charge and the number of patients has been up 30%.[38]

As a strategic issue, customer service can be considered a major dimension of competitiveness. In the most exhaustive study in its history, the American Management Association surveyed over 3000 international respondents:[39] 78% identified improving quality and service to customers as *the* key to competitive success, and 92% indicated that providing superior service is one of their key responsibilities, regardless of position. To say that your competitive edge is price is to admit that your products and services are commodities.

After being viewed as a manufacturing problem for most of the past decade, quality has now become a service issue as well. Total Quality Management relates not only to the product but to all the services that accompany it as well.

In many ways, defining and controlling quality of service is more difficult than quality assurance of products. Unlike manufacturing, service industries share unique characteristics that make the process of quality control less manageable but no less important. Moreover, the level of quality expected is less predictable. Service company operations are affected by several characteristics, including the intangible nature of the output and the inability to store the output. Other distinguishing characteristics include the following:

- The behavior of the service provider becomes a factor in service delivery.
- The service recipient has the final say regarding quality.
- Service requires contact (directly or via telephone) between the service provider and the service recipient.
- The image of the organization shapes the perception of customers.
- The customer is present during the production process and performing the final inspection.
- The measure of output is difficult to define.
- Quality can mean different things to different people given the same experience.
- Quality is defined in the context of the totality of the experience.

However, the most significant problem with the delivery of services is that it is typically measured at the customer interface — the one-on-one, face-to-face interaction between supplier and customer. If a problem exists, it is already too late to fix it.[40]

Wall Street Journal, March 4, 1993

Vice President Gore's sphere expanded yesterday with the announcement that he will lead the latest White House effort to answer the call for change in Washington: a task force that will supposedly examine each federal agency for ways to cut spending and improve services. The "total quality management" effort is an idea borrowed from industry.

EXERCISES

1-1 Give one or more examples of products made in Japan or Western Europe that are superior in quality to U.S.-made products. How do you explain this difference?

1-2 Illustrate how the TQM concept can integrate design, engineering, manufacturing, and service.

1-3 Explain why quality should be better by following the TQM concept than in a system that depends on final inspection.

1-4 What common elements or principles can you identify among (1) the Baldrige criteria and (2) Deming, Juran, and Crosby?

1-5 Describe how increased market share and profitability might result from improved quality.

1-6 Select one staff department (e.g., accounting, finance, marketing services, human resources) and describe how this department can deliver quality service to its *internal* customers.

ILLUSTRATIVE CASE

The West Babylon School District in Long Island, New York, is among the growing number of school systems, colleges, and universities that are adopting the quality philosophy and acting upon it. West Babylon began its quest for TQM by formally defining the educational philosophy of continuous improvement and quality. Total quality management became known to all employees of the district as total quality education (TQE) and was based on business terms such as customer satisfaction and mass inspection. The 14 points of TQE were the same as Deming's *universal 14 points.*

Questions

- How has an educational institution with which you are familiar failed to operate according to Deming's principles?
- How could these principles be put into practice in such an institution?

ENDNOTES

1. J. M. Juran, "Strategies for World Class Quality," *Quality Progress,* March 1991, p. 81.
2. Armand V. Feigenbaum, "America on the Threshold of Quality," *Quality,* Jan. 1990, p. 16. Feigenbaum, a pioneer and current expert in quality, estimates that TQM could mean a 7% increase in the country's gross national product. See Armand V. Feigenbaum, "Quality: An International Imperative," *Journal for Quality and Participation,* March 1991, p. 16.
3. Ronald M. Fortuna, "The Quality Imperative," *Executive Excellence,* March 1990, p. 1. It is expected that competition from Europe may become more severe than that from Japan. Despite the surface congeniality at the G-7 meetings in London in mid-July 1991, it is apparent that the European Economic Community plans a united front against the United States. German Chancellor Helmut Kohl stated, "Europe's return to its original unity means that the '90s will be the decade of Europe, not Japan. This is Europe's hour." (Peter Truell and Philip Revzin, "A New Era Is at Hand in Global Competition: U.S. vs. United Europe," *Wall Street Journal,* July 15, 1991, p. 1.)
4. American Society for Quality Control, *Quality: Executive Priority or Afterthought?* Milwaukee: ASQC, 1989.

5. American Society for Quality Control, *'88 Gallup Survey: Consumers' Perceptions Concerning the Quality of American Products and Services,* Milwaukee: ASQC, 1988, p. iv. This survey and the survey cited in endnote 4 were conducted by the Gallup Organization.
6. American Society for Quality Control, *Quality: Everyone's Job, Many Vacancies,* Milwaukee: ASQC, 1990. This survey was conducted by the Gallup Organization. Other findings include the following: (1) Where quality improvement programs exist, the level of participation among employees is actually higher in small companies and service companies; and (2) people who participate in quality improvement activities are more satisfied than non-participants with the rate of quality improvement their companies have been able to achieve.
7. A survey conducted by Integrated Automotive Resources, Wayne, PA.
8. David Mercer, "Total Quality Management: Key Quality Issues," in *Global Perspectives on Total Quality,* New York: Conference Board, 1991, p. 11. See also Walter E. Breisch, "Employee Involvement," *Quality,* May 1990, pp. 49–51; John Hauser, "The House of Quality," *Harvard Business Review,* May/June 1988, pp. 63–73; W. F. Wheaton, "The Journey to Total Quality: A Fundamental Strategic Renewal," *Business Forum,* Spring 1989, pp. 4–7.
9. Claude S. George, Jr., *The History of Management Thought,* Englewood Cliffs, N.J.: Prentice Hall, 1972, p. 53.
10. David A. Garvin, "Competing on the Eight Dimensions of Quality," *Harvard Business Review,* Nov./Dec. 1987, pp. 101–109. This same author has provided an excellent description of the background of quality developments in his book *Managing Quality* (New York: Free Press, 1988).
11. See Joel E. Ross and David E. Wegman, "Quality Management and the Role of the Accountant," *Industrial Management,* July/Aug. 1990, pp. 21–23.
12. Yunum Kathawala, "A Comparative Analysis of Selected Approaches to Quality," *International Journal of Quality and Reliability Management,* Vol. 6, Issue 5, 1989, pp. 7–17. Other writers and researchers with less visibility than those mentioned here have contributed to the literature. For a review of this rapidly expanding literature, see Jayant V. Saraph, P. George Benson, and Roger G. Schroeder, "An Instrument for Measuring the Critical Factors of Quality Management," *Decision Sciences,* Vol. 20, 1989. The research described in this article identified 120 prescriptions for effective quality management, which were subsequently grouped into eight categories that are quite similar to the Baldrige Award criteria: (1) the role of management leadership and quality policy, (2) the role of the quality department, (3) training, (4) product/service design, (5) supplier quality management, (6) process management, (7) quality data and reporting, and (8) employee relations.
13. W. Edwards Deming, *Quality, Productivity, and Competitive Position,* Cambridge, Mass.: Center for Advanced Engineering Study, Massachusetts Institute of Technology, 1982. See also W. Edwards Deming, *Out of the Crisis,* Cambridge, Mass.: Center for Advanced Engineering Study, Massachusetts Institute of Technology, 1982.

14. Juran's early approach appears in J. M. Juran, *Quality Control Handbook,* New York: McGraw-Hill, 1951. For more recent contributions, see (all by Juran) "The Quality Trilogy," *Quality Progress,* Aug. 1986, pp. 19–24; "Universal Approach to Managing Quality," *Executive Excellence,* May 1989, pp. 15–17; "Made in USA — A Quality Resurgence," *Journal for Quality and Participation,* March 1991, pp. 6–8; "Strategies for World Progress," *Quality Progress,* March 1991, pp. 81–85.
15. "Dueling Pioneers," *Business Week,* Oct. 25, 1991. This is a special report and bonus issue of *Business Week* entitled "The Quality Imperative."
16. Philip Crosby, *Quality Is Free,* New York: McGraw-Hill, 1979.
17. From *Quality,* a promotional brochure by Philip Crosby Associates, Inc.
18. Sara Jackson, "Calling in the Gurus," *Director (UK),* Oct. 1990, pp. 95–101. In this article, the author reports that in the U.K., it is not the quality gurus as much as government initiatives that have been responsible for raising quality awareness.
19. Marc Wortman, "Commencement," *Yale Alumni Magazine,* Summer 1991, p. 61. One of the authors happens to be a Yale graduate, although not of the class of 1928.
20. Daniel M. Stowell, "Quality in the Marketing Process," *Quality Progress,* Oct. 1989, pp. 57–62. For R&D, see Michael F. Wolff, "Quality in R&D — It Starts With You," *Marketing News,* Oct. 15, 1990, pp. 16–22.
21. Stanley Blacker, "Data Quality and the Environment," *Quality,* April 1990, pp. 38–42.
22. Carolyn Burstein and Kathleen Sediak, "The Federal Quality and Productivity Improvement Effort," *Quality Progress,* Oct. 1988, pp. 38–41.
23. General Dynamics and McDonnell Douglas are among the defense contractors that have achieved improvement through TQM. See Bruce Smith and William B. Scott, "Douglas Tightens Controls to Improve Performance," *Aviation Week & Space Technology,* Jan. 4, 1990, pp. 16–20. See also Glenn E. Hayes, "Three Views of TQM," *Quality,* April 1990, pp. 19–24.
24. Edwin L. Coate, "TQM at Oregon State University," *Journal for Quality and Participation,* Dec. 1990, pp. 90–101.
25. Jennifer Jordan, "Everything You Wanted to Know About TQM," *Public Manager,* Winter 1992–93, pp. 45–48.
26. Karen Matthes, "A Look Ahead for '93," *HR Focus,* Jan. 1993, pp. 1, 4. See also Richard Y. Chang, "When TQM Goes Nowhere," *Training and Development,* Jan. 1993, pp. 22–29.
27. Robert D. Buzzell and Bradley T. Gale, *The PIMS Principles: Linking Strategy to Performance,* New York: The Free Press, 1987, p. 87. PIMS is the acronym for Profit Impact of Market Strategy. A PIMS study in Canada reached a similar conclusion. See William Band, "Quality Is King for Marketers," *Sales and Marketing Management in Canada,* March 1989.
28. Robert D. Buzzell and Bradley T. Gale, *The PIMS Principles: Linking Strategy to Performance,* New York: The Free Press, 1987, p. 107.
29. A good example is the "Perdue Chicken" produced by Perdue Farms. Owner Frank Perdue set out to differentiate his chicken by color, freshness, availability, and meat-to-bone ratio. These criteria of quality, as defined by the customer, led the company to growth and improved market share and profitability. See Diane Feldman, "Building a Better Bird," *Management Review,* May 1989, pp. 10–14.

30. Tom Peters reported in *Thriving on Chaos* (New York: Knopf, 1987) that experts agree that poor quality can cost about 25% of the people and assets in a manufacturing firm and up to 40% in a service firm.
31. See Joel E. Ross and David Georgoff, "A Survey of Productivity and Quality Issues in Manufacturing: The State of the Industry," *Industrial Management,* Jan./Feb. 1991.
32. U.S. Bureau of Labor Statistics, *Monthly Labor Review,* Nov. 1989. Reported in *U.S. Statistical Abstract,* 1990, p. 395.
33. For example, Campbell USA has targeted the administrative and marketing side of the corporation in its latest quality program, "Quality Proud." See Herbert M. Baum, "White-Collar Quality Comes of Age," *Journal of Business Strategy,* March/April 1990, pp. 34–37.
34. David Garvin, "Quality Problems, Policies, and Attitudes in the United States and Japan: An Exploratory Study," *Academy of Management Journal,* Dec. 1986, pp. 653–673.
35. Frank K. Sonnenberg, "Service Quality: Forethought, Not Afterthought," *Journal of Business Strategy,* Sep./Oct. 1989, pp. 56–57.
36. American Society for Quality Control, *Quality: Executive Priority or Afterthought?* Milwaukee: ASQC, 1989, p. 8. In this survey conducted by the Gallup Organization, 57% of service company executives rated service quality as extremely critical (10 on a scale of 1 to 10), while only 50% of industrial company executives gave service quality the same rating.
37. Tom Peters, "Total Quality Leadership. Let's Get It Right," *Journal for Quality and Participation,* March 1991, pp. 10–15.
38. "Hospital Delivers: Emergency Room Guarantees Care in 20 Minutes," Associated Press, July 15, 1991.
39. Reported in Eric R. Greenberg, "Customer Service: The Key to Competitiveness," *Management Review,* Dec. 1990, p. 29. Reported fully in AMA Research Report, *The New Competitive Edge,* New York: American Management Association, 1991.
40. Lawrence Holpp, "Ten Steps to Total Service Quality," *Journal for Quality and Participation,* March 1990, pp. 92–96. The major steps referred to in the title include: (1) creating an awareness and a philosophy of constant improvement, (2) making the vision of the organization a personal vision for every employee, (3) empowering employees to act, (4) surveying customers personally, (5) measuring meaningful information, and (6) adopting a performance management system that rewards teamwork, improvement, and new behaviors consistent with interdepartmental cooperation.

REFERENCES

Crosby, Philip B., "Illusions About Quality," *Across the Board,* June 1996, pp. 38–41.

Masters, Robert, "Overcoming the Barriers to TQM's Success," *Quality Progress,* May 1996, pp. 53–55.

Mooney, Marta, "Deming's Real Legacy: An Easier Way to Manage Knowledge," *National Productivity Review,* Summer 1996, pp. 1–8.

Reddy, Allan C. and Claude R. Superville, "Measuring Total Quality Marketing, A New Approach," *International Journal of Management,* March 1996, pp. 108–116.

Shelley, G. C., "The Search for the Universal Management Elixir," *Business Quarterly,* Summer 1996, pp. 11–13.

Woodruss, David M., "Ten Essentials for Being the Low-Cost, High-Quality Producer," *Management Quarterly,* Winter 1995–96, pp. 2–7.

2

LEADERSHIP

> Getting quality results is not a short-term, instant-pudding way to improve competitiveness; implementing total quality management requires hands-on, continuous leadership.
>
> **Armand V. Feigenbaum**

The story is told of three executives traveling on the same flight to an international conference. One executive was British, one Japanese, and one American. They were hijacked by terrorists and immediately before execution were offered an opportunity to make a last request. The Englishman asked to sing a verse of "God Save the Queen." The Japanese executive wanted to give a lecture on Japanese management. Upon hearing this, the American said: "Let me be the first one to be shot. I simply can't take another lecture on Japanese management."

The point of this story is that many U.S. managers are growing weary of such comparisons, in which they appear to be second best. One such comparison involved a visit to several Japanese companies by seven Leadership Forum executives. The experience left them with a profound belief that the reason why Japanese companies are beating U.S. companies has little to do with trade barriers, culture, cost of capital, sympathetic unions, or a supportive government. They found that the primary reason is simply that the United States is being outled and outmanaged. With some notable exceptions, U.S. firms are lagging behind because they lack clear, consistent, and persistent leadership from the top. Joseph Jaworski, chairman of the American Leadership Forum, is among the many CEOs who suggest that quality depends upon a vision of excellence and that a vision becomes reality through excellent, compelling leadership.[1]

Some principles and practices of total quality management (TQM) may differ among firms and industries, but there is unanimous agreement as to the importance of leadership by top management in implementing TQM. Such leadership is a prerequisite to all strategy and action plans. According to Juran, it cannot be delegated.[2] Those firms that have succeeded in making total quality work for them have been able to do so because of strong leadership.[3] A U.S. General Accounting Office (GAO) study concluded, "Ultimately, strong visionary leaders are the most important element of a quality management approach."[4]

Dr. Curt Reimann, director of the Malcolm Baldrige National Quality Award, has reviewed hundreds of applications, including those of the award winners. His review of key excellence indicators of quality management is insightful and helpful for an award applicant or anyone using the Baldrige criteria as a benchmark to evaluate the quality of management. He summarizes the characteristics of excellent leadership as follows:[5]

- ***Visible, committed, and knowledgeable*** — They promote the emphasis on quality and know the details and how well the company is doing. Personal involvement in education, training, and recognition. Accessible to and routine contact with employees, customers, and suppliers.
- ***A missionary zeal*** — The leaders are trying to effect as much change as possible through their suppliers, through the government, and through any other vehicle that promotes quality in the U.S. They are active in promoting quality outside the company.
- ***Aggressive targets*** — Going beyond incremental improvements and looking at the possibility of making large gains, getting the whole work force thinking about different processes — not just improving processes.
- ***Strong drivers*** — Cycle time, zero defects, six sigma, or other targets to drive improvements. Clearly defined customer satisfaction and quality improvement objectives.
- ***Communication of values*** — Effecting cultural change related to quality. Written policy, mission, guidelines, and other documented statements of quality values, or other bases for clear and consistent communications.
- ***Organization*** — Flat structures that allow more authority at lower levels. Empowering employees. Managers as coaches rather than bosses. Cross-functional management processes and focus on internal as well as external customers. Interdepartmental improvement teams.
- ***Customer contact*** — CEO and all senior managers are accessible to customers.

Two of the many companies that have received a great deal of visibility for their TQM programs are Westinghouse and IBM, both with divisions that have won the Baldrige Award. Westinghouse committed significant capital resources to support the quality improvement efforts of all Westinghouse divisions, including the creation of the first corporate-sponsored Productivity and Quality Center in the U.S. The company's Total Quality Model (Figure 2-1) was developed for use by all division managers. Note that it is built upon a foundation of management leadership. The framework of IBM's corporate-wide quality program, called "Market Driven Quality," is shown in Figure 2-2. Again, note that the input or "driver" of the system is leadership.

David Kearns, chairman and CEO of Xerox, explains how the company's "Leadership through Quality" process achieves commitment at every level: "Training begins with our top-tier family work group — my direct reports and me. It then cascades through the organizations led by senior staff, gradually spreading worldwide to some 100,000 employees."[6] This "cas-

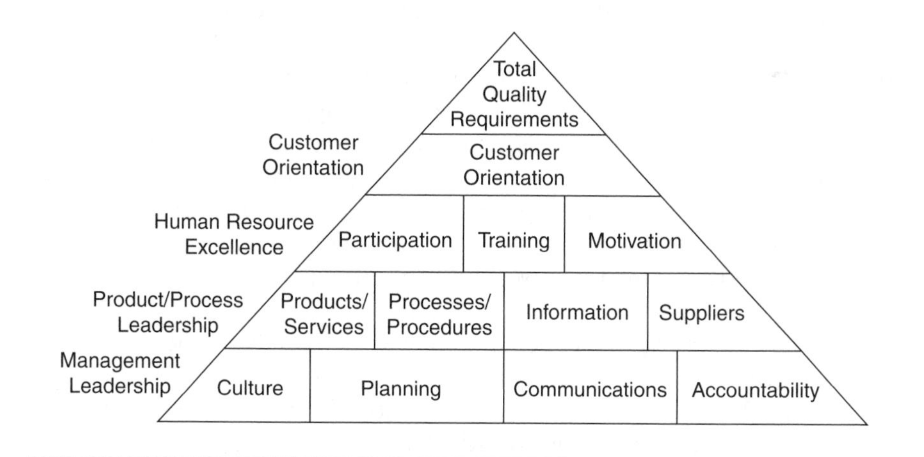

Figure 2-1 The Westinghouse Total Quality Model

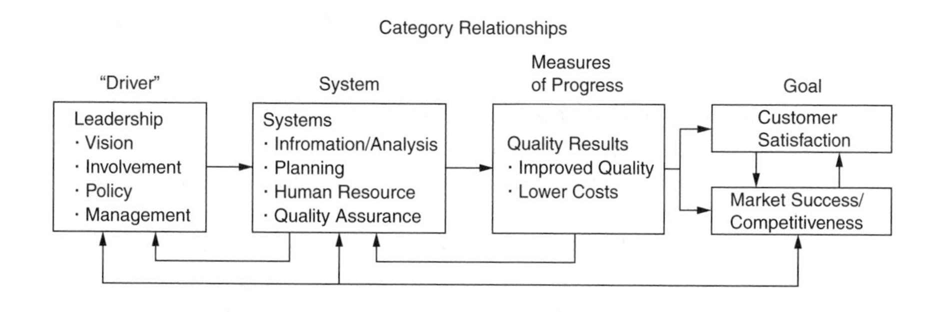

Figure 2-2 Framework of IBM's Market Driven Quality Program

cading" reflects the leverage effect of good leadership at all levels. As one executive remarked, "it goes up, down, and across the organization chart."

ATTITUDE AND INVOLVEMENT OF TOP MANAGEMENT

It would not be unfair to say that there has been a tendency among U.S. managers to focus on technology and hard assets rather than soft assets such as human resources and organizational competence.[7] The tendency has been to emphasize the organizational chart and the key control points within it. Many managers place priority on the budget and the business plan (to many, these are the same) and assume that rational people will get on board and perform according to standard. This popular perception does not fit with leadership and a philosophy of quality.

It is axiomatic that organizations do not achieve quality objectives; people do. If there is a big push for quality or a new program, each employee is justifiably skeptical (the BOHICA syndrome — bend over, here it comes again). According to A. Blanton Godfrey, chairman and CEO of the Juran Institute, top management should be prepared to answer the specific question that may be posed by each member of the organization: "What do you want me to do tomorrow that is different from what I am doing today?"[8] Thus, top managers need to be ambidextrous. They must balance the need for the *structural* dimension (e.g., hierarchy, budgets, plans, controls, procedures) on the one hand with the *behavioral* or personnel dimension on the other. The two dimensions need not be in conflict.

> At 3M Company, the leadership climate is proactive rather than reactive, externally focused rather than internally focused, and the quality perception views the totality of the business rather than just one aspect of it. In order to identify the gaps between its existing position and its vision of the future, 3M has developed "Quality Vision 2000" and implemented it through a process called Q90s which involves the total management system, making the process broader and deeper across the company worldwide.[9]

The commitment and involvement of management need to be demonstrated and visible. Speaking about his military experience, Dwight Eisenhower said: "They never listened to what I said, they always watch what I do."

Many managers send mixed signals. They endorse quality but reward bottom line or production. They insist on cost reduction even if it means canceling quality training. Still worse, some executives perceive the workers to be the cause of their quality problems.[10] This is hardly behavior that encourages individual involvement in decision making and personal "ownership" of the improvement process. Employee buy-in is unlikely in such a climate, where worker empowerment is talked about but not operationalized.

COMMUNICATION

Communication is inextricably linked in the quality process, yet some executives find it difficult to tell others about the plan in a way that will be understood. An additional difficulty is filtering. As top management's vision of quality gets filtered down through the ranks, the vision and the plan can lose both clarity and momentum. Thus, top management as well as managers and supervisors at all levels serve as translators and executors of top management's directive. The ability to communicate is a valuable skill at all levels, from front-line supervisor to CEO. Employees remain convinced that senior management knows something it is not telling the staff. Whether or not this assertion is actually true is not quite as important as the perception that it is true. A certain degree of transparency is necessary if department heads and managers are to disabuse the minds of their employees about this notion.

Quality-conscious companies are interested in the cost of poor communication in terms of both employee productivity and customer perception of product and service quality. More important than what is written or said is the recipient's perception of the message. Limited or inaccurate facts parceled out to employees may demoralize workers and lead to rumors.[11]

How Employees Receive Information

The culture of an organization can sometimes define how the employees receive information. The following represents the ways in which employees get their information:

- Rumor mill
- Monthly town meeting between the CEO and staff
- Monthly departmental meeting
- Email
- Members of the inner circle
- Company newsletter
- Memos
- External customers who call with questions
- Voice mail
- Verbal and/or written feedback from a manager or superior

Some of these techniques are grossly inappropriate and might actually lead to misinformation. When it comes to information, there is sometimes the notion that, to be among the "powerful," you must know something that nobody else knows, or possess information that no one else possesses. This can lead to a culture that encourages employees to withhold valuable information from other employees in order to preserve or validate their "power and influence."

According to Peter Drucker, a true guru of management thought and practice, "The communications gap within institutions and between groups in society has been widening steadily — to a point where it threatens to become an unbridgeable gulf of total misunderstanding."[12] Having said that, he provides an easily understood and simple approach to help communicate the strategy, vision, and action plans related to TQM.

Communication is defined as the *exchange of information and understanding* between two or more persons or groups. Note the emphasis on exchange and understanding. Without understanding between sender and receiver concerning the message, there is no communication. The simple model is as follows:

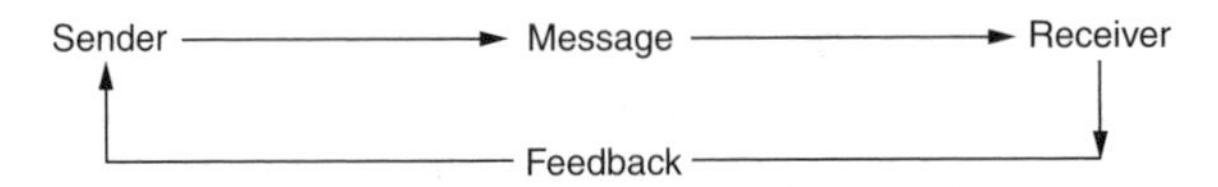

Unless the sender gets feedback that the receiver understands the message, no communication takes place. Yet most of us send messages with no feedback to indicate that the recipient (or percipient) has understood the message.

Despite the sorry state of communication, Drucker concludes that we do know something about communication in organizations and calls it "managerial communications." Communication is an extremely complex process. Many universities provide a doctoral program in the topic. At the risk of oversimplifying both communication theory and Drucker's approach, the essence of his principles can be paraphrased:

- One can only communicate in terms of the recipient's language and perception, and therefore the message must be in terms of individual experience and perception. If the employee's perception of quality is "do a better job" or "keep the customer happy," it is unlikely that the message of TQM will be understood. Measures of quality are needed to ensure agreement on the meaning of the message.
- Only the recipient can communicate — the communicator cannot. Thus, management systems (including training) should be designed

from the point of view of the recipient and with a built-in mechanism for feedback. Feedback and thus the exchange of information should be based on some measure, target, benchmark, or standard.

- All information is encoded, and prior agreement must be reached on the meaning of the code. Quality must be carefully defined and measures agreed upon.
- Communication downward cannot work because it focuses on what we want to say. Communication should be upward.
- Employees should be encouraged to set measurable goals.

Larry Appley, chairman emeritus of the American Management Association, has developed a company-wide productivity improvement program that has the model in Figure 2-3 as a centerpiece. Note that the direction of communication is *upward.* Recipient (subordinate) becomes sender, and sender (boss) becomes recipient. The message is specific and measurable, and the subordinate has ownership because he or she originated the message. Both parties can henceforth communicate about a message on which there is prior agreement. The Appley approach is therefore consistent with Drucker's ideas[13] and sound principles of communication. A modification tailored for a specific firm may be used as a vehicle for TQM implementation.

These concepts of effective communication can provide a practical approach for communicating about quality in the organization. It only remains to encode the message(s) in terms of recipient understanding.

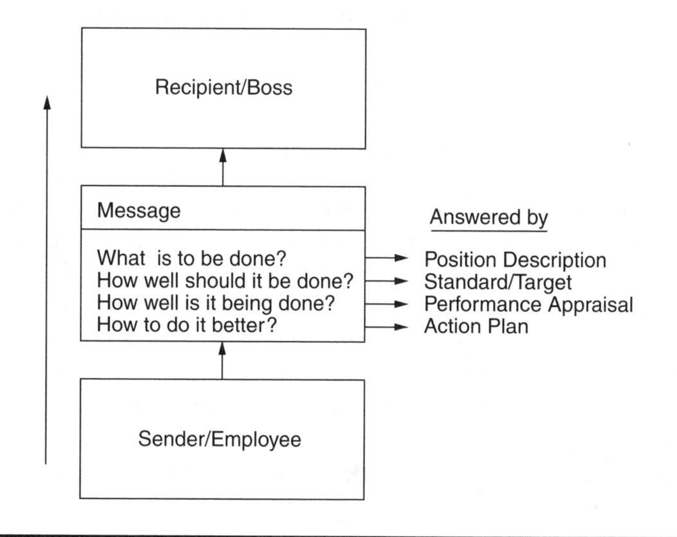

Figure 2-3 Effective Communication

The vehicles for communicating about quality are selected components of the TQM system:

- Training and development for both managers and employees. Managers must understand the processes they manage as well as the basic concept of systems optimization. Employee training should focus on the integration and appropriate use of statistical tools and problem-solving methods.
- Participation at all levels in establishing benchmarks and measures of process quality. Involvement is both vertical in the hierarchy as well as horizontal by cross-functional teams.
- Empowerment of employees by delegating authority to make decisions regarding process improvement within individual areas of responsibility, so that the individual "owns" the particular process step.
- Quality assurance in all organization processes, not only in manufacturing or operations but in business and supporting processes as well. The objective throughout is continuous improvement.
- Human resource management systems that facilitate contributions at all levels (up and down and across) the organizational chart.

The Digital Switching and Customer Service Division of Northern Telecom Canada Ltd. has received awards and international recognition for its quality systems and procedures. Continually communicating the importance of quality to its 5000 employees is considered vital by division management. Three internal communications specialists generate daily newsletters, monthly newspapers, and videos.[14]

CULTURE

Culture is the pattern of shared beliefs and values that provides the members of an organization rules of behavior or accepted norms for conducting operations. It is the philosophies, ideologies, values, assumptions, beliefs, expectations, attitudes, and norms that knit an organization together and are shared by employees.[15]

For example, IBM's basic beliefs are (1) respect for the individual, (2) best customer service, and (3) pursuit of excellence. In turn, these beliefs are operationalized in terms of strategy and customer values. In simpler terms, culture provides a framework to explain "the way things are done around here."

Other examples of basic beliefs include the following:

Ford	Quality is job one
Southwest Airline	Removing barriers
3M	Product innovation
Lincoln Electric	Wages proportionate to productivity
Caterpillar	Strong dealer support; 24-hour spare parts support around the world
McDonald's	Fast service, consistent quality

Institutionalizing strategy requires a culture that supports the strategy. For most organizations, a strategy based on TQM requires a significant if not sweeping change in the way people think. Jack Welch, head of General Electric and one of the most controversial and respected executives in America, states that cultural change must be sweeping — not incremental change but "quantum." His cultural transformation at GE calls for a "boundary-less" company where internal divisions blur, everyone works as a team, and both suppliers and customers are partners. His cultural concept of change may differ from Juran, who says, "When it comes to quality, there is no such thing as improvement in general. Any improvement is going to come about project by project and no other way."[16] The acknowledged experts agree on the need for a cultural or value system transformation:

- Deming calls for a transformation of the American management style.[17]
- Feigenbaum suggests a pervasive improvement throughout the organization.[18]
- According to Crosby, "Quality is the result of a carefully constructed culture, it has to be the fabric of the organization."[19]

It is not surprising that many executives hold the same opinions. In a Gallup Organization survey of 615 business executives, 43% rated a change in corporate culture as an integral part of improving quality. The needed change may be given different names in different companies. Robert Crandall, CEO of American Airlines, calls it an innovative environment,[20] while at DuPont it is "The Way People Think"[21] and at Allied Signal, "Workers attitudes had to change."[22] Xerox specified a five-year cultural change strategy called Leadership through Quality.[23] Tom Peters even adds what he calls "the dazzle factor."[24]

Successful organizations have a central core culture around which the rest of the company revolves. It is important for the organization to have a sound basis of core values into which management and other employees will be drawn. Without this central core, the energy of members of the organization will dissipate as they develop plans, make decisions, communicate, and carry on operations without a fundamental criteria of relevance to guide them. This is particularly true in decisions related to quality. Research has shown that quality means different things to different people and levels in the organization. Employees tend to think like their peers and think differently from those at other levels. This suggests that organizations will have considerable difficulty in improving quality unless core values are embedded in the organization.[25]

Commitment to quality as a core value for planning, organizing, and control will be doubly difficult if a concern for the practice is lacking. Research has shown that many U.S. supervisors believe that a concern for quality is lacking among workers and managers.[26] Where this is the case, the perceptions of these supervisors may become a self-fulfilling prophecy.

Embedding a Culture of Quality

It is one thing for top management to state a commitment to quality but quite another for that commitment to be accepted or embedded in the company. The basic vehicle for embedding an organizational culture is a teaching process in which desired behaviors and activities are learned through experiences, symbols, and explicit behavior. Once again, the components of the total quality system provide the vehicles for change. These components as well as other mechanisms of cultural change are summarized in Table 2-1. Above all, demonstration of commitment by top

Table 2-1 Cultural Change Mechanisms

Focus	*From traditional*	*To quality*
Plan	Short-range budgets	Future strategic issues
Organize	Hierarchy — chain of command	Participation/empowerment
Control	Variance reporting	Quality measures and information for self-control
Communication	Top down	Top down and bottom up
Decisions	Ad hoc/crisis management	Planned change
Functional management	Parochial, competitive	Cross-functions, integrative
Quality management	Fixing/one-shot manufacturing	Preventive/continuous, all functions and processes

management is essential. This commitment is demonstrated by behaviors and activities that are exhibited throughout the company. Categories of behaviors include the following:

- ***Signaling*** — Make statements or take actions that support the vision of quality, such as mission statements, creeds, or charters directed toward customer satisfaction. Publix supermarkets' "Where shopping is a pleasure" and JC Penney's "The customer is always right" are examples of such statements.
- ***Focus*** — Every employee must know the mission, his or her part in it, and what has to be done to achieve it. What management pays attention to and how management reacts to crisis is indicative of this focus. When all functions and systems are aligned and when practice supports the culture, everyone is more likely to support the vision. Johnson and Johnson's cool reaction to the Tylenol scare is such an example.
- ***Employee policies*** — These may be the clearest expression of culture, at least from the viewpoint of the employee. A culture of quality can be easily demonstrated in such policies as the reward and promotion system, status symbols, and other human resource actions.

Executives at all levels could learn a lesson from David T. Kearns, chairman and chief executive officer of Xerox Corporation. In an article for the academic journal *Academy of Management Executive,* he describes the change at Xerox: "At the time Leadership-Through-Quality was introduced, I told our employees that customer satisfaction would be our top priority and that it would change the culture of the company. We redefined quality as meeting the requirements of our customers. It may have been the most significant strategy Xerox ever embarked on."[27]

Among the changes brought about by the cultural change were the management style and the role of first-line management. Kearns continues: "We altered the role of first-line management from that of the traditional, dictatorial foreman to that of a supervisor functioning primarily as a coach and expediter."

Using a modification of the Ishikawa (fishbone) diagram, Xerox demonstrated (Figure 2-4) how the major component of the company's quality system was used for the transition to TQM.

MANAGEMENT SYSTEMS

No matter how comprehensive or lofty a quality strategy may be, it is not complete until it is put into action. It is only rhetoric until it has been

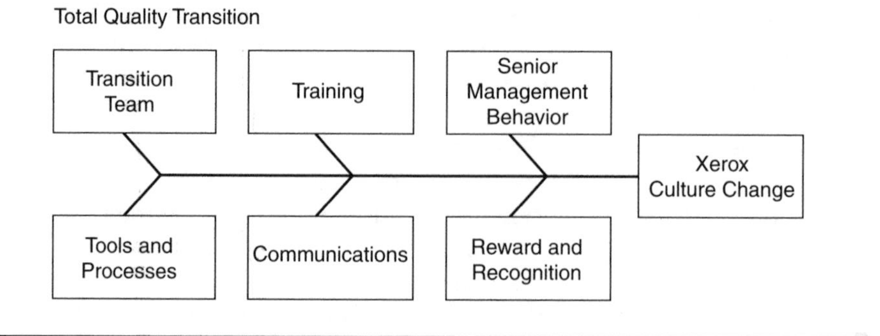

Figure 2-4 Transition to a Quality Culture at Xerox

implemented. Quality management systems are vehicles for change and should be designed to integrate all areas, not only the quality assurance department. They must be expanded throughout the company to include white-collar activities ranging from market research to shipping and customer service. They are directed toward achievement and commitment to purpose through four universal processes: (1) the specialization of task responsibilities through structure, (2) the provision of information systems that enable employees to know what they need to do in order to achieve goals, (3) the necessary achievement of results through action plans and projects, and (4) control through the establishment of benchmarks, standards, and feedback.

Each of these subsystems is the subject of a separate chapter in this book, but the implementation of each can only proceed from a base of clearly established goals. It is the specific task of top management to ensure that these goals are defined, disseminated, and implemented. Objectives in the areas of quality and productivity must be operationalized by establishing specific subobjectives for each function, department, or activity. Only then can courses of action be selected and plans implemented.

The problem, or conversely the opportunity, is to identify those *key* objectives and activities that are necessary in order to achieve a given strategy — in this case *quality*. The number of activities and processes in the typical organization is so large that a start-up quality management program cannot address all of them in the initial stages. Ultimately, every activity should be analyzed, its output evaluated in terms of value to both external and internal customers, and quality measures established.[28] Notwithstanding this longer-term need, it is desirable to begin by setting goals only for those activities that are critical to achieving the mission statement and strategy.

What are these activities and processes that are critical to the mission of quality? The answer lies in identifying the key success factors that must be well managed if the mission or objective is to be achieved; that is, the limited number of areas in which results, if satisfactory, will ensure successful competitive performance for the organization.[29] Each activity or process can then be rated as to its importance. Advertising is a key success factor for Coca-Cola but not for McDonnell Douglas; design is critical to a high-tech electronics firm but not to a bank.

This process can be used for any major objective, but it is also useful for providing a clear picture of things that must be done to implement a successful TQM program. Identification of key success factors emerges from three dimensions: (1) the drivers of quality such as cycle time reduction, zero defects, or six sigma; (2) operations that provide opportunities for reducing cost or improving productivity; and (3) the market side of quality, which relates to the salability of goods and services. These are converted to specific goals and targets, which form the basis for subsequent programs and the universal processes identified earlier. Some U.S. managers have adopted ideas and language from Japanese companies, many of whom call the process *policy deployment*.[30]

CONTROL

The classical control process will require significant change if TQM is to be successful. Traditionally, control systems have been directed to the end use of preparation of financial statements. Focus has been on the components of the profit-and-loss statement. Quality control has historically followed a three-step process consisting of (1) setting standards, (2) reporting variances, and (3) correcting deviations. One source has defined control as "to review, to verify, to compare with standards, to use authority to bring about compliance, and to restrain."[31] In an organization that perceives control systems in this way, there is the danger that the system will become the end rather than the means. This is not to say that classical control does not have a place in quality management.

If a company is in a declining industry and its generic strategy (see Chapter 4) is low cost, or if its product or service is a near-commodity for which differentiation by quality is difficult, then the management system should be directed toward tight cost control, frequent detailed control reports, structured organization and responsibility, and incentives based on meeting strict quantitative targets.[32] If, on the other hand, a company has chosen TQM as a strategy and culture, significant changes in traditional control may be needed. The central idea is to meet the needs of people so that they can be productive. These needs are both personal and job related, and a system of control should be based on both. If employees "buy in" to quality, the control system should not be perceived as domination, but rather as a means toward self-control. The danger of classical control has been summarized by Peter Drucker:

> A system of controls which is not in conformity with this true, this only effective, this ultimate control of the organization which lies in its people decisions will therefore at best be ineffectual. At worst it will cause never-ending conflict and push the organization out of control.[33]

The difference between TQM control and traditional control is the difference between self-control and control by variance report, between continuous control and historical control, between feedforward and feedback.

Consider historical feedback control, as depicted in Figure 2-5. Assuming that there is a measure of output (a doubtful assumption for most activities), the standard is compared to output, and variances are reported *after the fact*. The deviation has occurred and no amount of effort can change it. Typically, each period the first-line supervisor receives an after-the-fact statement of the quality control results for the entire plant. The worker receives nothing. This feedback is historical control by the numbers.

TQM control should be feedforward and predictive. Instead of measuring output after the fact, input is monitored by the individual or activity concerned, and output is forecast. If a deviation is predicted, action is taken to return to standard *before* the deviation occurs. There is no deviation because action is taken to avoid it before the fact. The concept is depicted in the bottom portion of Figure 2-5. The notion is fundamental for process control and continuous improvement of processes.

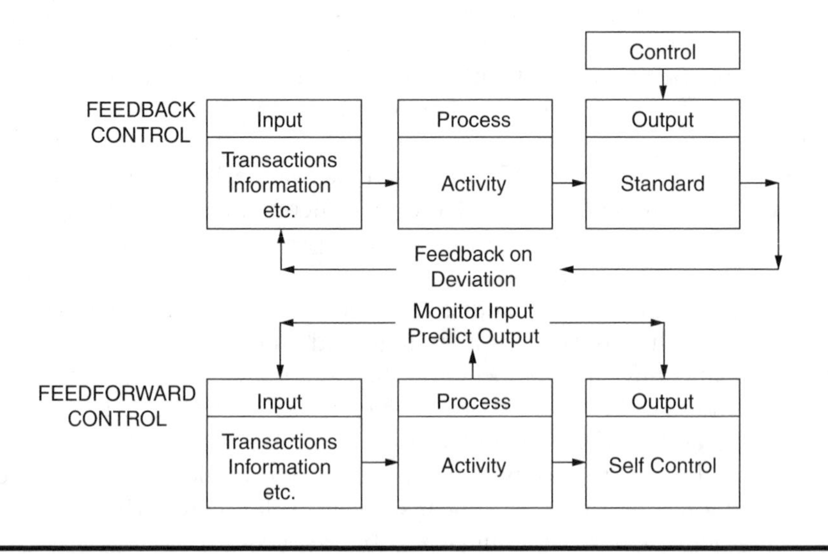

Figure 2-5 Feedback and Feedforward Control

EXERCISES

2-1 List the characteristics of excellent leadership for TQM.

2-2 Describe how leadership by top management is the *driver* of quality.

2-3 How can top management communicate the need for quality throughout the organization?

2-4 Describe how setting targets for quality improvement helps to establish a culture and climate.

2-5 Give an example of a company culture as reflected in a statement of basic beliefs. Would such a statement help to institutionalize a quality culture? If so, how?

2-6 How would an organization's commitment to quality facilitate or improve the following:

- The planning process
- Organization
- Control

2-7 Choose a manufacturing company and a service company. Identify a key activity for improving quality in each.

2-8 Discuss the pros and cons of each of the approaches used by some organization to communicate with its employees (as presented in the chapter).

ILLUSTRATIVE CASES

For years, Simplon Devices, a hi-tech manufacturer in the telecommunications industry, had chosen its suppliers based on a combination of price and quality as criteria, with an emphasis on price. A major supplier was Abbot Machine Tool Company. A sudden increase in demand for Simplon's products began growing faster than the company's ability to keep up. This, in turn, caused the company to put more pressure on its suppliers, particularly Abbot. Both companies were in a state of turmoil and both lost a number of sales to competitors. Both CEOs decided to work together on a longer-term basis in a partnering relationship to improve quality and avoid the ups and downs of changing demands. Both realized that changing to a TQM philosophy was necessary and that they themselves should be change agents, instead of reacting to customer pressure or the embarrassment caused by a competitor.

Questions

- What is a partnering arrangement? Discuss the pros and cons of entering into such an arrangement.
- How would you go about changing your company's philosophy and culture to one of TQM?

Las Vegas is a city with almost 100 casino/hotels offering various forms of entertainment and services. Mirage Resorts is aware that in order to compete in this highly competitive environment, it must attract and maintain a highly competent work force, the primary factor for success in such a service industry. The company spends a considerable amount of time and money in hiring and training the most qualified and the friendliest employees it can find. Employees are selected based on an estimate of whether they are likely to enjoy their jobs and contribute to Mirage's efforts to create and nurture customer goodwill. Following selection, employees are trained and motivated to provide quality customer service by offering them attractive benefits, including free meals, health insurance, education assistance, paid time off, bonuses for perfect attendance, and retirement plans.

Question

- Do you think the "material" benefits listed will result in motivation to provide quality customer service?

ENDNOTES

1. See Richard C. Whiteley, "Creating a Customer Focus," *Executive Excellence,* Sep. 1990, pp. 9–10.
2. J. M. Juran, "Made in USA — A Quality Resurgence," *Journal for Quality and Participation,* March 1991, pp. 6–8.
3. Thomas C. Gibson, "Helping Leaders Accept Leadership of Total Quality Management," *Quality Progress,* Nov. 1990, pp. 45–47.
4. U.S. General Accounting Office, "Quality Management Scoping Study," Washington, D.C.: U.S. General Accounting Office, Dec. 1990, p. 25.
5. Curt W. Reimann, "Winning Strategies for Quality Improvement," *Business America,* March 25, 1991, pp. 8–11.
6. David T. Kearns, "Leadership through Quality," *Academy of Management Executive,* Vol. 4 No. 2, 1990, p. 87.
7. Michael Moacoby, "How to Be a Quality Leader," *Research-Technology Management,* Sep./Oct. 1990, pp. 51–52. See also Brian L. Joiner and Peter R. Scholtes, "The Quality Manager's New Job," *Quality Progress,* Oct. 1986, pp. 52–56.

8. A. Blanton Godfrey, "Strategic Quality Management," *Quality,* March 1990, pp. 17–22.
9. Doug Anderson, "The Role of Senior Management in Total Quality," *Global Perspectives on Total Quality,* Conference Board Report Number 958, New York: Conference Board, 1991, p. 17. The author is Director, Corporate Quality Services, 3M Company.
10. Lance Arrington, "Training and Commitment: Two Keys to Quality," *Chief Executive,* Sep. 1990, pp. 72–73.
11. Dianna Booher, "Link between Corporate Communication & Quality," *Executive Excellence,* June 1990, pp. 17–18.
12. Peter Drucker, *Management: Tasks, Responsibilities, Practices,* New York: Harper & Row, 1973, p. 481.
13. The author had the pleasure and learning experience of working with Larry Appley on the development of this program, which had great success in a number of companies.
14. Bruce Van-Lane, "Good as Gold," *PEM: Plant Engineering & Maintenance (Canada),* April 1991, pp. 26–28.
15. Ralph H. Kilmann, Mary J. Saxton, and Roy Serpa, "Issues in Understanding and Changing Culture," *California Management Review,* Winter 1986, p. 89.
16. See "Jack Welch Reinvents General Electric — Again," *The Economist (UK),* March 30, 1991. See also Joseph M. Juran, *Juran on Leadership for Quality: An Executive Handbook,* New York: The Free Press, 1989.
17. W. Edwards Deming, "Transformation of Today's Management," *Executive Excellence,* Dec. 1987, p. 8.
18. Armand V. Feigenbaum, "Seven Keys to Constant Quality," *Journal for Quality and Participation,* March 1989, pp. 20–23.
19. Philip B. Crosby, *The Eternally Successful Organization,* New York: McGraw-Hill, 1988.
20. Aaron Sugarman, "Success through People: A New Era in the Way America Does Business," *Incentive,* May 1988, pp. 40–43.
21. Thomas C. Gibson, "The Total Quality Management Resource," *Quality Progress,* Nov. 1987, pp. 62–66.
22. Syed Shah and George Woelki, "Aerospace Industry Finds TQM Essential for TQS," *Quality,* March 1991, pp. 14–19.
23. U.S. General Accounting Office, "Quality Management Scoping Study," Washington, D.C.: U.S. General Accounting Office, Dec. 1990, p. 64.
24. Tom Peters, "Total Quality Leadership: Let's Get It Right," *Journal for Quality and Participation,* March 1991, pp. 10–15.
25. Frederick Derrick, Harsha Desai, and William O'Brien, "Survey Shows Employees at Different Organizational Levels Define Quality Differently," *Industrial Engineering,* April 1989, pp. 22–26.
26. David A. Garvin, "Quality Problems, Policies, and Attitudes in the United States and Japan: An Exploratory Study," *Academy of Management Journal,* Dec. 1986, pp. 653–673.
27. David T. Kearns, "Leadership through Quality," *Academy of Management Executive,* Vol. 4 No. 2, 1990, p. 87.

28. The need to set activity output and productivity measures (productivity being the ratio of output to input) has long been recognized. The accounting profession, recognizing the need to focus not only on cost but also on quality and productivity, has embraced a method known as *activity-based accounting*. See H. Thomas Johnson, "A Blueprint for World-Class Management Accounting," *Management Accounting,* June 1988, pp. 23–30. A very insightful and useful treatment of the shortcomings of traditional accounting systems is contained in H. Thomas Johnson and Robert S. Kaplan, *Relevance Lost,* Boston: Harvard Business School Press, 1991. Also included are excellent prescriptions for the improvement of accounting systems for managerial decision making.
29. Joel K. Leidecker and Albert V. Bruno, "Identifying and Using Critical Success Factors," *Long Range Planning,* Vol. 17 No. 1, 1984, pp. 23–32.
30. John E. Newcomb, "Management by Policy Deployment," *Quality,* Jan. 1989, pp. 28–30.
31. Andrew D. Szilagyi, Jr. and Marc J. Wallace, Jr., *Organizational Behavior and Performance,* 5th ed., Glenview, Ill.: Scott, Foresman, 1990, p. 620.
32. Michael E. Porter, *Competitive Strategy,* New York: The Free Press, 1980, p. 40.
33. Peter Drucker, *Management: Tasks, Responsibilities, Practices,* New York: Harper & Row, 1973, p. 504.

REFERENCES

Chang, Fred and Henry Wiebe, "The Ideal Culture Profile for Total Quality Management: A Competing Values Perspective," *Engineering Management Journal,* June 1996, pp. 19–26.

Crosby, Philip B., "The Leadership and Quality Nexus," *Journal for Quality and Participation,* June 1996, pp. 18–19.

Grant, Robert M. and Renato Cibin, "The Chief Executive as Change Agent," *Planning Review,* Jan./Feb. 1996, pp. 9–11.

McLagan, Patricia and Christo Nel, "A New Leadership Style for Genuine Total Quality," *Journal for Quality and Participation,* June 1996, pp. 14–16.

Peters, Tom, "Brave Leadership," *Executive Excellence,* Jan. 1996, pp. 5–6.

3

INFORMATION AND ANALYSIS

> Since quality programs are dependent on good information systems, chief information officers have the opportunity to plan an integral and highly visible role in shaping the quality of the corporation.
>
> **Curt Reimann, Director**
> *Malcolm Baldrige Award*

Information is the critical enabler of total quality management (TQM). More and more successful companies agree that information technology and information systems serve as keys to their quality success. Conversely, this component of TQM is frequently the roadblock to improvement in many firms. In these firms, better quality and productivity may not be the issue; rather, the real issue may be better quality of information. Dr. Curt W. Reimann, director of the Malcolm Baldrige National Quality Award, suggests that the critical constraint for many companies in applying for the award is the lack of a proper information system for tracking and improving areas in the remaining award categories.[1]

ORGANIZATIONAL IMPLICATIONS

John Sculley, former chairman of Apple Computer, concludes that information systems and technology can no longer be regarded as staff or service functions for management. Moreover, information systems will become the most important means for companies to create distinctive quality and unique service at the lowest possible cost.[2] At a 1988 symposium in Washington, D.C., for some 175 chief executive officers of major

U.S. corporations, the main topic was quality improvement and the information systems to support that effort. Designing the product, the plant configuration, and even the organizational structure is less challenging than designing the information system, which is the central component of TQM that allows the process to function.[3] It may be that the rigor of the production process is not matched by that of the information system, and the cause may lie in the increased complexity and breadth of the latter. Information is critical to *all* functions, and *all* functions need to be integrated by information.

The natural progression of information systems (used interchangeably with management information systems) in the past has frequently resulted in temporary fixes or "islands of mechanization," as applications such as inventory control, production scheduling, and sales reporting were designed without much regard for integration among each other or among other functions and activities within the organization. In recent years, additional and more sophisticated applications have emerged, such as quality function deployment, Taguchi methods, statistical process control, and just-in-time. These are now considered basic to the TQM process.[4] The challenge remains the same: to integrate these techniques and principles into a structured approach that includes related decision-making requirements across the board.

In Chapter 6, the argument is made for designing processes for continuous improvement in quality and productivity. A natural accompaniment is the design of the information systems to facilitate decisions related to these improvements. Indeed, these modern processes are all but impossible without sophisticated information systems.

Historically, companies have automated the easy applications: payroll, financial accounting, production control, and so on. Today, the concept of *re-engineering* is emerging. Rather than automating tasks and isolating them into discrete departments, companies are attempting to integrate the related activities of engineering, manufacturing, marketing, and support operations. Actions proceed in parallel, rather than in sequential order. Cycle time is reduced and products get to market faster with fewer defects. In short, the process is reengineered, and computer power is applied to the new process in the form of information systems. The focus is changing from buying information technology in order to automate paperwork to a focus designed to improve the process.

Information Technology

Systems design may be a constraint, but information technology (IT) is not. The geometric acceleration of developments is well known and can only be described as dramatic and spectacular.[5] If industry is capable of absorbing the technology, a further increase in the sophistication and importance of information will occur. Capital and direct labor will continue to be sources of value added, but the proportion contributed by intellectual and information activity will increase. Indeed, information can be considered to be a substitute for other assets because it can increase the productivity of existing capital and reduce the requirement for additional expenditures. It should be exploited.[6]

In 1990, Federal Express spent more than $243 million on IT. Then CEO Fred Smith stated that IT is absolutely the key to the organization's operations and that the entire quality process depends on statistical quantification which, in turn, depends on IT. Information is generated for both employees and customers.

Decision Making

The ability to make decisions quickly has always been critical to management at all levels, and information is essential to the process. It has emerged as a crucial competitive weapon.[7] Yet middle managers, who are the real change agents, spend most of their time exchanging information with subordinates, peers, or the boss, leaving little time for customers or for innovation and change. In the jargon of information systems, they need a decision support system.

Information Systems in Japan

In what continue to be customary comparisons between the United States and Japan, it is useful to examine how IT and information systems are perceived in Japan. Japanese executives believe that customer satisfaction drives the development of new services and products and that IT can be a vital means to facilitate strategies and operations to this end.[8] In true Japanese fashion, this view is apparently promoted by the national government as well. To build a foundation for future technicians and managers, the Ministry of Education has implemented national education policies for the full-scale use of computers in education.[9] There is also a national policy on software. The Ministry of International Trade and Industry has launched the Sigma Project, which calls for computerizing the software process and industrializing and computerizing software production.[10]

The Deming Prize is awarded each year to Japanese companies that demonstrate outstanding improvement in quality control. Yokogawa Hewlett-Packard, a joint venture of Hewlett-Packard and Yokogawa Electric Works, was awarded the prize for an information systems approach that yielded dramatic increases in profit, productivity, and market share.

STRATEGIC INFORMATION SYSTEMS

The integration of management information systems (MIS) with strategic planning has been suggested as a necessary prerequisite to strategy formulation and implementation. If we assume that the basic requirement of a strategy is environmental positioning in order to meet customer requirements, and if we further assume that the ultimate purpose of each function and process within an organization is to contribute to strategy, the role of information becomes clear.

As is discussed in Chapter 9, the value chain is a useful concept for determining the structure and processes needed by an organization in order to achieve a competitive advantage, keeping in mind that competitiveness is decided neither by the industry nor by the company, but rather by the customer.

Beginning with the customer, integration of processes and information can proceed as follows:

- Identify the market segment in which you want to compete.
- Use data collection and analysis to define the customer requirements in the chosen segment.
- Translate these requirements into major design parameters to develop, produce, deliver, and service the product that meets the customers' requirements. These are the primary functions and activities (processes) of the value chain.
- Complement the primary processes with support activities such as planning, finance and accounting, MIS, personnel, and so on.
- Subdivide or "explode" the organization design parameters into the processes (functions, activities, and so on) that are necessary to achieve the quality differentiation.
- Design the information requirements necessary to manage each process and to integrate all processes horizontally.

The support activities are sometimes taken for granted and their linking potential is often overlooked. Moreover, their potential contribution to differentiation may not be realized. Marketing services, for example, when combined with the customer's expertise, can generate differentiated product and service opportunities. The customer will place high value on a supplier who delivers the right information quickly. Engineering services, usually perceived as a commodity product, can also differentiate a firm. In both cases, the information systems support is cost effective.

Environmental Analysis

Strategy formulation requires an analysis of the different environments: general, industry, and competitive (see Chapter 4 for further discussion). One study found that small business owners spend over one-fourth of the day in external information search activities.[11] Competitive information is particularly valuable but is difficult to obtain.[12] In general, the minimum information needed about competitors can be related to how they stand on the key success factors for a market segment. These may differ by industry and segment but usually include the following:[13]

- Market share
- Product line breadth
- Proprietary advantages
- Age and location of facility
- Experience curve effects
- R&D advantage and position
- Growth rate
- Distribution effectiveness
- Price competitiveness
- Capacity and productivity
- Value added
- Cash throw-off

Porter has identified the information needed for positioning in an industry and in a chosen market segment, and his system is widely used. His categories are:

- Intensity of rivalry
- Bargaining power of buyers
- Bargaining power of suppliers
- Threat of substitution
- Threat of new entrants.[13]

Each category includes a number of elements or subtopics that should be determined and tracked with some type of information system.

Central to all information relating to strategy formulation and implementation is the need to *define and measure* the concept of quality of product and service — as determined by the customer. This step is fundamental to positioning and subsequent follow-up.

SHORTCOMINGS OF ACCOUNTING SYSTEMS

Financial information is perhaps the most widespread indicator of performance, and for many firms is the only indicator. Critics of accounting systems claim that they do not really support the operations and strategy of the company, two dimensions in which quality plays a dominant role. Despite the widely held conclusion that we are in the information age, management accounting would probably be labeled inadequate by managers who seek to support company operations and strategy through quality improvement. This is increasingly evident in the "new" manufacturing environment, which is characterized by the trends and implications listed in Table 3-1.

Accountant bashing is becoming increasingly popular in the management literature. The trend is summarized by Harvard Business School Professor Robert Kaplan in his popular book *Relevance Lost.*[14] He concludes that today's accounting information provides little help in reducing costs and improving quality and productivity. Indeed, he suggests that this information might even be harmful. Peter Drucker, another critic, describes some of the shortcomings that are generally recognized:[15]

1. Cost accounting is based on a 1920s reality, when direct labor was 80% of manufacturing costs other than raw material. Today it is 8 to 12%, and in some industries (e.g., IT) it is about 3%.
2. Non-direct labor costs, which can run up to 90%, are allocated in proportion to labor costs, an arbitrary and misleading system. Benefits of a process change are allocated in the same way.
3. The cost system ignores the costs of *non-producing,* whether this be downtime, stockouts, defects, or other costs of non-quality.
4. The system cannot measure, predict, or justify change or innovation in product or process. In other words, accounting measures direct or real costs and not benefits.
5. Accounting-generated information does not recognize linkages between functions, activities, or processes.
6. Manufacturing decisions cannot be made as *business* decisions based on the information provided by accounting. The system confines itself to measurable and objective decisions and does not address the intangibles.

Efforts are underway to make accounting a true management and business system. For example, Computer-Aided Manufacturing-International (CAM-I) is a cooperative effort by automation producers, multinational manufacturers, and accountants to develop a new cost accounting system. Even internal auditors are examining their new role in TQM.[16]

Table 3-1 New Manufacturing Environment

Trend	*Implication for quality*
Focus on manufacturing strategy	Quality rapidly becoming the central competitive edge of strategy
Production of high-quality goods	Quality directly related to market share, growth, profits
Reduction of inventory levels	Reduction of costs associated with excess inventory by just-in-time inventory
Tight schedules	Improves availability to customer, another competitive edge perceived as quality by the customer
Product mix and variety	Allows focus on strategy and market segmentation
Equipment automation	Provides justification for quality and productivity improvement
Shortened product life cycle	Provides opportunity to expedite market shifts and incorporate new technologies into the product, but imposes additional stress on the quality management program
Organizational changes	Responsibility for quality delegated to strategic business units and product managers
Information technology	Allows greater control of cost of quality, quality management, and cross-functional integration

ORGANIZATIONAL LINKAGES

The importance of data linkages is illustrated by data on service calls, a primary source of measuring product field performance. These are an important source of information for design, engineering, manufacturing, sales, and service. One research study reported that in some cases among air conditioner manufacturers, the aggregate data on failure rates were of little use because of organization barriers:

> The *service tracking report* at American Express monitors performance for all centers worldwide. For the credit card division, for example, performance is measured against 100 service measures, including how long it takes to process an application,

> authorize charges, bill card members, and answer customer billing inquiries. Each measure is based on customer expectations, the competition, the economy, and legislation. Application processing time has been reduced by 50 percent and the bottom line has been increased by $70 million.[17]

This example illustrates the widespread need for organization linkages and cross-functional MIS and the need to track a process on a continuous basis. Figure 3-1 shows how each step in the life cycle of a product involves related processes as cross-functional lines.

Each step in the product life cycle involves a number of processes at these cross-functional lines in a continuous flow from design to preproduction planning to vendor management to incoming material to in-process control to finished goods to customer service. The steps along the flow should be accompanied by appropriate information.[18] Thus, the linkage concept may focus on internal customers (those who use products in a later step of the process) as well as external customers:

> Federal Express, the first service sector company to earn the Baldrige Award, integrates a variety of internal measurement systems into the core of its business. The objective is "zero service defects." The system, SQI (service quality indicators), measures twelve critical points at which failure can occur in the service process and continually reinforces how employees are doing compared to their goals.

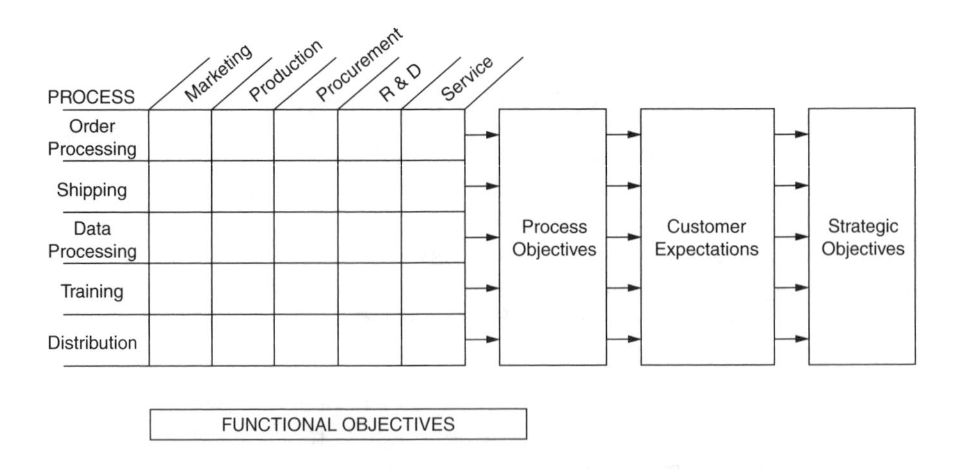

Figure 3-1 Cross-Functional Lines in the Life Cycle of a Product

White-Collar Measures

The large number of white-collar and service personnel in the typical manufacturing firm was noted earlier. These activities not only comprise a large (perhaps major) share of total costs, but are essential to a systems approach to TQM. If TQM is to work, it must address the involvement of employees in developing measurement systems that will need to be in place and accessible to all levels.

The characteristics of white-collar work make it more difficult to measure than work in the manufacturing process. The American Productivity Center in Houston has developed a continuous performance improvement process for white-collar work called IMPACT. *Measures development* is the fourth step of the six-step process. To quote:

> IMPACT provides a "family of measures" that allows each pilot group to track its progress from "Where are we?" to "Where do we want to go?" The family of measures provides the pilot group the tools it needs to measure progress, to give feedback, and to know when to take additional corrective actions. In addition, current measures are inventoried and used along with new measures. It is most effective if both the customer and the supplier participate in this phase.[19]

Sara Lee Hosiery launched a company-wide quality program that focuses on customer service and providing training designed to help information systems, networking, and other service departments understand how to apply total quality concepts to their work. The key is to focus on tools to enhance those processes that are based on customer needs and expectations.

ADVANCED PROCESSES/SYSTEMS

SPC, QFD, CAD, CIM, MRP — one gets the impression of "alphabet management." These and other basic applications represent the major systems of TQM. None stand alone and there are overlaps among them. Some advocates promote one or more as the "total system." Most if not all of these processes depend upon IT and a sophisticated information system design. Because systems design begins with the objective of the process, it is useful to list the objectives of the major processes (Table 3-2).

At Motorola's Automotive and Industrial Electronics Group in Arcade, New York, over 1000 employees were trained in SPC. Operators then began doing their own inspections and plotting hourly control charts to

Table 3-2 Objectives of Major Processes in Systems Design

Process/system	*Objective*
Statistical quality control (SQC)	Build in the control limits of a process that spots and identifies causes of variations
Statistical process control (SPC)	Provide information on how productivity and quality can be *continuously improved* through problem identification (it has been estimated that U.S. firms invest 20–25% of their operating budgets in finding and fixing mistakes[20])
Just-in-time (JIT)	Reduce inventory cost, production time, and space requirements
Computer-integrated manufacturing (CIM)	Lower cost, shorten lead time, and improve quality based on information sharing by linking management and financial information systems, departmental computing, process management systems, and factory systems for controlling machinery and manufacturing processes
Quality function deployment (QFD)	Integrate the three dimensions of (1) company-wide quality, (2) focus on customer requirements, and (3) translation of quality perceptions into product characteristics and then into the manufacturing process

control their own projects. Quality control inspectors were transferred out. Improvements included (1) achieving 10:1 goal of improvement, (2) significant increases in yields, and (3) reduction in scrap. The facility received the Q1 quality award as a supplier to Ford.

Donald Bell, general manager of Monsanto's Fibers Division, envisioned the "Plant of the 1990s." The scheme is a three-tiered approach encompassing human resources planning, total quality concepts, and computer-integrated manufacturing. Productivity gains of 40 to 50% have already been achieved. The program emphasizes the needs of internal customers — those who use products in a later step of the manufacturing process. Computer training has enabled greater acceptance of these concepts.[21]

INFORMATION AND THE CUSTOMER

According to examiners who visit companies that apply for the Baldrige Award, most companies lack the processes that ensure efficient flow of information on customer demands and related information throughout the organization.[22] In other words, most companies do not devote the same attention to the customer that they do to the internal processes of shipping, inventory, just-in-time, manufacturing, and so on. This is unfortunate because the operating processes cannot be managed according to the principles of TQM unless the loop is closed with customer feedback. Information systems should be extended beyond the plant into the marketplace. Some companies tend to define quality in terms of customer satisfaction or some other non-specific term and then relax after shipment is made, overlooking the competitive success that accompanies after-the-sale service, spare parts, or distribution.[23]

Why do information systems directly related to customer satisfaction frequently take a back seat to what otherwise might be acceptable or excellent information systems in support of quality and process control? The answer may be that it is difficult to specify information needs for an elusive system to measure customer requirements and satisfaction, which in themselves are difficult to define. Or it may be that the pressures of crisis management and internal information exchange leave little time for the customer.[24] Whatever the cause, it is a good idea to design a system that measures the pulse of the market and the customer base. It is estimated that failure to do so will cost twice as much as poor internal quality.

> The First National Bank of Chicago found that quality can be the difference between acquiring and keeping customers. Because competitive pricing varies by only a few pennies, the customer must be enlightened as to the benefits of strong quality. The bank measures customer satisfaction by how often inaccurate information is given. In 1982, the error rate was 1 in 4000 transactions; in 1990 it dropped to 1 in 810,000 transactions.[25]

Information Needs

After the objective of an information system is established, the next step is to determine the information needs. This is the most difficult step in designing an MIS for customer satisfaction. Everything else is detail and technique. Manager/user involvement is essential here.

If there is one fundamental principle of TQM, it is that *quality is what the buyer defines it to be,* not what the company defines it to be. Ford learned this lesson in the late 1970s, when the company definition of DQR (durability, quality, and reliability) was found to be presented in terms (engineering design tolerances and specifications) understandable only within the company, rather than in terms that represented quality to the customer. Only after reassessing quality in terms understandable to the customer was Ford able to adopt a policy called "Ford Total Quality Excellence" and achieve organization-wide commitment to continuous improvement and customer focus.

The first step, then, is to define quality as perceived by the customer by viewing it *externally* from the customer's perspective. By profiling how customers make purchase decisions, it is possible to determine which product attributes are most important and to determine how customers rate each attribute. As discussed in Chapter 8, this process forms the basis of *benchmarking*.[26]

Market research methods ranging from focus groups to shopper surveys are means for profiling customers and defining quality as perceived by customers. The information system can then be designed to provide the input for decisions regarding the operating plan, organizational implications, and follow-up control.

THE INFORMATION SYSTEMS SPECIALISTS

It would be an understatement to say that the power of IT and computers has exploded and will continue to do so. Computer power is estimated to double every five years, while the cost continues to decrease. This expansion of IT and computer power has been accompanied by a growth in "knowledge workers" or "information workers" — people who control the quality of streams of information. These streams flow into a business from customers and the external environment, then flow through a business from product development to manufacturing and distribution, and flow out in the form of sales effort and service follow-up.

It is unfortunate that the growth of the white-collar sector has outpaced its productivity. Still worse, while investment in IT tripled from 1978 to 1988, output per hour among 80% of the total work force remained virtually stagnant. This represents a major opportunity to improve the productivity and quality of all workers by providing better decision-making information for process improvement throughout the company. Achieving this goal requires training in the techniques of systems design and use.

The Chief Information Officer

> General Dynamics was awarded the Premier 100 ranking as the top aerospace firm by the magazine *Computerworld*. To General Dynamics, TQM means near perfect products and an information system that meets the continuous process improvements.

Information systems is a growth industry, particularly for companies aspiring to TQM. Therefore, it is likely that quality information systems as described in this book will become more important and more voluminous than financial systems (the traditional and widespread source of operating data). This trend is reflected in the growing number of organizations with positions titled CIO.[27]

The CIO position is still evolving, but the ideal job description would have the individual responsible for developing IT/information systems planning and tying it into the strategic plan of the business (something that accounting does not generally do). Additionally, the CIO's function would focus on performance measures based on customer satisfaction and would then apply productivity tools to improve the related processes.[28] These functions would be in addition to the normal duties of quality assurance, cost–benefit analysis, software development, technology transfer, and technology forecasting. Providing quality output within the department to internal customers is a given because it sets a climate and provides a role model for others who deal with external customers. Service quality can only be built from the inside out, and how the information systems function delivers its services to internal customers can influence the way external customers are served.

SYSTEMS DESIGN

After reviewing hundreds of applications for the Baldrige Award, Curt Reimann, director of the award, concluded that the area of information and analysis represents a serious national problem.[29] Many firms have failed to design individual applications to fit an overall master plan. The result has been a temporary solution with little integration between functions and activities.

A master plan should be centered around corporate goals and the critical success factors and cost–performance drivers related to these goals. In a manufacturing firm, data from engineering, production, and field service are used to improve product design and manufacturing techniques. If reducing cycle time in bringing a product to market is a critical success factor (as it is), a good deal of this information will flow sideways and across departmental lines, rather than upward and vertically, as in the traditional model.

The individual manager/user has the job of designing his or her own system requirements and fitting these into the overall master plan. This is not easy. In discussions with dozens of system analysts, they almost always report that their number one difficulty in system design is the inability or unwillingness of the user to define information needs. This definition is not the job of the analyst — it is the job of the individual user. Before design can proceed, two critical steps must be taken: define *system objectives* and *information needs.*

Surprisingly, many users cannot define an objective. They will define it as "having the right part at the right place at the right time" or "preparing a field service report." Statements such as these are elusive, not quantifiable, and unsuitable for conversion to information needs. On the other hand, when objectives are stated in more specific terms (such as "reduce final inspection in the production process to the point of elimination" or "reduce throughput time to six days"), the designer has a benchmark from which to proceed.

The next step is to define *information needs,* another requirement that users have difficulty defining. The question is: "What information do I need to achieve the objective?" If performance measures are established, the determination of both objective and information needs will become more apparent. Successful companies benchmark their performance against world-class quality leaders. For example, Xerox measured its performance in about 240 key areas of product, service, and business performance. This process is discussed further in Chapter 8.

EXERCISES

3-1 Describe how lack of information can be a roadblock to implementing one or more TQM actions.

3-2 How do traditional accounting systems provide inadequate information for control of processes in an industry with low labor content?

3-3 Choose two functions or activities (market research, R&D, design, production planning, procurement, human resources) and show how information can serve to integrate them across functional lines.

3-4 How does information technology affect organizational structure? Give an example of how information technology can facilitate TQM.

3-5 How would you go about designing an MIS for getting customer input for quality improvement?

3-6 How does market segmentation influence information needs?

ILLUSTRATIVE CASE

The Engine and Foundry Division of Navistar International Transportation had set quality and productivity goals and decided to develop supporting information systems. The desired systems would be for real-time quality monitoring and control that was easy to use and maintain. The company chose an outside consulting group for statistical process control. Consultants helped Navistar work out specifications to recalculate control limits, to evaluate and graph non-normal distributions, and to support a customized-gage interface that supports multiple inputs and multiple recipes from any terminal. The result was an expansion of 1,000% in the amount of data that the company collects and evaluates. Plants can collect data from hundreds of gages and more than 10,000 measurement points and still enjoy real-time evaluation.

Questions

- Would it have been more appropriate to use in-house personnel rather than consultants? Which system design steps are appropriate for a statistical process control system?
- Would a multidisciplinary team been appropriate? If so, which functions should be represented?
- Is the huge expansion of data collected justified? Can you have too much data?

ENDNOTES

1. Telephone interview with Curt W. Reimann.
2. John Sculley, "The Human Use of Information," *Journal for Quality and Participation,* Jan./Feb. 1990, pp. 10–13.
3. Elizabeth A. Haas, "Breakthrough Manufacturing," *Harvard Business Review,* March/April 1987, pp. 75–81. It is estimated here that companies adopting integrated strategies may succeed in increasing productivity by 10 to 15% annually. See also Julian W. Riehl, "Planning for Total Quality: The Information Technology Component," *Advanced Management Journal,* Autumn 1988, pp. 13–19.
4. Nael A. Aly, Venetta J. Maytubby, and Ahmad K. Elshennawy, "Total Quality Management: An Approach & A Case Study," *Computers and Industrial Engineering,* Issues 1–4, 1990, pp. 111–116.
5. For a description of what lies ahead, see Robb Wilmot, "Computer Integrated Management — The Next Competitive Breakthrough," *Long Range Planning,* Vol. 21 No. 6, 1988, pp. 65–70.
6. James Heskett, "Lessons in the Service Sector," *Harvard Business Review,* March/April 1987, pp. 118–126.
7. Kathleen M. Eisenhardt, "Speed and Strategic Choice: How Managers Accelerate Decision Making," *California Management Review,* Spring 1990, pp. 39–54.
8. Dennis Normile, "Japan Inc. Bows to the Customer," *CIO,* Aug. 1990, pp. 91–93. A major benefit of information systems is increasing the speed from product concept to marketing, an improvement that translates into customer satisfaction.
9. Takashi Yamagiwa, "Computer Use in the Japanese Educational System," *Business Japan,* March 1988, pp. 38–39.

10. Ryozo Hayashi, "National Policy on the Information Service Industry," *Business Japan,* March 1990, pp. 49–61.
11. J. Lynn Johnson and Ralph Kuehn, "The Small Business Owner/Manager's Search for External Information," *Journal of Small Business Management,* July 1987, pp. 53–60.
12. The pioneering book is Frank J. Aguilar, *Scanning the Business Environment,* New York: Macmillan, 1967.
13. For a comprehensive treatment of competitive information and its sources, see Michael E. Porter, *Competitive Advantage,* New York: The Free Press, 1985. Based on research data collected from more than 3000 strategic business units, the Strategic Planning Institute, through its PIMS program, has identified the following characteristics of the most profitable companies in an industry: (1) higher market share, (2) higher quality, (3) higher labor productivity, (4) higher capacity utilization, (5) newer plant and equipment, (6) lower investment intensity per sales dollar, and (7) lower direct cost per unit. See Robert D. Buzzell and Bradley T. Gale, *The PIMS Principles: Linking Strategy to Performance,* New York: The Free Press, 1987.
14. H. Thomas Johnson and Robert S. Kaplan, *Relevance Lost: The Rise and Fall of Management Accounting,* Boston: Harvard Business School Press, 1991.
15. Peter Drucker, "The Emerging Theory of Manufacturing," *Harvard Business Review,* May/June 1990, pp. 94–102. See also Robert S. Kaplan, "The Four Stage Model of Cost Systems Design," *Management Accounting,* Feb. 1990, pp. 22–26; James M. Reeve, "TQM and Cost Management: New Definitions for Cost Accounting," *Survey of Business,* Summer 1989, pp. 26–30.
16. Fred J. Newton, "A 1990s Agenda for Auditors," *Internal Auditor,* Dec. 1990, pp. 33–39.
17. David A. Garvin, *Managing Quality,* New York: The Free Press, 1988, pp. 167–169. In this study, the best plants maintained sophisticated systems to track data and report it back to interested departments and functions.
18. See Raymond G. Ernst, "Why Automating Isn't Enough," *Journal of Business Strategy,* May/June 1989, pp. 38–42. The author argues that companies too often attempt to improve manufacturing by making large investments in automation without improving their business processes. He estimates that the savings of 10 to 20% that can be derived from automating can be increased to 70% when improvements are made to existing business processes as well. The process improvements can be achieved through a product-information flow.
19. Jackie P. Comola, "Designing a New Family of Measures," in *Total Quality Performance,* New York: The Conference Board, 1988, pp. 59–64. Mr. Comola is Vice President, White-Collar Productivity, at the American Productivity Center. Measures are developed through a nominal group technique in which teams accomplish the following steps: state the problem, list possible measures, collect possibilities round-robin, edit nominations, vote and rank, discuss, and reach consensus. The six phases of IMPACT are (1) planning, (2) assessment, (3) direction setting, (4) development of measures, (5) service (re)design and implementation, and (6) results review and recycle.
20. Otis Port, "The Push for Quality," *Business Week,* June 8, 1987, pp. 130–135.

21. For an excellent description of how quality function deployment is implemented, see John R. Hauser and Don Clausing, "The House of Quality," *Harvard Business Review,* May/June 1988, pp. 63–73. See also Chia-Hao Chang, "Quality Function Deployment (QFD): Processes in an Integrated Quality Information System," *Computers and Industrial Engineering,* Vol. 17 Issues 1–4, 1989, pp. 311–316.
22. Peter Burrows, "Commitment to Quality: Five Lessons You Can Learn from Award Entrants," *Electronic Business,* Oct. 15, 1990, pp. 56–58.
23. Morris A. Cohen and Hau L. Lee, "Out of Touch with Customer Needs? Spare Parts and After Sales Service," *Sloan Management Review,* Winter 1990, pp. 55–66.
24. Robert W. Wilmot, "Computer Integrated Management — The Next Competitive Breakthrough," *Long Range Planning,* Dec. 1988, pp. 65–70. This author has found that typical middle managers spend less than 10% of their time with customers and a tiny fraction sponsoring innovation and orchestrating change.
25. John Goodman and Cynthia J. Grimm, "A Quantified Case for Improving Quality Now," *Journal for Quality and Participation,* March 1990, pp. 50–55.
26. Bradley T. Gale and Robert D. Buzzel, "Market Perceived Quality: Key Strategic Concept," *Planning Review,* March/April 1989, pp. 6–15.
27. A survey by the Healthcare Financial Management Association found increasing use of a CIO in large hospitals and a growing emphasis on strategic planning. See John J. May and Ed H. Bowman, "Information Systems for the Value Management Era," *Healthcare Financial Management,* Dec. 1986, pp. 70–74. Insurance companies and banks represent other industries that are adopting this position.
28. Herbert Z. Halbrecht, "What's Good For the Boss..." *Computerworld,* Aug. 21, 1989, p. 78.
29. Curt W. Reimann, "Winning Strategies for Quality Improvement," *Business America,* March 25, 1991, pp. 8–11.

REFERENCES

Aiken, Milam, Bassam Hasan, and Mahesh Vanjani, "Total Quality Management: A GDSS Approach," *Information Systems Management,* Winter 1996, pp. 73–75.

Betts, Mitch, "Dallas' Qualitative Edge," *Computerworld,* April 18, 1994, p. 122.

Kumar, Ram and Connie Crook, "Educating Senior Management on the Strategic Benefits of Electronic Data Interchange," *Journal of Systems Management,* March/April 1996, pp. 42–47.

Pollalis, Yannis A., "A Systematic Approach to Change Management: Integrating IS Planning, BPR, and TQM," *Information Systems Management,* Spring 1996.

4

STRATEGIC QUALITY PLANNING

The basics of total quality management (TQM) can effectively govern executive-level strategic management and goal-setting.

Executive
Academy of Management

Ford's slogan, "Quality Is Job 1," has caught on with increasing segments of the car-buying public. The company's North American Automobile Group is gaining market share among U.S. manufacturers.[1] Things were not always this way. Between 1978 and 1982, market share slipped to 16.6% and sales fell by 49%, with a cumulative loss in excess of $3 billion. Ford was losing $1000 on every car it sold. The company sought advice from W. Edwards Deming. Reports John Betti, at that time a senior executive at Ford, "I distinctly remember some of Dr. Deming's first visits. We wanted to talk about quality, improvement tools, and which programs work. He wanted to talk to us about management, cultural change, and senior management's vision for the company. It took time for us to understand the profound cultural transformation he was proposing."[2] The company's subsequent turnaround is a classic example of the results that can be obtained from a strategic change based on quality. The major changes responsible for reversing the company's fortune were as follows:

- Emphasize quality and review new product planning and design.
- Keep investing in new products and processes.
- Make employee relations a source of competitive advantage.[3]

3M's approach to quality is so highly regarded that executives from leading U.S. companies travel to St. Paul to attend monthly briefings sponsored by 3M. In *Thriving on Chaos,*[4] Tom Peters described 3M as the only truly excellent company today. *Forbes* chose 3M as one of America's three most highly regarded companies. Its TQM implementation strategy includes:

- Defining 3M's quality vision
- Changing management perceptions through specialized training
- Empowering employees to focus on and satisfy customer expectations
- Sustaining the process through an ongoing culture change

One executive of the company explained it as follows: "How do you meet such a wide variety of expectations in a coherent way? I think you do it with a corporate philosophy on what constitutes a total quality process…a philosophy that you can apply across the company…to all your operations."[5]

These comments reflect the importance that successful companies place on the strategy issue. In the American Management Association's survey of over 3000 international managers, the key to competitive success was defined as the improvement of quality. There is little doubt that a strategy based on quality begins with strategic planning and is implemented through program and action planning.[6]

STRATEGY AND THE STRATEGIC PLANNING PROCESS

What is strategy and what is the strategic planning process? The answers to these questions are important because evidence suggests that those companies with strategies based on TQM have achieved stunning successes.[7]

Most of these successful companies will attribute their progress to a quality-based strategy that was developed through a formal structured approach to planning. The Commercial Nuclear Fuel Division of Westinghouse, another Baldrige winner, has discovered that the total quality concept must be viewed as a pervasive operating strategy for managing a business every day:

> Total Quality begins with a *strategic decision* — a decision that can only be made by top management — and that decision, simply put, is the decision to compete as a world-class company. Total Quality concentrates on quality performance — in every facet of the business — and the primary strategy to achieve and maintain competitive advantage. It requires taking a sys-

> tematic look at an organization — looking at how each part interrelates to the whole process. In addition, it demands continuous improvement as a "way of life."[8]

Major contributors to the development of the strategic concept and to the planning process include Professors Andrews, Christensen, and others in the Policy group at the Harvard Business School.[9] A recent definition by this group is contained in their highly regarded text on the subject:

> Corporate strategy is the pattern of decisions in a company that (1) determines, shapes, and reveals its objectives, purposes, or goals; (2) produces the principal policies and plans for achieving these goals; and (3) defines the business the company intends to be in, the kind of economic and human organization it intends to be, and the nature of the economic and non economic contribution it intends to make to its shareholders, employees, customers, and communities.[10]

Michael Porter is perhaps the most highly regarded and certainly the most popular writer on the subject of strategy.[11] He describes the development of a competitive strategy as "a broad formula for how a business is going to compete, what its goals should be, and what policies will be needed to carry out those goals."

Strategic Planning is a deliberate process used by organizations to develop a mission, vision, guiding values, strategic objectives, and specific strategies for achieving the objectives. Prior to embarking on this process, It is often helpful to conduct a SWOT analysis. SWOT is the acronym for strengths, weaknesses, opportunities, and threats. Through the process of SWOT analysis, an organization is able to answer the following questions:

- What are the organization's strengths?
- What are the organization's weaknesses?
- What opportunities exist out there for the organization to consider?
- What factors (external and internal) constitute a threat to the organization?

The value of a SWOT analysis lies in its ability to provide basic information that would bring clarity to a strategic plan. Following a SWOT analysis, the strategic plan will involve a process to search for the answers to the following questions:

- Who are we?
- What are we known for?

- What do we do better than 90% of our competitors?
- What do our competitors beat us on?
- What do we wish to be known for?
- Where are we headed as an organization?
- Where do we wish to be headed?
- How would we get there?
- What would it take to get us there?

STRATEGIC QUALITY MANAGEMENT

This pattern of goals, policies, plans, and human organization is not something to be taken lightly. It is likely to be in place over a long period of time and therefore affects the organization in many different ways. The culture that guides members of the organization and other stakeholders, the position that it will occupy in an industry and market segments, and determining particular objectives and allocating resources to achieve them all follow from the decision processes determined by strategy. It is easy to see how pervasive a strategy based on quality can become. It provides the basis upon which plans are developed and communications achieved. A basic rule of strategic planning is that *structure follows strategy*. Although the process of formulation and implementation may require staff input, the ultimate decision is fundamental to the job of the chairman or CEO. It cannot be delegated.

The pervasive role that quality plays in strategic planning can best be understood by examining the components of a strategy:

- Mission, vision, and guiding values
- Product/market scope
- Competitive edge (differentiation)
- Supporting policies
- Objectives
- Organizational culture

These components are developed through a process of strategy formulation, the outline of which is shown in Figure 4-1. Note that the process involves positioning yourself against forces in the environment in such a way that action plans can minimize your weaknesses and take advantage of your strengths relative to the competition. Quality is the means of differentiation for the satisfaction of customer needs. Research that includes over 300 U.S. companies indicates that firms with superior quality address quality offensively, as a distinct competitive advantage, while firms with inferior quality treat it defensively (e.g., eliminate defects, lower cost of product failure).[12]

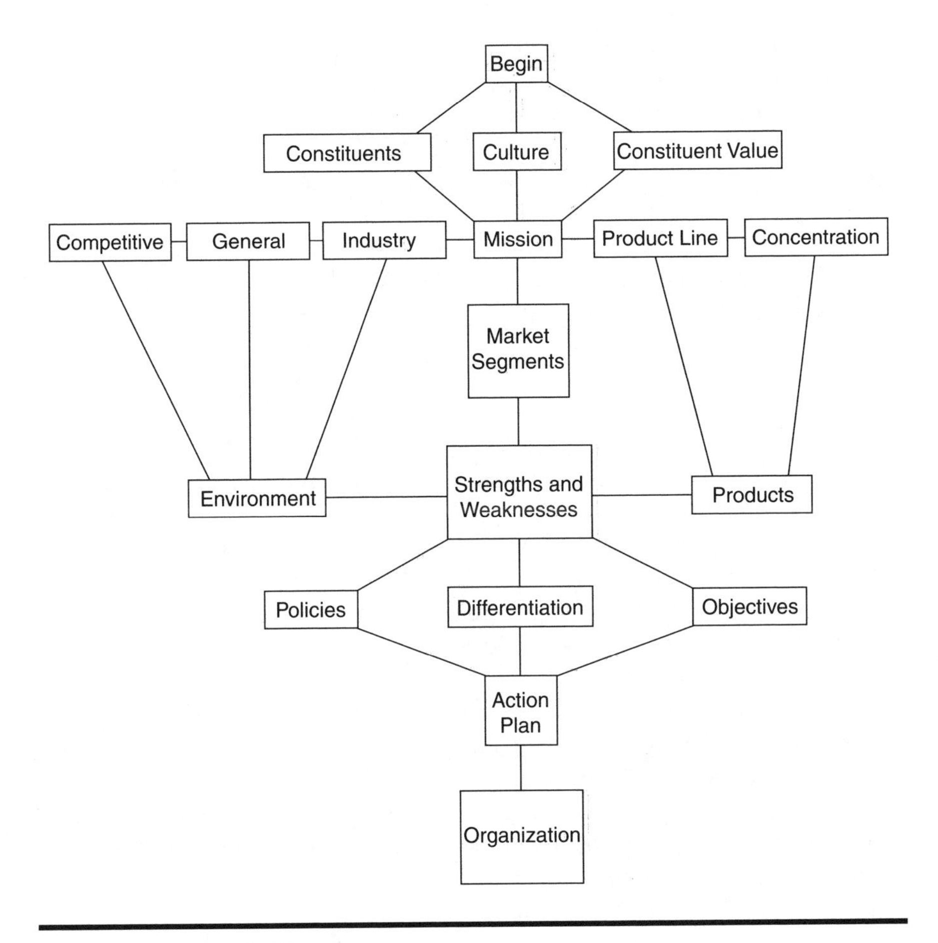

Figure 4-1 Strategic Planning

Mission

The mission is the primary overall purpose of an organization and its expressed reason for existence. The simplest statement of mission might be to "meet the needs/values of constituents."

> The mission of NCR is stated simply: "Create Value for Our Stakeholders." Stakeholders are identified as employees, shareholders, suppliers, communities, and customers.[13] The mission can be operationalized by statements of how it will be implemented for each stakeholder.
>
> At Goodyear, every employee carries a credit-card-sized mission statement: "Our mission is constant improvement in products

> and services to meet our customers' needs. This is the only means to business success for Goodyear and prosperity for its investors and employees."[14]

Southwest Airlines has the following for its mission statement:[14a]

> The mission of Southwest Airlines is dedication to the highest quality of Customer Service delivered with a sense of warmth, friendliness, individual pride, and Company Spirit.
>
> **To Our Employees**
>
> We are committed to provide our Employees a stable work environment with equal opportunity for learning and personal growth. Creativity and innovation are encouraged for improving the effectiveness of Southwest Airlines. Above all, Employees will be provided the same concern, respect, and caring attitude within the organization that they are expected to share externally with every Southwest Customer.

The mission statement includes the value that is being added and the direction the company intends to move. Because a mission can only be achieved by the people in an organization, it should have the commitment of the entire organization. Deming's first and what he considers his most important point of management obligation is to "create constancy of purpose for improvement of product and service with a plan to become competitive and stay in business."

This consistency must be achieved by a mission that can be operationalized and implemented. Consider the following examples:

- All employees at Motorola consistently strive for a six sigma target.
- 3M's mission focuses on innovation. To ensure consistency of purpose, the company established a requirement that 25% of each profit center's sales must come from products less than five years old.
- Ford spent more than a year defining its mission. The real test of consistency and commitment came when the company withheld releasing a new Thunderbird, a "sure bet" for car of the year, because the car's quality was not yet suitable for a production model.

A vision reflects where the organization is headed or wishes to be. It is like a destination dreamed up by the organization. Every decision made by the organization must be informed by its vision. An organization's vision must come from top management, and must be well articulated and understood by all. The guiding values reflect the beliefs that shape and mold the decisions and choices an organization makes.

Environment

The major determinant of a mission is the environment in which the firm plans to operate: the general environment, the industry environment, and the competitive environment. Strategy is essentially the process of positioning oneself in that environment as trends and changes unfold. Thus, it is necessary to identify trends in the environment and how they affect the strategy of the firm. Figure 4-2 illustrates how a major U.S. manufacturer of computer equipment and software documented the major changes in that industry. The impact on strategy, as these issues relate to quality, is illustrated in Figure 4-3.

Product/Market Scope

This answers the questions: What am I selling and to whom am I selling it? The answers are more complex than they appear. What is Domino's Pizza selling: dough and tomato sauce or reliable delivery? What is a physician selling: surgery and diagnosis or patient involvement? Wal-Mart and Bloomingdale's are both in the retail business, but are their products

Information Industry Shift

Technology		
· Predictable Cycles · Custom Design · Proprietary	→	· Fast Paced · Off-the-Shelf · Open Architecture
Customers		
· DP Professionals · Large Companies · Mostly U.S.	→	· End Users, Consumers · All Sizes · International
Offerings		
· Individual Products · Hardware	→	· Complete Business Solutions · Software/Services
Competitors		
· Few, Large U.S. Fully Integrated · Independent	→	· Many, Diverse, Global Specialized · Alliances

Figure 4-2 Defining the Environment

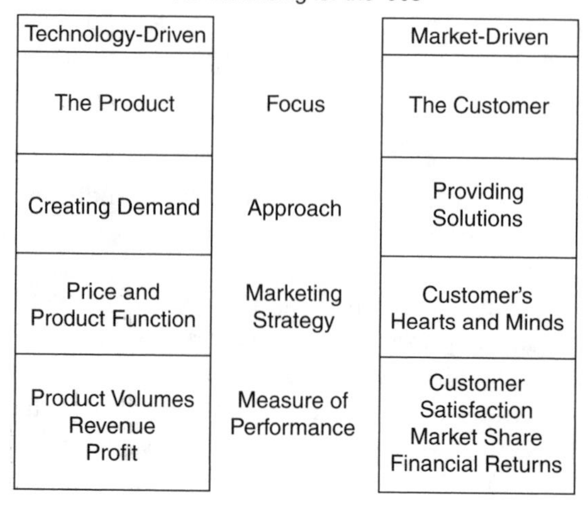

Figure 4-3 Impact of Changes in the Environment on Strategy

simply what is on the shelves and racks in their stores? A company does not simply sell shoes or soap or banking services. It sells value to a particular segment of the market. The answers to these questions should be clear, as well as the role of quality in customer value.

What is value? It is, of course, what the customer — not the company — says it is. Timex sells watches, but does Rolex sell jewelry and prestige? Canada Dry sells sparkling water, but does Perrier sell snob appeal? Thom-McCann sells loafers, but what does Gucci sell? This does not mean that Timex, Canada Dry, and Thom-McCann do not sell on the basis of quality. Indeed, they do. However, quality is defined differently for a different segment of the market. Each company must define its market segment and customer value in that segment. Ford's product mix includes the Lincoln Town Car and the Escort, but each is targeted at a different market segment, and quality (value) is different for each segment.

Every purchase decision is a function of price and quality. Price is generally known, but quality is in the mind of the individual customer. General Electric is aware of this and has broadened its perspective from "product quality" to "total customer satisfaction." The "product" is now defined by the customer.[15] It only remains to define customer satisfaction, perception, or expectation.

To repeat, in today's heightened competitive environment, a product or service is not simply sold to anyone who will buy it. To be effective, value must be sold to a particular market or customer segment. Strategic planning involves the determination of these strategy components, and quality plays a major role in this process.

Differentiation

Differentiation, frequently called the competitive edge, answers the question: Why should I buy from you? Michael Porter, in his landmark book *Competitive Strategy,* identified two generic competitive strategies: (1) overall cost leadership and (2) differentiation.[16] Cost leadership in turn can be broad in market scope (e.g., Ivory Soap, Emerson Electric, Black & Decker) or market segment focused (e.g., La Quinta Motels, Porter Paint). The second strategy involves differentiating the product or service by creating something that is perceived by the buyer as unique. Differentiation can also be broad in scope (American Airlines in on-time service, Caterpillar for spare parts support) or focused (e.g., Godiva chocolates, Mercedes automobiles). Thus, there are four generic strategies, but each depends on something different — something unique or distinguishing. Even an effective cost leadership strategy must start with a good product.

Selecting a strategy and recognizing quality as the competitive dimension is important for strategic purposes. Product and service quality has become widely recognized as a major force in the competitive marketplace and in international trade.[17]

Research indicates that eight out of ten customers consider quality to be equal to or more important than price in their purchase decisions.[18] This is a doubling of buyer emphasis in ten years and the trend is expected to continue. The message here is that whether a cost leadership or differentiation strategy is chosen, quality must be a competitive consideration in either case.

Differentiation can command a premium price or allow increased sales at a given price. Moreover, differentiation is one of two types of competitive advantage, the other being price. Price, however, should not be the sole basis of differentiation unless the product is perceived to be a commodity. Even if the product is a commodity or near-commodity, it can still be differentiated by such service characteristics as availability or cycle time.

The several sources of differentiation are not well understood. Many managers perceive their uniqueness in terms of the physical product or in their marketing practices rather than in terms of value to the customer. They may waste money because their uniqueness does not provide real value to the buyer. Why spend money on extra tellers or checkout lines to reduce waiting time to one minute if the customers are willing to wait two minutes? On the other hand, buyers frequently have difficulty estimating value and how a particular firm can provide it. This incomplete knowledge can become an opportunity if the firm can adopt a new form of differentiation and educate the buyers to value it.

DEFINITION OF QUALITY

The concept and vocabulary of quality are elusive. Different people interpret quality differently. Few can define quality in measurable terms that can be operationalized. When asked what differentiates their product or service, the banker will answer "service," the healthcare worker will answer "quality healthcare," the hotel or restaurant employee will answer "customer satisfaction," and the manufacturer will simply answer "quality product." When pressed to provide a specific definition and measurement, few can do so.[19] There is an old maxim in management that says, "If you can't measure it, you can't manage it," and so it is with quality. If the strategic management system and the competitive advantage are to be based on quality, every member of the organization should be clear about its concept, definition, and measurement as it applies to his or her job. As will be discussed, it may be entirely appropriate for quality to be defined or perceived differently in the same company, depending on the particular phase of the product life cycle.

Harvard professor David Garvin, in his book *Managing Quality,*[20] summarized five principal approaches to defining quality: transcendent, product based, user based, manufacturing based, and value based.

People from around the world travel to view the *Mona Lisa* or Michaelangelo's *David*, and most would agree that these works of art represent quality. But can they define it? Those who hold the *transcendental* view would say, "I can't define it, but I know it when I see it." Advertisers are fond of promoting products in these terms. "Where shopping is a pleasure" (supermarket), "We love to fly and it shows" (airline), "The great American beauty ... It's elegant" (automobile), and "It means beautiful eyes" (cosmetics) are examples. Television and print media are awash with such undefinable claims, and therein lies the problem: quality is difficult to define or to operationalize. It thus becomes elusive when using the approach as a basis for competitive advantage. Moreover, the functions of design, production, and service may find it difficult to use the definition as a basis for quality management.

Product-based definitions are different. Quality is viewed as a quantifiable or measurable characteristic or attribute. For example, durability or reliability can be measured (e.g., mean time between failure, fit, and finish), and the engineer can design to that benchmark. Quality is determined objectively. Although this approach has many benefits, it has limitations as well. Where quality is based on individual taste or preference, the benchmark for measurement may be misleading.

User-based definitions are based on the idea that quality is an individual matter, and products that best satisfy their preferences (i.e., perceived quality) are those with the highest quality. This is a rational approach but leads to two problems. First, consumer preferences vary widely, and it is difficult to aggregate these preferences into products with wide appeal. This leads to the choice between a niche strategy (see later) or a market aggregation approach which tries to identify those product attributes that meet the needs of the largest number of consumers.

Another problem concerns the answer to the question: "Are quality and customer satisfaction the same?" The answer is probably not. One may admit that a Lincoln Continental has many quality attributes, but satisfaction may be better achieved with an Escort. One has only to recall the box office success of recent motion pictures that suffer from poor quality but are evidently preferred by the majority of moviegoers.

Manufacturing-based definitions are concerned primarily with engineering and manufacturing practices and use the universal definition of "conformance to requirements." Requirements, or specifications, are established by design, and any deviation implies a reduction in quality. The concept applies to services as well as products. Excellence in quality is not necessarily in the eye of the beholder but rather in the standards set by the organization. Thus, both Cadillac and Cavalier possess quality, as do Wal-Mart and Bloomingdale's, as long as the product or service "conforms to requirements."

This approach has a serious weakness. The consumer's perception of quality is equated with conformance and hence is internally focused. Emphasis on reliability in design and manufacturing tends to address cost reduction as the objective, and cost reduction is perceived in a limited way — invest in design and manufacturing improvement until these incremental costs equal the costs of non-quality such as rework and scrap. This approach violates Crosby's concept of "quality is free" and is examined further in Chapter 11.

Value-based quality is defined in terms of costs and prices as well as a number of other attributes.[21] Thus, the consumer's purchase decision is based on quality (however it is defined) at an acceptable price. This approach is reflected in the popular *Consumer Reports* magazine, which ranks products and services based on two criteria: quality and value. The highest quality product is not usually the best value. That designation is assigned to the "best-buy" product or service.

Which Approach(es)?

Which definition or concept of quality should be adopted? If each function or department in the company is allowed to pursue its own concept, potential conflicts may occur:

Function	*Quality concerns*
Marketing	Performance, features, service, focus on customer concerns User-based concerns that raise costs
Engineering	Specifications Product-based concerns
Manufacturing	Conformance to specifications Cost reduction

Adopting a single approach could lead to cost increases as well as customer dissatisfaction. Each function has a role to play, but it cannot be played in isolation. A blend is needed to coordinate meeting each of the concerns listed.

Market Segmentation (Niche) Quality

Quality means different things to different people. In terms of strategic quality management, this means that the firm must define that segment of the industry, that generic strategy, and that particular customer group which it intends to pursue. This can be called a segmented quality strategy. The big three automobile manufacturers have wide product lines, each of which is marketed to a different part of the market and each with differing quality attributes.

Recent efforts to codify the concepts of quality and provide baselines for measurement have yielded the characteristics listed in Table 4-1. None of these dimensions stands alone. Differentiation may depend on one or more or a combination, but the point is that when differentiating based on quality, quality must be defined in terms that meet customer expectations, even if this is only what the customer perceives as quality.

A survey of purchasers of consumer products by the American Society for Quality Control summarized the factors influencing decisions to purchase (on a scale of 1 to 10) (Table 4-2).

Objectives

Management statesman Peter Drucker has said, "a company has but one objective: to create a customer." Following this statement, he proceeded to popularize the concept of management by objectives (MBO) and identified eight key areas for which objectives must be set: (1) marketing, (2) innovation, (3) human organization, (4) financial resources, (5) physical resources, (6) productivity, (7) social responsibility, and (8) profit requirements.[22] These types of objectives have been widely adopted by industry.

Table 4-1 Measurement of Quality

Category	*Example*
Performance	On-time departure of aircraft Acceleration speed of automobile
Features	Remote control for stereo Double coupons at the supermarket
Reliability	Absence of repair during warranty Thirty-minute pizza delivery
Conformance	Supplier conforms to specifications Cost of performance failures
Durability	Maytag's ten-year transmission warranty Mean time between failures
Serviceability	Consumer "hot line" for repair information Time to answer the telephone for reservation or complaint
Aesthetics	Restaurant ambiance Perfume fragrance
Perceived quality	Japanese vs. American automobiles Doctor A is better than Doctor B

Table 4-2 Factors in Decisions to Purchase

Factor	*Mean*
Performance	9.37
Lasts a long time	9.03
Easy to repair	8.80
Service	8.62
Warranty	8.13
Price	8.11
Ease of use	8.09
Appearance	7.54
Brand name	6.09

Within these eight broad areas, a company can set more specific objectives to identify the ends it hopes to achieve by implementing a strategy. Marketing becomes market share, innovation becomes new products, financial resources becomes capital structure, productivity becomes output per employee, profitability becomes return on investment or earnings per share, and so on.

Here, the question of quality becomes blurred. Is it a mission or an objective? It hardly matters if it is woven into the fabric of company strategy. If quality is chosen as central to a mission, other objectives begin to fall into place. For example, cycle time reduction, cost reduction, competitive standing, and return to shareholders can be related to the central mission.

> Digital Equipment Corporation launched a TQM program in order to tie together various efforts scattered throughout the company. The goal is to have a consistent company vision and language. Included is a six sigma program motivated by a desire to improve competitive position.[23]

Many quality improvement programs were started in the 1980s and 1990s in reaction to the increasing importance of quality and the need to compete for market share. Many companies failed, often because they had no action plan for implementing a strategy that was based on objectives, a prerequisite for follow-on operational planning.[24]

Supporting Policies

Policies are guidelines for action and decision making that facilitate the attainment of objectives. Taken together, a company's policies delineate its strategy fairly well. Tell me your policies and I can tell you your strategy.

The role of policies as a critical element of strategy is displayed in Figure 4-4, which can be called the *policy wheel*. In the center are the mission (the purpose of the organization), the differentiation (how to compete in the market), and the key objectives of the business. The spokes of the wheel represent the functions of the business. Each function requires supporting policies (functional strategies) to achieve the *hub*. If the firm's strategy calls for competing on quality, then this becomes the impetus for policy determination. Each functional policy supports this central strategy and the objectives that are determined during the planning process.

A firm's policy choices are essential as drivers of differentiation. They determine what activities to perform and how to perform them. Grey Poupon's advertising policy for its premium mustard sets the product

Figure 4-4 Policy Wheel

apart. Bic Pen's manufacturing policy of low-cost automation supports its low price. Avon's door-to-door distribution policy sets it apart. McDonald's policy of strict franchisee training and control allows it to retain its quality image. An airline's policy of "answering the phone on the third ring" reinforces a competitive edge of service.

Testing for Consistency of Policies

Assuming that the company has decided to make quality the central focus of its strategy, objectives are then set for profitability, growth, market share, innovation, productivity, etc. The test for consistency of supporting policies for a hypothetical firm is provided in Table 4-3. Of course, each policy is related to the hub and radiates from it. Like a wheel, the spokes must be connected.

CONTROL

The propensity of the U.S. manager to focus on short-term financial goals is well known. In its simplest and most prevalent form, the control system consists of setting financial standards (the budget), getting historical feedback on performance (the variance report), and trying to meet targets after deviations have occurred.

Table 4-3 Consistency of Supporting Policies

Function	*Illustration of policy*
Target market	Map the industry and seek out those segments where we have the advantage
Product line	Product line breadth is confined to those products where our value chain is appropriate for focus segment
Marketing	Market research to be directed toward defining customer expectations
Sales	Sales force hired and trained to promote our competitive edge
Distribution	Select distributors that complement our quality edge
Manufacturing	Invest in automation for improvement of quality and productivity
Supplier	Select suppliers that have applied for Baldrige Award Make life contracts
Human resources	Require skill and experience level for new hires Partnership relations with union
Research and development	Percentage of budget devoted to quality improvement Products designed for ease of repair
Finance	Service procedures in billing activity Financial arrangements with suppliers

Much has been written about the shortcomings of this approach. The major problem is the lack of focus on productivity (absolute, not financial measures), quality, and other strategic issues.

A system to control quality objectives, as distinct from quality on the shop floor, requires measures and standards designed for that purpose. Indeed, Juran suggests that the traditional control process may be put on hold while increasing the emphasis on quality planning and improvement.[25] Thus, planning and control of quality come together in an integrated system. The focus is on quality improvement set out in the planning process. The difference between traditional dollar accounting budgeting and the control of quality objectives is the participation of those who set standards and targets. Each function, department, or individual sets targets and provides real-time feedback as operations unfold.

SERVICE QUALITY

The differences between service and product quality are discussed in Chapter 1. This topic is examined further in Chapter 7 ("Customer Focus and Satisfaction"). It is both more difficult and yet simpler to plan and control service quality than it is to plan and control product quality. It is more difficult because measurement is elusive and production is frequently one-on-one. Like product quality, service quality should live up to expectation, but this can be a pitfall if too much service is promised.

Service quality may be more easily planned, provided objectives are defined and people committed. In any case, the payoff can be years away, and no service can overcome other weaknesses in a business.[26] The system for quality service also requires new approaches, such as restructuring incentives. In any case, a good beginning approach can be based on the Baldrige Award criteria, which are the same for both product and service. Process control in service industries is discussed further in Chapter 6.

SUMMARY

Quality has taken center stage as the main issue in both national and corporate competitive strategies. Those organizations that adopt quality as a differentiation and a way of organizational life will, over the longer term, pull ahead of competition. Achieving this goal is not easy. It is more than just issuing pronouncements and engaging in company promotion.

When an organization chooses to make quality a major competitive edge, it becomes the central issue in strategic planning — from mission to supporting policies. An essential idea is that the product is customer value rather than a physical product or service. Another concept that is basic to the process is the need to develop an organizational culture based on quality. Finally, no strategy or plan can be effective unless it is carefully implemented.

EXERCISES

4-1 Assume that an airline, a hotel, and a hospital have chosen quality for differentiation. Identify two or more measures of quality for a firm in each of these industries.

4-2 Illustrate a definition of:

Transcendental quality
Product-based quality
User-based quality
Value-based quality

4-3 Choose an industry and a product or service within that industry. Show how quality may differ for different segments or customer groups within that industry.

4-4 Is the objective of cost reduction in conflict with quality improvement? If so, illustrate how.

4-5 How can quality be reflected in the following?

Distribution policy
Human resources
Sales
Suppliers

4-6 Illustrate how trends in an industry can change a company's strategy.

4-7 What characteristics would be used to evaluate quality for the following products?

Ceiling fan
Bathing soap
Toothbrush

ILLUSTRATIVE CASES

Wawa Food Markets took pride in its old-fashioned values but realized that change was needed to meet the increasing competition in the convenience store market and the customer challenges in that industry segment. To stand out from the competition and develop a competitive advantage, Wawa Stores adopted the following goals: improve customer service and satisfaction, develop an organization culture that supports continuous improvement, reduce costs and bolster return on equity, prioritize strategic plans, and enable employees to contribute to the company's bottom line. The company adopted a customer-focused quality strategy that focuses on five vital issues, namely, strategic planning, team projects, education, supplier alliances, and customer information.

Questions

- Are the goals and issues too general in nature? Should they be quantified? Are they sufficiently specific to provide differentiation or a competitive advantage?
- How would you go about developing an organization culture that supports continuous improvement?

Following the deregulation of the trucking industry in the 1980s, the competition became more intense, and trucking firms adopted a number of strategies to survive. Southeast Freight Systems identified employee performance as a major area that needed improvement. The company wanted to transform its sales force into one that could attract business that was both of high quality and a good geographical fit and could provide exceptional service to all of Southeast's existing accounts. The company was convinced that achieving these goals for its sales professionals would help generate a competitive advantage. To operationalize this advantage, the company created career planning and professional incentive programs.

Questions

- How do you think a trucking company can differentiate its service in order to provide a sustained competitive advantage? Would the actions taken by Southeast provide such advantage?
- Which of the policies from Figure 4-4 would support such differentiation?

ENDNOTES

1. U.S. General Accounting Office, "Quality Management: Scoping Study," Washington, D.C.: U.S. General Accounting Office, Dec. 1990, p. 67.
2. U.S. General Accounting Office, "Quality Management: Scoping Study," Washington, D.C.: U.S. General Accounting Office, Dec. 1990, p. 15.
3. The details of Ford's transformation are contained in HBS Case 390-083, available from HBR Publications, Harvard Business School, Boston, MA 02163. See also Richard T. Pascale, *Managing on the Edge,* New York: Simon & Schuster, 1990.
4. Tom Peters, *Thriving on Chaos: Handbook for a Management Revolution,* New York: Knopf, 1987.
5. Remarks of A. F. Jacobson at the Conference Board Quality Conference in Dallas, April 2, 1990.
6. Eric Rolf Greenberg, "Customer Service: The Key to Competitiveness," *Management Review,* Dec. 1990, pp. 29–31.
7. J. M. Juran, "Made in USA — A Quality Resurgence," *Journal for Quality and Participation,* March 1991, pp. 6–8.
8. "Performance Leadership through Total Quality," a presentation made to the Conference Board Quality Conference, April 2, 1990. Two other Westinghouse divisions were runners-up for the Baldrige Award in 1989 and 1990.
9. Kenneth R. Andrews, *The Concept of Corporate Strategy,* New York: Dow Jones-Irwin, 1971.

10. Joseph L. Bower, Christopher A. Bartlett, C. Roland Christensen, Andrall E. Pearson, and Kenneth R. Andrews, *Business Policy: Text and Cases,* 7th ed., Homewood, Ill.: Irwin, 1991, p. 9.
11. Michael Porter, *Competitive Strategy: Techniques for Analyzing Industries and Competitors,* New York: The Free Press, 1980. See also his *Competitive Advantage: Creating and Sustaining Superior Performance,* New York: The Free Press, 1985, and *The Competitive Advantage of Nations,* New York: The Free Press, 1990.
12. Joel Ross and David Georgoff, "A Survey of Quality Issues in Manufacturing: The State of the Industry," *Industrial Management,* Jan./Feb. 1991.
13. Company brochure entitled "NCR Mission."
14. U.S. General Accounting Office, "Quality Management: Scoping Study," Washington, D.C.: U.S. General Accounting Office, Dec. 1990, p. 23. T. Boone Pickens, the quintessential LBO raider, was not very charitable to Goodyear. In a speech to the Strategic Planning Institute in Boston on October 23, 1989, he used chairman Robert Mercer as an example of corporate America in the early 1980s: "bloated, uncompetitive, bureaucratic and barely accountable to anyone…what I call the BUBBA syndrome."
14a. Southwest.com, 2004.
15. Elyse Allan, "Measuring Quality Costs: A Shifting Perspective," a presentation made to the Conference Board Quality Conference, April 2, 1990. *Global Perspectives on Total Quality,* Report Number 958, New York: The Conference Board, 1990, p. 35.
16. Michael Porter, *Competitive Strategy: Techniques for Analyzing Industries and Competitors,* New York: The Free Press, 1980, pp. 35–37.
17. J. M. Juran, "Strategies for World-Class Quality," *Quality Progress,* March 1991, pp. 81–85.
18. Armand V. Feigenbaum, "How to Implement Total Quality," *Executive Excellence,* Nov. 1989, pp. 15–16.
19. Y. K. Shetty and Joel Ross, "Quality and Its Management in Service Businesses," *Industrial Management,* Nov./Dec. 1985, pp. 7–12; Joel Ross and Y. K. Shetty, "Making Quality a Fundamental Part of Strategy," *Long Range Planning (UK),* Feb. 1985, pp. 53–58.
20. David A. Garvin, *Managing Quality,* New York: The Free Press, 1988, pp. 40–46.
21. In a survey of consumers' purchasing decisions conducted by the Gallup Organization, consumers were asked to rank (on a scale of 1 to 10) the importance of selected factors in the decision to purchase; 42% ranked price as 10. Other factors ranked as 10 were performance (72%), lasts a long time (58%), easily repaired (52%), service (50%), warranty (48%), ease of use (37%), appearance (28%), and brand name (15%). See *'88 Gallup Survey of Consumers' Perceptions Concerning the Quality of American Products and Services,* Milwaukee: American Society for Quality Control, 1988, p. 9.
22. Peter F. Drucker, *Management: Tasks, Responsibilities, Practices,* New York: Harper & Row, 1973, p. 100.
23. Rick Whiting, "Digital Strives for a Consistent Vision of Quality," *Electronic Business,* Nov. 26, 1990, pp. 55–56.

24. A. Blanton Godfrey, "Strategic Quality Management," *Quality,* March 1990, pp. 17–22.
25. J. M. Juran, "Universal Approach to Managing for Quality," *Executive Excellence,* May 1989, pp. 15–17. See also Bradley Gale and Donald J. Swmre, "Business Strategies that Create Wealth," *Planning Review,* March/April 1988, pp. 6–13. Traditional strategic planning based on financial measures is being called into question because it does not look beyond more important measures such as quality.
26. David Eva, "The Myth of Customer Service," *Canadian Business,* March 1991, pp. 34–39.

REFERENCES

Cruz, Clarissa, "Quality Program Softens Boundaries at Champion," *Purchasing,* Jan. 11, 1996, pp. 73–76.

Dean, James W. and Scott Snell, "The Strategic Use of Integrated Manufacturing," *Strategic Management Journal,* June 1996, pp. 459–480.

Feurer, Rainer, "Analysis of Strategy Formulation and Implementation at Hewlett-Packard," *Management Decision,* Vol. 4 Issue 3, Dec. 15, 1995.

Slater, Stanley F., "The Challenge of Sustaining Competitive Advantage," *Industrial Marketing Management,* Jan. 1996, pp. 79–86.

Voss, Bristol, "The Total Quality Corporation: How 10 Major Companies Turned Quality and Environmental Challenges to Competitive Advantage in the 1990s," *Journal of Business Strategy,* March/April 1996, p. 63.

5

HUMAN RESOURCE DEVELOPMENT AND QUALITY MANAGEMENT

At the heart of Total Quality Management (TQM) is the concept of intrinsic motivation. Empowerment — involvement in decision making — is commonly viewed as essential for assuring sustained results.

Healthcare Forum

The effective management of human resources is at the heart of any successful quality management process. The following questions underscore this point:

- What is the organization's record of success at finding the right people who would support or promote a quality culture?
- Is the organization able to retain the right people?
- Is the organization investing a sufficient amount of resources in professional development and training for staff?
- Are hiring and firing decisions and functions (at all levels) linked to the organization's mission, vision, and guiding principles?
- Does the organization value employee input and participation?
- Does the organization handle employee reward and recognition in a manner that complements the organization's mission, vision, and guiding principles?

Kaizen is a Japanese concept that means *continuous improvement.* Despite the perception of many U.S. managers that kaizen is not appropriate for American firms, there is abundant evidence that the concept is entirely in keeping with American values and norms. The approach offers a substantial potential for improvement if accompanied by an appropriate human resources effort. Indeed, it is becoming a maxim of good management that *human factors* are the most important dimension in quality and productivity improvement. People really do make quality happen.

Chief executive officers of some of America's most quality-conscious companies are quick to point out that the best way to achieve organization success is by involving and empowering employees at all levels. Some even say that employee empowerment is a revolution that will turn top-down companies into democratic workplaces.

> The whole employee involvement process springs from asking all your workers the simple question, "What do you think?"
>
> **Donald Peterson**
> *Former Chairman of Ford*

> To get every worker to have a new idea every day is the route to winning in the '90s.
>
> **John Welch, Chairman**
> *General Electric*

> The teams at Goodyear are now telling the boss how to run things. And I must say, I'm not doing a half-bad job because of it.
>
> **Stanley Gault**
> *Chairman*

Recall W. Edwards Deming's 14 points discussed in Chapter 1. The basis of his philosophy is contained in the following principles:

- Institute training on the job.
- Break down barriers between departments to build teamwork.
- Drive fear out in the workplace.
- Eliminate quotas on the shop floor.
- Create conditions that allow employees to have pride in their workmanship and abolish annual reviews and merit ratings.
- Institute a program of education and self-improvement.

Total quality management (TQM) has far-reaching implications for the management of human resources. It emphasizes self-control, autonomy, and creativity among employees and calls for greater active cooperation rather than just compliance.

INVOLVEMENT: A CENTRAL IDEA OF HUMAN RESOURCE UTILIZATION

> Back in 1987, the Ames Rubber Corporation decided to adopt a TQM strategy as a major change for implementing its determination to become more competitive. The executive committee identified its best and brightest managers and asked them to reorganize around functional processes. By 1992, every employee was assigned to an *involvement* group or team.

The human resource professional magazine *HR Focus* asked over 1000 readers to rate the key issues they faced in 1993. Employee involvement was rated as one of the top three concerns by 46% of the respondents. Customer service followed with 39% and TQM with 34%.[1]

At the heart of TQM is the concept of intrinsic motivation-involvement in decision making. Employee involvement is a process for *empowering* members of an organization to make decisions and to solve problems appropriate to their levels in the organization. The logic is that the people closest to a problem or opportunity are in the best position to make decisions for improvement if they have ownership of the improvement process. Empowerment is equally effective in service industries, where most frequently the customer's perception of quality stands or falls based on the action of the employee in a one-on-one relationship with the customer.

At Federal Express, the driver represents the company. He or she *is* the company and must deal directly with customer problems. Quality in an airline is represented not by CEOs and pilots, but by counter personnel and flight attendants.

> One of the more successful efforts to *empower* employees was the Astronautics Groups at Martin Marietta's Denver, Colorado operation (MMAG). The group instituted a TQM process. To build employee support, the group dropped its pyramid hierarchy of management in favor of a flatter structure and a more participative management approach. High-performance work teams were organized to *empower* people closest to the work to make decisions about how the work is performed. Aside from the substantial production area savings, less tangible benefits included improved morale.

Quality improvement can result from a reduction in cost or cycle time, an increase in throughput, or a decrease in variation within the process. In the past, the focus in achieving such improvement was frequently the *system* — traditional techniques and methods of quality control. Such a focus may overlook the fact that operation of the system depends on people, and no system will work with disinterested or poorly trained employees. The solution is simple: Coordinate the system and the people.

Contrast two production management styles in manufacturing industries. The "buffered" approach is characterized by large stocks of inventory and narrowly specialized workers. "Lean" systems, utilizing just-in-time (JIT) techniques, operate with small inventory stocks, multiskilled workers, and a team approach to work organization. Lean plants are more productive because they do not have valuable resources tied up in idle inventory. Plants are smaller and more efficient, with increased communication among departments, and workers tend to have a view of the organization as a whole.

Two examples of the lean approach involving worker participation are General Motors' New United Motor Manufacturing (NUMMI) plant (a joint venture with Toyota) and Dynatech's automotive test division. In both companies, *internalization* of the JIT philosophy and worker participation have increased worker pride and involvement on the shop floor. At GM, productivity levels are 40% higher than typical GM plants, and the plant has the highest quality levels GM has ever known. At Dynatech, cycle time was reduced by as much as 90% and setup by 67 to 100%.[1a]

ORGANIZING FOR INVOLVEMENT

Human resource professionals generally agree that a major shortcoming of human resource programs is a failure to match employee talent with organizational effectiveness. A strategy of empowerment must be operationalized through some organizational vehicle. A suggestion system is certainly not the total answer, despite the fact that many companies consider it to be an employee involvement program, and in many cases it is the only program.

Properly organized and administered small groups and teams are an effective motivational device for improving productivity and quality. They can reduce the overlap and lack of communication in a functionally based classical structure characterized by chain of command, territorial battles, and parochial outlooks. The danger always exists that functional specialists may pursue their own interests at the expense of the overall company mission or strategy. Team membership, particularly in a cross-functional team, reduces many of these barriers and encourages an integrative systems approach to achievement of common objectives — those that are common to both the company and the team or group. Consider the following success stories:

- Globe Metallurgical, the first small company to win the Baldrige Award, had a 380% increase in productivity that was attributed to self-managed work teams.
- Ford increased productivity by 28% by using a "partnering" concept that required a new corporate culture of participative management.
- At Decision Data Computer Corporation, middle management was trained to support "Pride Teams."
- Martin Marietta Electronic and Missiles Group achieved success with performance measurement teams.[1b]

Quality circles are perhaps the most widespread form of employee involvement teams. They are defined as a small group of employees doing similar or related work who meet regularly to identify, analyze, and solve product quality and production problems and to improve general operations. The concept has had some success in white-collar operations, but the major impact has been among "direct labor" employees in manufacturing, where concerns focus primarily on quality, cost, specifications, productivity, and schedules. Few cross-functional problems are considered because problem solving is generally confined to similar work areas.

Quality circles have not met the expectations that were set for them. As many as 50% of Fortune 500 companies have disbanded their circles. The major reason has been a general lack of commitment to the concept of participation and involvement and the lack of interest by management. Many middle managers perceived quality circles as a threat to their power and authority.

Task teams are a modification of the quality circle concept. The major differences are that the task teams can exist at any level and the goal is given to the team, whereas quality circles are generally free to choose the problems they will address.

Self-managing work teams are also an extension of the quality circle concept but differ in one major respect: Members are empowered to exercise control over their jobs and optimize the effectiveness of the total process rather than the individual steps within it. Team members perform all the tasks necessary to complete an entire job, such as setting up work schedules and making assignments to team members.

Cross-functional teams represent an attempt to modify the classic hierarchical form of an organization based on a vertical chain of command. They include horizontal coordination in order to plan and control processes that flow laterally. If no lateral coordination is achieved, the organization becomes a collection of islands of specialization, without integration of business processes that flow horizontally across the organizational chart. The concept of linking cross-functional processes is shown in Figure 3-1. Note that a cross-functional approach achieves the objectives of customer, functions, processes, and the total organization.

TRAINING AND DEVELOPMENT

Increased involvement means more responsibility, which in turn requires a greater level of skill. This must be achieved through training. Baldrige Award winners place a great deal of emphasis on training and support it with appropriate provision of resources. Motorola allocates about 2.5% of payroll costs or $120 million annually to training, 40% of which goes to quality training. The company calculates the training return at about $29 for each dollar invested. Additional benefits include (1) improved communications, (2) change in corporate culture, and (3) demonstration of management's commitment to quality. (Xerox has extended quality training to 30,000 supplier personnel.)[1c]

> Since the early 1980s, Hughes Aircraft has made quality one of its chief operating philosophies. The cornerstone of the company's TQM thrust is continuous measurable improvement (CMI). Recently, the firm has championed a unique "trickle-down" training system to sustain its quality and productivity improvements. Under CMI (Cascaded Measurable Input), the managers responsible for achieving improvement teach the philosophy and principles of CMI leadership throughout the organization.[2]

Although the type of training depends on the needs of the particular company and may or may not extend to technical areas, the one area that should be common to all organization training programs is *problem solving*. Problem solving should be institutionalized and internalized in many, if not most, companies. This would be a prerequisite to widespread empowerment.

Training usually falls into one of three categories: (1) reinforcement of the quality message[3] and basic skill remediation, (2) job skill requirements, and (3) knowledge about principles of TQM. The latter typically covers problem-solving techniques, problem analysis, statistical process control, and quality measurement — areas that go beyond typical job skills. If groups or teams are utilized, training in the group process and group decision making is included. According to a survey conducted by the Conference Board, top companies commonly address the following topics in quality training curricula:

- Quality awareness
- Quality measurement (performance measures/quality cost benchmarking, data analysis)
- Process management and defect prevention
- Team building and quality circle training
- Focus on customers and markets
- Statistics and statistical methods
- Taguchi methods

> Research Testing Laboratories, Inc., a TQM company providing clinical research services, encourages employees to make changes in processes in order to minimize and eliminate errors early in the work process. The goal is 100% customer satisfaction. To achieve this goal, employees are provided with a 25-hour training program in which they learn (1) effective interactive skills, (2) the problem-solving process, and (3) the quality improvement process.

Managerial training may take the form of the third item above (TQM principles). In addition, programs often are directed toward sensitizing individuals to the strategic importance of quality, the cost of poor quality, and their role in influencing the quality of products and services.

The International Quality Study was conducted among 584 companies representing four industries. The use of quality tools in the American auto industry is expected to increase 1.5- to 6-fold over the next three years. Quality training was found to have the greatest impact when coupled with other practices, such as measurement and reward systems.[4]

SELECTION

Selection is choosing from a group of potential employees (or placement from existing employees) the specific person to perform a given job. In theory, the process is simple: Decide what the job involves and what abilities are necessary, and then use established selection techniques (ability tests, personality tests, interviews, assessment centers) as indicators of how the candidate will perform.

The process is not so simple, however, when TQM enters the picture. The job requirements for a typist, a machinist, or even a manager can be determined by job analysis, and the qualifications of a candidate can be compared to these requirements. When a company commits to TQM, an entirely new dimension is introduced. The skills and abilities required for a specific job can usually easily be identified and then matched with an individual. People well suited for operating in a quality climate may require additional characteristics, such as attitude, values, personality type, and analytical ability.

Persons working in a quality environment need sharp problem-solving ability in order to perform the quantitative work demanded by statistical process control, Pareto analysis, etc. Because of the emphasis on teams and group process, personnel must function well in group settings. Motorola shows applicants videotapes of problem-solving groups in action and asks them how they would respond to a particular quality issue. Presumably this technique encourages *self-selection*.

What is perhaps different in the selection process in a TQM environment is the emphasis on a *quality-oriented organization culture* as the desired outcome of the selection process.[5]

PERFORMANCE APPRAISAL

The purpose of performance appraisal is to serve as a diagnostic tool and review process for development of the individual, team, and organization. Appraisals are used to determine reward levels, validate tests, aid career development, improve communication, and facilitate understanding of job duties.[6]

Deming cites *traditional* employee evaluation systems as one of seven deadly diseases confronting U.S. industry. He states that *individual* performance evaluations encourage short-term goals rather than long-term planning. They undermine teamwork and encourage competition among people for the same rewards. Moreover, the overwhelming cause of nonquality is not the employee but the system; by focusing on individuals, attention is diverted from the root cause of poor quality: the system.

Many TQM proponents, like Deming, argue that traditional performance appraisal methods are attempts by management to pin the blame for poor organization performance on lower level employees, rather than focusing attention on the system, for which upper management is primarily responsible.

Should individual performance appraisal be eliminated, as Deming suggests?[7] This is unlikely in view of the historical and widespread use of this human resource management tool. What, then, can be done to relate individual and group performance to a total quality strategy?

Performance appraisals are most effective when they focus on the objectives of the company and therefore of the individual or group. Because the eventual outcome of all work is quality and customer satisfaction, it follows that appraisal should somehow relate to this outcome — to the objectives of the company, the group, and the individual. In other words, a performance appraisal system should be aligned with the principle of shared responsibility for quality. This can be accomplished by focusing on development of the skills and abilities necessary to perform well and, as such, directly support collective responsibility.

> In a model used by the Hay Group (a consulting organization), individuals are evaluated for base pay on such variables as ability to communicate, customer focus, and ability to work as a team. Managers are rated on employee development, group productivity, and leadership. Variable pay for both is based on what is accomplished. Because customer focus is a critical part of any TQM effort, a three-category rating system that involves (1) not meeting customer expectations, (2) meeting them, and (3) far exceeding them is easy to implement.[8]

Answering Deming and the other critics is not easy. The integration of total quality and performance appraisal is necessary. One should reinforce the other. One approach might be to modify existing systems in accordance with the following principles:

- Customer expectations, not the job description, generate the individual's job expectation.
- Results expectations meet different criteria than management-by-objectives statements.
- Performance expectations include behavioral skills that make the real difference in achieving quality performance and total customer satisfaction.
- The rating scale reflects actual performance, not a "grading curve."
- Employees are active participants in the process, not merely "drawn in."

Regardless of which specific system is adopted, there seems to be little question that performance management practices need to be in line with and supportive of TQM.

COMPENSATION SYSTEMS

This may be one of the most elusive and controversial of all systems that support TQM. Historically, compensation systems have been based on (1) pay for performance or (2) pay for responsibility (a job description). Each of these is based on individual performance, which creates a competitive atmosphere among employees. In contrast, the TQM philosophy emphasizes flexibility, lateral communication, group effectiveness, and responsibility for an entire process that has the ultimate outcome of customer satisfaction. No wonder research and writing have offered little in the way of new approaches that are more in tune with the needs of TQM.

> Shawnee Mission (Kansas) Medical Center attempted to set up an infrastructure to push TQM ideals throughout the organization. In 1992 the center operationalized its new evaluation system based on personal development, education, and teamwork. Everyone received the same raise.

Both training and performance appraisal are desirable components of a TQM implementation strategy, but compensation is an equally necessary dimension. Employees may perceive the system as a reflection of the company's commitment to quality.

Individual or Team Compensation?

A company's infrastructure, specifically its reward and compensation systems, provides an accurate picture of its strategic goals. If compensation criteria are focused exclusively on individual performance, a company will find that initiatives promoting teamwork may fail. A TQM vision and the principles supporting it are unlikely to take hold unless the values on which they are based are built into the underlying structure.

> Target Stores is among the growing number of companies in the retail industry that are going beyond logistics-specific performance measures and are tying pay into the effectiveness of TQM programs. Throughout the logistics field, pay for performance and pay for quality appear to be becoming more intertwined.

There are several compensation plans in U.S. industry, including gain sharing, profit sharing, and stock ownership. These are among the systems designed to create a financial incentive for employees to be involved in performance improvements. Gain sharing is one of the most rapidly growing compensation and involvement systems in U.S. industry. It is a system of management in which an organization seeks higher levels of performance through the involvement and participation of its people. Employees share financially in the gain when performance improves. The approach is a team effort in which employees are eligible for bonuses at regular intervals on an operational basis. Gain sharing reinforces TQM, partially because it contains common components, such as involvement and commitment.[9]

The jury is still out on the effectiveness of these plans, but evidence suggests that effectiveness is a function of strong communication programs and widespread employee involvement.

Summary

Many reasons have been offered as the cause of poor performance in organizations:

- System failure
- Misunderstanding of job expectations
- Lack of awareness about performance
- Lack of time, tools, or resources to succeed
- Lack of necessary knowledge or skills
- Lack of appropriate consequences for performance
- Bad fit for the job

Although a compensation system supportive of TQM is not the only remedy, combined with other human resource management systems it will go a long way toward improvement of performance and development among individuals, groups, and the organization.

TOTAL QUALITY ORIENTED HUMAN RESOURCE MANAGEMENT

Human resource executives are faced with both a challenge and an opportunity. They are not generally perceived with the same regard as line managers. Philip Crosby describes the human resource department as behind the times and the human resource executive as his or her own worst enemy. On the other hand, the department can play a critical role in the implementation of a holistic quality environment in support of a strategic initiative. To accomplish this role, the function should not only be designed to support TQM throughout the organization, but should make sure that good quality management practices are followed within the processes of the function itself. This means continuous improvement as a way of department life. Bowen and Lawler suggest putting the following principles of TQM to work *within* the human resource department:[10]

- Quality work the first time
- Focus on the customer
- Strategic holistic approach to improvement
- Continuous improvement as a way of life
- Mutual respect and teamwork

It is evident that some modification of traditional human resource management practices is required if the function is to support the TQM program throughout the company. Planning is the first step. The 1993 Baldrige Award criteria describe human resource planning:[11]

> Human resource plans might include the following: mechanisms for promoting cooperation such as internal customer/supplier techniques or other internal partnerships; initiatives to promote labor–management cooperation, such as partnerships with unions; creation and/or modification of recognition systems; mechanisms for increasing or broadening employee responsibilities; creating opportunities for employees to learn and use skills that go beyond current job assignments through redesign of processes; creation of high performance work teams; and education and training initiatives. Plans might also include forming partnerships with educational institutions to develop employees or to help ensure the future supply of well-prepared employees.

EXERCISES

5-1 Would a quality improvement program based on process control be more appropriate for employee involvement than a system based on traditional production methods? If so, explain why.

5-2 What effect does employee involvement have on motivation? Explain the effect in terms of motivational theory.

5-3 What is the impact of low employee retention on an organization's quality?

5-4 Contrast the benefits of the different types of small groups or teams. Which would be more appropriate for achieving integration across organizational functions or departments?

5-5 A Deming principle advises to "create conditions that allow employees to have pride in their workmanship." What are these conditions and how can they be implemented?

5-6 Assume that a company has just committed to change from a traditional style of management to one based on TQM. What topics would you include for:

Shop floor employees
Front-line supervisors
Middle-level managers

5-7 Describe how training in problem solving would improve:

Process control
Employee motivation

ILLUSTRATIVE CASES

Norton Manufacturing Company contracted with a local vocational school to have an instructor based at its tool-and-die plant to offer about two hours of instruction each week to each of the three shifts. Classes include machine-shop math, basic blueprint reading, and statistical process control. Another firm, Taco, Inc. (a heating and cooling equipment manufacturer), believes that there can be a payoff from a training program if it helps create a bond between employer and employee and encourages valued employees to stay. The company provides training in job skills as well as in areas such as art appreciation, gardening, and aerobics.

Question

- Contrast the "hard skills" training such as statistical process control with "soft" training provided by Taco. Should training include both as well as strategic or philosophical training in the concepts of TQM?

The Telecommunications Products Division (TPD) was formed by Corning Incorporated to commercialize its revolutionary optical-fiber product and process technology. TPD was one of only two 1995 Baldrige Award winners (the other was the Building Products Division of Armstrong World Industries). TPD's commitment to its human resource development and management system includes four basic components:

1. ***Planning*** — Develop and maintain a human resource strategy as a fully integrated functional element of the overall business strategy plan.
2. ***High-performance work systems*** — Includes cross-functional teams and employee-designed work teams empowered to make decisions to resolve customer concerns and encouraged to take initiative in preventing and solving problems. Compensation reward and recognition systems complement work teams.
3. ***Education, training, and development*** — Tools and programs that provide for development of competencies required for currently assigned jobs; those required to accomplish division, unit, and work group objectives; and those required for future growth in responsibilities for cultural change.
4. ***Employee satisfaction and well-being*** — Ensuring a safe and healthful environment, providing an array of employee support services, and measuring and continuously improving employee satisfaction.

Question

- Compare TPD's human resource system with that of a company with which you are familiar.

ENDNOTES

1. *HR Focus,* Jan. 1993, pp. 1, 4.

1a. Cameron, Iain and Hilary Duckett, "New Systems Implementation — A Human Resource Approach," *Logistics Focus,* Jan./Feb. 1996, pp. 4–5.

1b. Galvin, Robert W., "Knowledge Makes the Difference at Motorola," *Strategy and Leadership,* March/April 1996, pp. 42–44.

1c. Kochanski, James and Donald Ruse, "Designing a Competency-Based Human Resources Organization," *Human Resource Management,* Spring 1996, pp. 19–33.

2. Judy Rice, "Cascaded Training at Hughes Aircraft Helps Ensure Continuous Measurable Improvement," *National Productivity Review,* Winter 1992/93, pp. 111–116.

3. Bernie Knill, "The Nitty-Gritty of Quality Manufacturing," *Materials Handling Engineering,* July 1992, pp. 40–42. In a Conference Board survey, training is first used to reinforce the quality message and then to build skills. Another finding of the survey is that leaders link TQM to performance review and compensation.

4. Trace E. Benson, "When Less Is More," *Industry Week,* Sep. 7, 1992, pp. 68–77.

5. David E. Bowen and Edward E. Lawler III, "Total Quality-Oriented Human Resource Management," *Organization Dynamics,* Spring 1992, pp. 29–41.

6. David E. Bowen and Edward E. Lawler III, "Total Quality-Oriented Human Resource Management," *Organization Dynamics,* Spring 1992, p. 36.

7. Some articles that treat performance appraisal in a TQM context include Kathleen A. Guinn, "Successfully Integrating Total Quality Management and Performance Appraisal," *Human Resource Professional,* Spring 1992, pp. 19–25; Mike Deblieux, "Performance Reviews Support the Quest for Quality," *HR Focus,* Nov. 1991, pp. 3–4; Jean B. Ferketish and John W. Hayden, "HRD & Quality: The Chicken or the Egg?" Jan. 1992, pp. 38–42.

8. Linda Thornburg, "Pay for Performance: What You Should Know (Part 1)," *HR Magazine,* June 1992, pp. 58–61.

9. Robert L. Masternak, "Gainsharing at B. F. Goodrich: Succeeding Together Achieves Rewards," *Tapping the Network Journal,* Fall/Winter 1991, pp. 13–16.

10. David E. Bowen and Edward E. Lawler III, "Total Quality-Oriented Human Resource Management," *Organizational Dynamics,* Spring 1992, p. 29.

11. "Malcolm Baldrige National Quality Award. 1993 Award Criteria," Gaithersburg, Md.: U.S. Department of Commerce, National Institute of Standards and Technology, 1993, p. 21.

REFERENCES

Konczak, Lee J., "Creating High Performance Organizations: Practices and Results of Employee Involvement and Total Quality Management in Fortune 1000 Companies," *Personnel Psychology,* Summer 1996, pp. 495–499.

Mohrman, Susan, Edward Lawler, and Gerald Ledford, "Do Employee Involvement and TQM Programs Work?" *Journal of Quality and Participation,* Jan./Feb. 1996, pp. 6–10.

Morris, Linda, "Training, Empowerment and Change," *Training and Development,* July 1996, p. 54.

6

MANAGEMENT OF PROCESS QUALITY

> A Deming-style "total quality management" approach to improving service quality is rooted in the unglamorous and never fashionable discipline of statistics. Using Dr. Deming's statistical approach to total quality management, we have reduced service expenses 35% over the past 12 months while improving service quality.
>
> **President**
> *Savin Copiers*

The need for top management to display leadership in setting the climate and culture for total quality management (TQM) is outlined in Chapter 2. Climate and culture, however, are not enough. It is unlikely that exhortations and slogans will be effective unless accompanied by action planning and implementation. A statement such as "We Are the Quality Company" convinces no one — not the employees and not the customers. The company should be organized for quality assurance in the context of modern quality management.

Assume that the criteria of the Baldrige Award fairly represent what is generally accepted as the national standard for management of process quality:

> The *Management of Process Quality* category examines systematic processes the company uses to pursue ever-higher quality and company operational performance. The key elements of process management are examined, including research and development, design, management of process quality for all work units and suppliers, systematic quality improvement, and quality assessment.

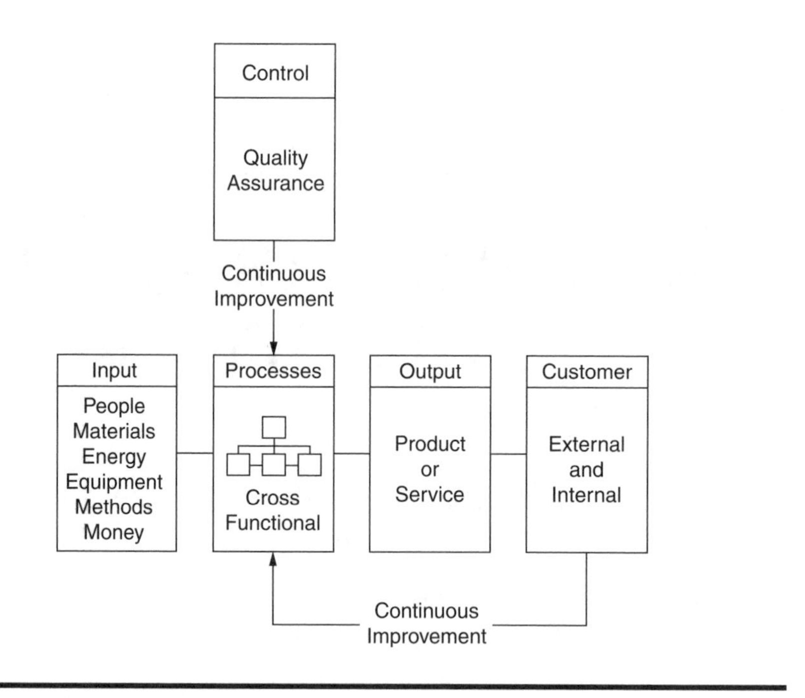

Figure 6-1 Management System

It is apparent that this definition is directly related to how well the *processes* are managed — *all* of the processes in the organization that contribute directly or indirectly to quality as the customer defines it. The concept is illustrated in Figure 6-1. Note that the control component (quality assurance) has moved from measuring output (the traditional control system) to managing the *continuous improvement of the process.*

The traditional approach to quality control was inspection of the final product, and this approach is still practiced by many firms. It is now abundantly clear that quality cannot be inspected into a product; it must be built into it. This chapter will introduce methods and techniques that are significantly more advanced and more effective than the practice of "final inspection," which has been used for so long. Although the concepts in this chapter are not the last word in modern TQM, they represent substantial potential for improving quality, cost, and productivity in almost any company.

A BRIEF HISTORY OF QUALITY CONTROL

Concern for product quality and process control is nothing new. Historians have traced the concept as far back as 3000 B.C. in Babylonia. Among the references to quality from the Code of Hammurabi, ruler of Babylonia, is the following excerpt: "The mason who builds a house which falls down and kills the inmate shall be put to death." This law reflects a concern for quality in antiquity.[1] Process control is a concept that may have begun with the pyramids of Egypt, when a system of standards for quarrying and dressing of stone was designed. One has only to examine the pyramids at Cheops to appreciate this remarkable achievement. Later, Greek architecture would surpass Egyptian architecture in the area of military applications. Centuries later, the shipbuilding operations in Venice introduced rudimentary production control and standardization.

Following the Industrial Revolution and the resulting factory system, quality and process control began to take on some of the characteristics that we know today. Specialization of labor in the factory demanded it. Interchangeability of parts was introduced by Eli Whitney when he manufactured 15,000 muskets for the federal government. This event was representative of the emerging era of mass production, when inspection by a skilled craftsman at a workbench was replaced by the specialized function of inspection conducted by individuals not directly involved in the production process.

Specialization of labor and quality assurance took a giant step forward in 1911 with the publication of Frederick W. Taylor's book *Principles of Scientific Management*.[2] This pioneering work had a profound effect on management thought and practice. Taylor's philosophy was one of extreme functional specialization and he suggested eight functional bosses for the shop floor, one of whom was assigned the task of inspection:

> The inspector is responsible for the quality of the work, and both the workmen and the speed bosses [who see that the proper cutting tools are used, that the work is properly driven, and that cuts are started in the right part of the piece] must see that the work is finished to suit him. This man can, of course, do his work best if he is a master of the art of finishing work both well and quickly.[3]

Taylor later conceded that extreme functional specialization has its disadvantages, but his notion of process analysis and quality control by inspection of the final product still lives on in many firms today. Statistical quality control (SQC), the forerunner of today's TQM or total quality control, had its beginnings in the mid-1920s at the Western Electric plant of the Bell System. Walter Shewhart, a Bell Laboratories physicist, designed the original version of SQC for the zero defects mass production of complex telephone exchanges and telephone sets. In 1931 Shewhart published his landmark book *Economic Control of Quality of Manufactured Product*.[4] This book provided a precise and measurable definition of quality control and developed statistical techniques for evaluating production and improving quality. During World War II, W. Edwards Deming and Joseph Juran, both former members of Shewhart's group, separately developed the versions used today.

It is generally accepted today that the Japanese owe their product leadership partly to adopting the precepts of Deming and Juran. According to Peter Drucker, U.S. industry ignored their contributions for 40 years and is only now converting to SQC.[5]

> The Willimatic Division of Rogers Corporation, an IBM supplier, uses just-in-time techniques along with X-bar and R charts for key product attributes to achieve statistical process control. Rework is reduced by 40%, scrap by 50%, and productivity is increased by 14%.[6]

PRODUCT INSPECTION VS. PROCESS CONTROL

> Structure follows strategy.
>
> Nothing happens until a sale is made.
>
> If you can't measure it, you can't manage it.

These statements are typical of the popular catchphrases adopted by particular functions (e.g., planning, sales, accounting) within the business. The popularity of the expression usually means that there is a measure of truth behind it. Truisms in the field of quality management include "don't inspect the product, inspect the process" and "you can't inspect it in, you've got to build it in."

There is sound thinking behind these two statements. In the previous discussion of the control process, the point was made that controlling the output of the system *after the fact* was historical action, and nothing could be done to correct the variation after it had already occurred. This is feedback control. The same is true of inspecting the product. The variation or the defect has already occurred. What is needed is a feedforward system that will prevent defects and variations. Better yet is a system that will improve the process. This is the idea behind process control (Figure 6-1).

What is the process? Does it begin with material inspection at the receiving dock and end with final inspection, or does it begin with design and end with delivery to the customer? Does it begin with market research and end with after-sale service? If we take the broader view, the process might begin with the concept of the product idea and extend through the life cycle of the product to ultimate maturity and phaseout. This definition matches the concept of TQM.

It is clear that in the philosophy of TQM, most (if not all) business functions and activities (i.e., processes) are interrelated and none stands alone — not purchasing, engineering, shipping, order processing, or manufacturing. Key business objectives and organization success are dependent on cross-functional processes. Moreover, these processes must change as environments change. The conclusion emerges that true process optimization requires the application of tools and methods in all activities, not just manufacturing.

Historically, there have been two major barriers to effective process control. The first has been the tendency to focus on volume of output rather than quality of output. Volume of production has been the major objective in the mistaken notion that more units of output means lower unit cost. Another barrier is the quality control system that measures products or service against a set of internal conformance specifications that may or may not relate to customer expectations. The result in many cases has been inferior quality products that are reworked or scrapped or, worse, products that customers *did not buy.* As will be discussed in Chapter 11, the cost of poor quality can amount to 25 to 30% of sales revenues. The profit potential in quality improvement is greater than simply improved production of inferior quality.

> Bytex Corporation manufactures electronic matrix switches for Citicorp, MasterCard, American Express, and others. The company has focused on understanding the process, concentrating on eliminating non-value-added transactions. Cycle time is down by 60%, inventory down by 43%, final assembly time down by 52%, and floor space down by 30%. The resulting product is superb.[7]

MOVING FROM INSPECTION TO PROCESS CONTROL

Process control may still require measurement that is determined by inspection, but the activity of inspection is now transformed into a diagnostic role. The objective is not merely to discover defects, but rather to identify and remove the cause(s) of defects or variations. Process control now becomes problem solving for *continuous improvement.* Moving from inspection to process control takes place in steps or phases:

Step	Action
1	Characterize process Define process requirements and identify key variables
2	Develop standards and measures of output Involve work force
3	Monitor compliance to standards and review for better control Identify any additional variables that affect quality
4	Identify and remove cause(s) of defects or variations (this requires a step-by-step documentation of the process and process control charting)
5	Achieve process control with improved stability and reduced variation

STATISTICAL QUALITY CONTROL

Statistical Quality Control (SQC) is the oldest and most widely known of the several process control methods. It involves the use of statistical techniques, such as control charts, to analyze a work process or its outputs. The data can be used to identify variations and to take appropriate actions in order to achieve and maintain a state of statistical control (predetermined upper and lower limits) and to improve the capability of the process. It is the best-known innovation among Deming's ideas.

Rigorously applied, SQC can virtually eliminate the production of defective parts.[8] By identifying the quality that can be expected from a given production process, control can be built into the process itself. Moreover, the method can spot the causes of variations — incoming materials, machine calibration, temperature of soldering iron, or whatever.

Despite the maturity of the method and its proven benefit, many firms do not take full advantage of it. One survey found that 49% of responding electronic manufacturers reported using SQC techniques, but 75% of them also continued to use traditional 100% inspection. This is in an industry where quality in the manufacturing process is essential.

> At Motorola, SQC has been integrated into the corporate culture and is being applied in all areas of the plant. Steps to place a process under statistical control include (1) characterizing the process, (2) controlling it, and (3) adjusting the process when non-random deviations are observed. Six sigma is the goal.

Statistical Process Control (SPC) is the companion to SQC. The term *statistical process control* can be misleading because it is so frequently confined to manufacturing processes, whereas the methods can be useful for improving results in other non-manufacturing areas such as sales and staff activities. Moreover, the methods can be used in many of the activities and functions of service industries. It is also worth noting that the only universal technique for SQC is logical reasoning applied to the improvement of a process. Thus it is a systematic way of problem solving.

A *process* is a set of causes and conditions and a set of steps comprising an activity that transforms inputs into outputs. Consider the number of processes involved in the airline industry: the process of taking and confirming a reservation, of baggage handling, of loading passengers, of meal service, etc. The process is any set of people, equipment, procedures, and conditions that work together to produce a result — an output.

The process is expected to add value to the inputs in order to produce an output. The ratio of output to input is called productivity and the objectives are to (1) increase the ratio of output to input and (2) reduce the variation in the output of the process. If the variation is too small or insignificant to have any effect on the usefulness of the product or service, the output is said to be within tolerance. Should the output fall outside the desired tolerance, the process can be improved and returned to tolerance by defining the cause of the change (the problem) and taking action to make sure that the cause does not recur.

BASIC APPROACH TO STATISTICAL QUALITY CONTROL[9]

SQC and its companion, *statistical process control* (SPC), were developed in the United States in the 1930s and 1940s by W. A. Shewhart, W. E. Deming, J. M. Juran, and others. These techniques (some call them philosophies) have been used for decades by some U.S. firms and many Japanese companies. Despite the proven effectiveness of the techniques, many U.S. firms are reluctant to use them.[10]

The approach is designed to identify the underlying causes of problems which cause process variations that are outside predetermined tolerances and to implement controls to fix the problems. The basic approach contains the following steps:

1. Awareness that a problem exists.
2. Determine the specific problem to be solved.
3. Diagnose the causes of the problem.
4. Determine and implement remedies to solve the problem.
5. Implement controls to hold the gains achieved by solving the problem.

The Deming Cycle

The Deming cycle is an approach that provides a systematic framework for continuously improving a process. The Deming cycle is often referred to as the PDSA (Plan-Do-Study-Act) cycle. First, a plan is developed, based on the careful definition of the problem, an understanding of the process, data collection, and analysis. At the end of the "Plan," an alternative solution for improvement is developed. In the "Do" phase, the plan is tested on a trial basis through designed experiments. The outcomes of the experiments are evaluated (study); and appropriate steps are taken on the process (act). These steps can lead to a modification of the plan in a never-ending cycle of improvement.

MANUFACTURING TO SPECIFICATION VS. MANUFACTURING TO REDUCE VARIATIONS

Among production managers who manufacture to specifications or those who depend upon final inspection, the common problem can be traced to the control loop. Defect statistics are generated by inspection, but appropriate action is not taken to define problems, determine cause(s), and correct variations. Companies continue to live with a reject rate that is considered to be "normal," as typified by statements such as "We can live with *X*% defectives" or "that's fairly common in the industry."

The benefit of manufacturing to reduce variations (process control) is generally recognized.[11] It is the purpose of SQC to *identify* and *reduce* variations from standard and *continuously improve* the process until a theoretical condition of "zero defects" is achieved.[12] The causes of variations are many and vary from industry to industry. Common sources include (1) material balance disturbances, (2) energy balance changes, (3) process instabilities, (4) equipment failure and wear, and (5) poor control loop performance.[13] SQC is used to develop control limits for each step within the process. Measuring sample parts and graphing trends leads to identification of the cause(s) of any erratic (non-random) behavior in the process.

The objective of process control is not only production of quality output, but reduction of costs as well. Quality is defined as the total acceptable variation divided by the total actual variation or *Cp* index. When used alone, this measure may be misleading because it assumes acceptable quality product design.[14] This, of course, is not always the case and suggests the need for the cross-functional process control mentioned earlier.

Data acquisition and monitoring is an essential step if the process is to remain in control. This tracking is generally accomplished by the operator concerned. In more sophisticated plants, particularly in unattended manufacturing, the goal is to have in-process measurement and correction in real time through the use of sensors or other measuring devices.[15] Devices such as bar code readers, vision systems, and counters are some of the tools available for collection of data. Of course, data alone is not enough. Data must be organized in such a way that process decisions can be made.

PROCESS CONTROL IN SERVICE INDUSTRIES

Examination of the U.S. Government Standards Industrial Classification of Industries suggests many industries in which the use of SPC would be appropriate. Use of the techniques is spreading to such industries as transportation,[16] healthcare, and banking.[17]

To some extent, the service process is more difficult to control than manufacturing because quality is typically measured at the customer interface, when it is already too late to fix the problem. Hence, "final inspection" will always be a part of the process; the customer serves as the inspector.

Service failures are analogous to bad parts in manufacturing, and measures of service may be compared to manufacturing tolerances or standards. SPC can be used to measure consistency of service and determine causes of deterioration from prescribed standards and the cause(s) of variations. In transportation, the cause may be missed appointments, refusals, or weekend closures.[18] At the First National Bank of Chicago, a number of processes are checked weekly against over 500 customer-sensitive measures.[19]

> L.L. Bean, a mailiorder company in Freeport, Maine, is known worldwide for its outstanding distribution system. It is the ideal company to benchmark for that function. The company analyzed all key activities and processes, including benchmarking competitors. It is ranked number 1 in virtually every product category in which it is evaluated by outside sources.[20]

Customer Defections: The Measure of Service Process Quality

Measures of output, as the customer defines them, are not too difficult to identify in service firms. An airline can measure on-time departures and the time it takes to make a reservation. A bank can measure the ratio of ATM downtime to total number of ATM minutes available and so on. Measures such as these are necessary, but the most important measure is *customer defections* or customers lost to the competition.

What is the cost of a customer defection? Conversely, what is the value of a customer retention? Defections have a substantial effect on profits and cost, more so than market share, economies of scale, or unit costs. Simply put, losing a customer costs money and retaining one makes money.

The initial cost of acquiring a new customer involves a number of one-time costs for prospecting, advertising, records, and such. Banks, attorneys, mutual funds, and credit card companies are examples of firms that spend to recruit a customer and establish an account. However, once a relationship is established, the marginal cost of each additional dollar of sales *diminishes* — provided the customer does not defect.

Improving the processes and reducing the process variations that reduce customer defections can be perceived not as a cost but as an investment. Consider the following examples:

- Taco Bell calculates that the lifetime value of a retained customer is $11,000.
- An auto dealer believes that the lifetime value of retaining a customer is $300,000 in sales.
- MBNA America, a credit card company, has found that a 5% improvement in defection rates increases its average customer value by 125%.

PROCESS CONTROL FOR INTERNAL SERVICES

Until it moved to Raleigh, North Carolina, IBM's personal computer assembly operation was located at its plant in Boca Raton, Florida. The general manager was committed to internal as well as external quality. In support of this commitment, the following policy was adopted, widely disseminated, and implemented through "Excellence Plus" groups:

Excellence Plus Commitment

IBM Boca Raton will deliver defect-free, competitive products and services, on time, to all customers. Quality will be the primary consideration in all decisions related to cost and delivery. *Likewise, each department will provide defect-free work to the next user of its output or service* (italics added).

An inventory of the many functions and activities in an organization will reveal that each activity is responsible for the operations of one or more processes where the customer is an *internal* user of its output or service. Many, if not most, of these processes lend themselves to process control methods.

AT&T's support services organization in Chicago is responsible for word processing and reprographics. Through SPC, a fivefold improvement in typing accuracy and a halving of turnaround time in reprographics was achieved. Most of the gain was attributed to better communications with customers.[21]

QUALITY FUNCTION DEPLOYMENT

For centuries, and even today, navies have built ships in the same process sequence:

Design → Build hull and launch → Outfit →
Trial run → Return to shipyard → Rework →
Operational check → Return → Fix → Operational

This sequence in modern construction of ships and other weapon systems almost always results in time and cost overruns and subsequent operational deficiencies. This is evidence of inadequate process control, which may change as a result of the Department of Defense's shift from testing the product to testing the process. This shift is a part of the Pentagon's TQM strategy.[22]

It is generally agreed that maybe nine out of ten new product developments end up as a design, manufacturing, or marketing failure. These failures may be more the fault of the organization than the market. Many firms lack a system to integrate the market demands with the organization processes. Most applicants for the Baldrige Award, according to examiners, lack management processes that ensure the efficient flow of customer demands throughout the organization.[23]

If quality definition (customer expectation) is not introduced early in the concept or design stage, there is the risk (indeed the probability) that design errors and product defects will only be discovered at later stages of production or final inspection. The worst scenario is discovery by the customer in the marketplace. Motorola estimates that whereas design accounts for only 5% of product cost, it accounts for 70% of the influence on manufacturing cost.

The major functions of the organization and the matching activities/processes are shown in Figure 6-2. Each is necessary throughout the life cycle of the product, but if the beginning of each process or activity must wait for the end of the preceding one, the time to market is lengthened and the product may be obsolete or overtaken by competition midway through

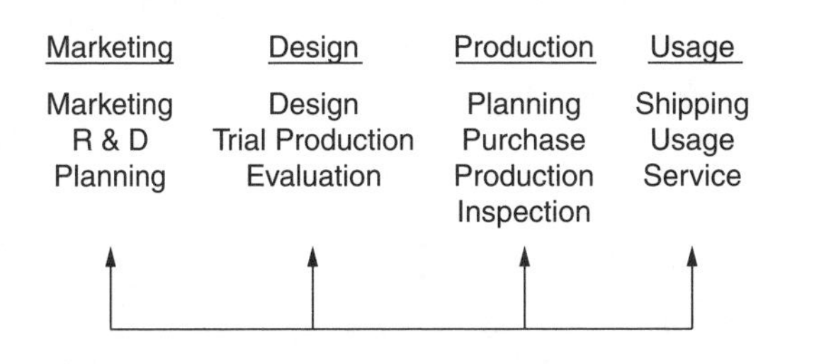

Figure 6-2 Quality Function Deployment Chart

the processes. A method is needed to integrate all processes and relate them to the customer.

Every chief executive officer would welcome a TQM system that would:

- Implement strategic quality management, including market segment differentiation based on customer expectations
- Communicate a culture of quality throughout the organization
- Translate technical requirements into process requirements and then to production planning
- Organize the potential for world-class competition
- *Integrate*
    1. The special interest functions of the company
    2. The stream of processes and provide a basis for process design and control
    3. Suppliers and customers
    4. Everyone in the process while promoting a team culture with interfunctional teams

This is a lot to ask of any method, but proponents of quality function deployment (QFD) suggest that this method has the potential to achieve many of these requirements. It has proven so effective as a competitive advantage in some companies (e.g., Ford, Digital Equipment, Black & Decker, Budd, Kelsey Hayes) that they are unwilling to talk about it.[24] In Japan, where the method was first used, companies have achieved dramatic improvement in the design-development process, including reductions of 30 to 50% in engineering changes and design-cycle time and 20 to 60% in start-up costs.[25]

QFD is a group of techniques for planning and communicating that coordinates the activities within an organization. It is a dynamic, iterative method performed by interfunctional teams from marketing, design, engineering, manufacturing engineering, manufacturing, quality, purchasing, and accounting, and in some cases suppliers and customers as well. Thus, a common quality focus is achieved across all functions: quality function deployment. The basic premise is that products should be designed to reflect the desires and tastes of customers. An additional benefit is improvement of the company's management processes.[26]

The primary technique is a visual planning matrix called the "House of Quality," which links customer requirements, design requirements, target values, and competitive performance in one easy-to-read chart. The concept, without the details, is illustrated in Figure 6-3.[27]

QFD unfolds in the following steps or phases. Note that step numbers for product planning and design processes are entered in the sections of the House of Quality in Figure 6-3.

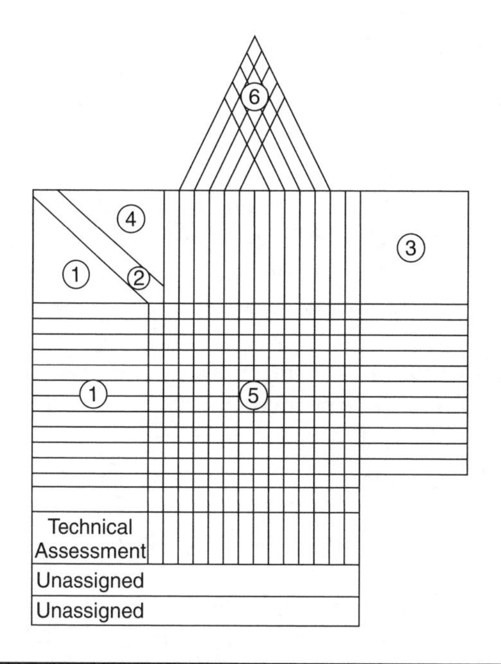

Figure 6-3 House of Quality Concept

Step 1 ***Product planning*** — This begins with customer requirements, defined by specific and detailed phrases that the customers in their own words use to describe desired product characteristics.

Eaton Corporation, a supplier to the automobile industry, selected a control device for a QFD pilot process. A matrix chart was prepared that related desired product features to part quality characteristics. Each quality characteristic was ranked. Through QFD, selling price and engineering expenses were reduced by 50%.[28]

Step 2 ***Prioritize and weight*** the relative importance that customers have assigned to each characteristic. This can be done on a scale (e.g., 1 to 5) or in terms of percentages that sum to 100%.[29]

Step 3 ***Competitive evaluation*** — For those who want to be world-class or meet or beat the competition, it is essential to know how their products compare. Specifically, will the characteristics identified in steps 1 and 2 provide a strategic competitive advantage? (See Chapter 8 for further information on benchmarking.)

Step 4 ***The design process*** — This is where the customer's product characteristics meet the *measurable* engineering characteristics that directly affect customer perceptions.

Step 5 ***Design (continued)*** — The central relationship matrix indicates the degree to which each engineering characteristic affects the customer's characteristics. Strengths of relationships are entered.

Step 6 ***Design (continued)*** — The roof of the "house" matrix encourages creativity by allowing changes between steps 4 and 5 in order to judge potential trade-offs between engineering and customer characteristics.

Step 7 ***Process planning*** — Output from the design process goes to process planning, where the key processes (e.g., cutting, stamping, welding, painting, assembly, etc.) are determined. This step may have its own matrix.

Step 8 ***Process control*** — Output from step 7 goes to process control, where the necessary process flows and controls are designed.

The entire QFD process is "deployed" as illustrated in Figure 6-4. The "hows" of one step become the "whats" of the next. Many of the statistical techniques mentioned previously can be used. Market research has particular methods for that function.

In all cases, the interfunctional teams are involved. This is necessary to avoid rework and redesign as well as overruns in cost and time. Questions need to be answered along the way: What does the customer really want? Can we design it? Can we make it? Is it competitive? Can we sell it at a profit? Do the processes support it? Hewlett-Packard estimates that quality programs have saved the company $400 million in warranty costs. Prior to implementing QFD and quality programs, the company estimated that non-quality costs added up to 25 to 30% of sales dollars.[30]

The essential prerequisite for QFD is the determination of customer requirements, as defined by specific and detailed phrases that customers, in their own words, use to describe desired product characteristics. To achieve this degree of specificity, it may be necessary to communicate with customers one-on-one or in focus groups. A less desirable method is to use surveys or other indirect means.

JUST-IN-TIME (JIT)

> By the third year of JIT implementation, Isuzu (a Japanese company) had reduced the number of employees from 15,000 to 9,900, reduced work in process from 35 billion yen to 11 billion yen, and decreased the defect rate by two-thirds.[31]

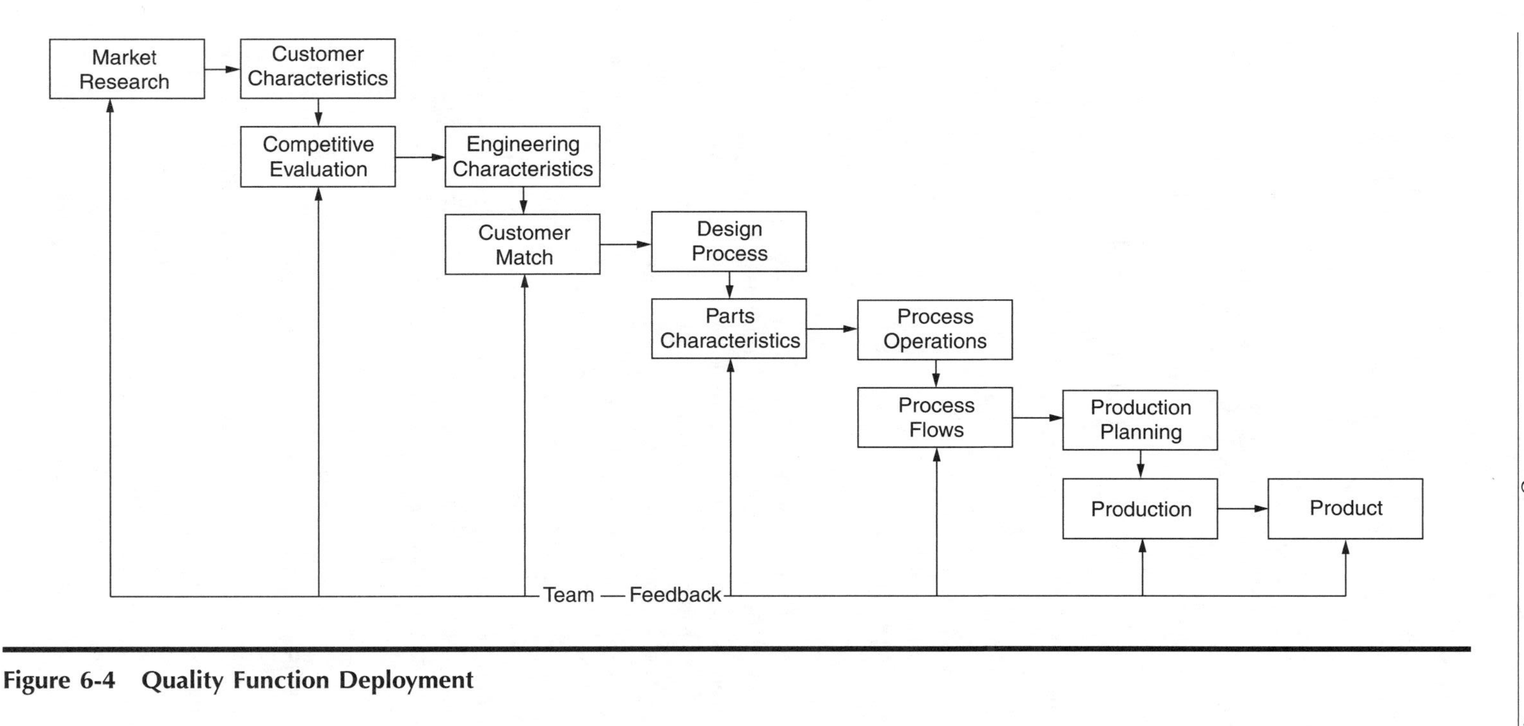

Figure 6-4 Quality Function Deployment

> Hewlett-Packard has spread JIT to all areas, including cost accounting, procurement, and engineering. At one plant where 290 pieces of equipment are hand-assembled, product reliability has improved sixfold and productivity is up considerably.[32]
>
> As part of its conversion to JIT, Westinghouse Electric's Asheville, North Carolina, plant was run as a number of miniplants. Cycle time has been reduced two to four times, on-time performance is up over 90%, and shop productivity is up by 70%. Employees are trained to perform multiple functions, and each will end up knowing how to build the complete product.

U.S. manufacturing has been characterized by mass production, high-volume output, and machine capacities that are pushed to the limit. This is changing as U.S. managers begin to discover a production method called just-in-time. Proponents say that it is more than a manufacturing system; they call it a philosophy and a way of approaching business goals that incorporates (1) producing what is needed when it is needed, (2) minimizing problems, and (3) eliminating production processes that make safety stocks necessary.[33]

Prior to the 1960s, the goal of production planning was cost optimization. In the early 1970s, it became requirements planning, and the technique of materials requirements planning (MRP) computed material needs to meet a sales forecast and production plan. MRP was, and is, an effort to balance the sometimes conflicting demands of safety stocks, inventory carrying costs, economic order quantity, and risk factors related to possible stockouts. Today, the modern corporation is turning to manufacturing as a crucial strategic resource and is adopting JIT as a basic component of manufacturing strategy. The view is that the expense and risk of maintaining inventory can be reduced so that lower costs becomes a way of improving both productivity and quality. Of course, inventory is not the only consideration of JIT. It involves all functions and all processes.

JUST-IN-TIME OR JUST-IN-CASE

JIT assumes that "less is best," while just-in-case (JIC) involves the use of buffer or safety stocks. Conventional reasons given to explain the need for buffer stock include avoiding risks of stockouts or failure of suppliers, getting a better price for volume purchasing, or avoiding an anticipated price increase. The presence of such "excess" inventory increased the risk of obsolescence and deterioration, increased the need for warehouse and shop floor space, and by "pushing" parts through the assembly process encouraged a number of wasteful practices. Operators were unconcerned with workstations other than their own. The attitude became "there's plenty more where that came from." If a defective part was discovered, the tendency was to blame it on a previous operation or assume that it would be corrected later in the process or at the rework area.

Shigeo Shingo, who is credited with designing Toyota's JIT production system, believes that the "push" process used in the United States generates process-yield imbalances and interprocess delays.[34] *Kanban,* as JIT is called in Japan, means "visible record." It is a means of pulling parts through the assembly process; production is initiated only when a worker receives a visible cue that assembly is needed for the next step in the process. The worker orders the product from the previous operation so that it arrives just when needed. If one of the key processes fails to produce a quality part, the production line stops. Individual operators are their own inspectors and are cross-trained for a number of tasks. The system is continuously being fine-tuned.

Benefits of JIT[35]

JIT is not just an inventory control method. It is a system of factory production that interrelates with all functions and activities. The benefits include:

- Reduction of direct and indirect labor by eliminating extraneous activities
- Reduction of floor space and warehouse space per unit of output
- Reduction of setup time and schedule delays as the factory becomes a continuous production process
- Reduction of waste, rejects, and rework by detecting errors at the source
- Reduction of lead time due to small lot sizes, so that downstream work centers provide feedback on quality problems
- Better utilization of machines and facilities
- Better relations with suppliers
- Better plant layout
- Better integration of and communication between functions such as marketing, purchasing, design, and production
- Quality control built into the process

THE HUMAN SIDE OF PROCESS CONTROL

One study found that a very small percentage of employees could define quality or could relate what their companies were doing to improve it.[36]

The problems of managing streams of processes are both methodological and organizational. Peter Drucker concludes that SQC has its greatest impact on the factory's social organization.[37] The essence of his argument relates to the way that the use of statistical tools in the production process places information and hence accountability in the hands of the machine operator rather than non-operators such as inspectors, expediters, repair crews, and supervisors. Each operator becomes his or her own inspector. Operators "own" the machines, which allows them to spot malfunctions and correct problems.[38]

If Drucker is right, the potential exists for significant improvement in quality, cost, and productivity. However, there is a down side. Strict adherence to rigid methods and procedures means that workers and teams may lose the autonomy they previously enjoyed, only to have it replaced by the regimentation necessitated by process control. By their very nature, SPC and JIT require a focus on the process as a whole, an environment that may be strange to an operator accustomed to the segmented approach previously in effect.[39]

It is almost universally accepted that control of any process rests upon measuring against some standard, measure, benchmark, or target. Yet in many organizations, workers and managers operate with two different sets of goals and in two different cultures. It becomes an "us vs. them" split culture. As we move from inspection to process control, it is essential that control measures become the property of the workers. SPC and JIT achieve this. Workers are involved in measures over which they have some control in monitoring continuous improvements. Control of measures alone, however, may not be enough. Understanding of and involvement in the system would enhance job satisfaction, which is a necessary dimension. Moreover, like any process or system, the people with hands-on involvement are a valuable resource for refinement and improvement.

Attention to the human resource dimension provides a basis for significant improvements in job development, job satisfaction, training, and morale. Suggested actions to improve the changes include:

- Like all major change, top management support is essential.
- Change the focus from production volume to quality, from speed to flow, from execution to task design, from performing to learning.
- Invest in training, a necessary prerequisite.

EXERCISES

6-1 Explain the difference between feedforward and feedback (final inspection) control. Why is feedforward more appropriate for TQM?
6-2 What are the steps in moving from a system of final inspection to process control?
6-3 Choose a non-manufacturing (service) process and show how statistical quality control would be appropriate.
6-4 How would a *sequential* approach to product design and introduction result in overruns in time, cost, and quality? How would quality function deployment improve the system?
6-5 Is customer defections a measure of service quality? If so, how can the measure be used to reduce customer defections?
6-6 Describe the Deming Cycle.
6-7 Explain the benefits of just-in-time.

ILLUSTRATIVE CASES

The Building Products Operations Division of Armstrong World Industries is one of two winners of the 1995 Baldrige Quality Award. The division's process management systems are labeled Managing Process Improvement (MPI) and are integrated into the approaches used for designing and improving processes and products as well as services. The six-step model of MPI is as follows:

1. Identify the process (scope, suppliers, customers).
2. Define the output (customer requirements).
3. Define the process (flow diagrams, measurement).
4. Define subprocesses.
5. Improve the process (goals and actions).
6. Continuously improve the process (refinement).

Services are designed and improved using the MPI framework. For example, architects, designers, and end users wanted fast answers to technical and installation questions, a service not available in the industry at the time. A Technical Elegance Feasibility Team was formed, customer requirements were defined, current procedures documented and measured, goals set, and actions taken. This included an analysis of competitors' technical services and benchmarking. The basis for continuous process improvement began based on industry best practices.

Question

- Show how the principles of process control are applicable to service processes. Select a service process (e.g., university registration, product service query) and describe how you might go about improving it.

A survey by the *Engineering Management Journal* (June 1996) suggested that the United States has met the Japanese competitively in many areas in global manufacturing but still lags in the quality areas that can affect manufacturing cycle time, such as product design.

Question

- How would the use of quality function deployment techniques reduce manufacturing cycle time?

ENDNOTES

1. Claude S. George, Jr., *The History of Management Thought,* Englewood Cliffs, N.J.: Prentice-Hall, 1972, p. 10. For an excellent summary of sources that have traced the history of the quality movement, see David A. Garvin, *Managing Quality,* New York: The Free Press, 1988, p. 251.
2. Frederick W. Taylor, *Principles of Scientific Management,* New York: Harper & Row, 1911.
3. Frederick W. Taylor, *Shop Management,* New York: Harper & Row, 1919, p. 101.
4. W. A. Shewhart, *Economic Control of Quality of Manufactured Product,* New York: E. Van Nostrand Company, 1931.
5. Peter Drucker, "The Emerging Theory of Manufacturing," *Harvard Business Review,* May/June 1990, p. 95.
6. Harry W. Kenworthy and Angela George, "Quality and Cost Efficiency Go Hand in Hand," *Quality Progress,* Oct. 1989, pp. 40–41.
7. Barbara Dutton, "Switching to Quality Excellence," *Manufacturing Systems,* March 1990, pp. 51–53.
8. Bob Johnstone, "Prophet with Honor," *Far Eastern Economic Review (Hong Kong),* Dec. 27, 1990, p. 50. A survey of Japanese automobile parts suppliers showed that 93% used SPC in their operations.
9. For a detailed treatment of SQC and SPC, see Kaoru Ishikawa, *Guide to Quality Control,* rev. ed., 1982 (available in the United States from UNIPUB [Tel. 800-274-4888]). See also J. M. Juran, *Quality Control Handbook,* 3rd ed., New York: McGraw-Hill, 1974. For application of process control charts, see such standard texts as J. M. Juran and Frank Gryna, Jr., *Quality Planning and Analysis,* New York: McGraw-Hill, 1980 and E. L. Grant and R. S. Leavenworth, *Statistical Quality Control,* 5th ed., New York: McGraw-Hill, 1980.

10. One research study of over 300 U.S. firms found that less than half believed that they had a state-of-the-art quality control program that includes SQC and SPC and utilizes a computer support system. Joel E. Ross and David Georgoff, "A Survey of Productivity and Quality Issues in Manufacturing: The State of the Industry," *Industrial Management,* Jan./Feb. 1991.
11. Ken Jones, "High Performance Manufacturing: A Break with Tradition," *Industrial Management (Canada),* June 1988, pp. 30–32.
12. Genichi Taguchi and Don Clausing, "Robust Quality," *Harvard Business Review,* Jan./Feb. 1990, pp. 65–75. This article provides a summary of the collection now known as Taguchi methods, a popular approach that is opposed to the zero defects concept, based on the conclusion that the concept promotes quality in terms of acceptable deviation from targets rather than a consistent effort to hit them. Zero defects, according to Taguchi, fixes design before the effects of the quality program are felt.
13. Kenneth E. Kirby and Charles F. Moore, "Process Control and Quality in the Continuous Process Industries," *Survey of Business,* Summer 1989, pp. 62–66.
14. Larry H. Anderson, "Controlling Process Variation Is Key to Manufacturing Success," *Quality Progress,* Aug. 1990, pp. 91–93.
15. Chester Placek, "CMMs in Automation," *Quality,* March 1990, pp. 28–38.
16. Ray A. Mundy, Russel Passarella, and Jay Morse, "Applying SPC in Service Industries," *Survey of Business,* Spring 1986, pp. 24–29.
17. Aleta Holub, "The Added Value of the Customer–Provider Partnership," in *Making Total Quality Happen,* Research Report No. 937, New York: The Conference Board, 1990, pp. 60–63.
18. John E. Tyworth, Pat Lemons, and Bruce Ferrin, "Improving LTL Delivery Service Quality with Statistical Process Control," *Transportation Journal,* Spring 1989, pp. 4–12.
19. Aleta Holub, Endnote 17.
20. Thomas C. Day, "Value-Driven Business = Long-Term Success," in *Total Quality Performance,* Research Report No. 909, New York: The Conference Board, 1988, pp. 27–29.
21. Laurence C. Seifert, "AT&T's Full-Stream Quality Architecture," in *Total Quality Performance,* Research Report No. 909, New York: The Conference Board, 1988, pp. 47–49.
22. Pam Nazaruk, "Commitment to Quality: Test Process Not Product, Orders Pentagon," *Electronic Business,* Oct. 15, 1990, pp. 163–164.
23. Peter Burrows, "Commitment to Quality: Five Lessons You Can Learn from Award Entrants," *Electronic Business,* Oct. 15, 1990, pp. 56–58.
24. Gary S. Vasilash, "Hearing the Voice of the Customer," *Production,* Feb. 1989, pp. 66–68.
25. Ronald Fortuna, "Beyond Quality: Taking SPC Upstream," *Quality Progress,* June 1988, pp. 23–28. The author is manager of the Chicago office of Ernst & Young and observes that the control charts of SPC are considered one of the "Seven Old Tools" in Japan, along with Pareto analysis, cause-and-effect diagrams, data stratification, histograms, checksheet, graphs, and scatter diagrams.
26. William Band and Richard Huot, "Quality & Functionality Equal Satisfaction," *Sales and Marketing Management in Canada (Canada),* March 1990, pp. 4–5.

27. For a more detailed and practical description of how to formulate and implement the method, see John R. Hauser and Don Clausing, "The House of Quality," *Harvard Business Review,* May–June 1988, pp. 63–73. Clausing, one of the authors, introduced QFD to Ford and its supplier companies in 1984, and the process was used in the successful design and introduction of the Taurus.
28. Dennis De Vera et al. "An Automotive Case Study," *Quality Progress,* June 1988, pp. 35–38.
29. At a Conference Board Quality Conference on April 2, 1990, A. F. Jacobson of 3M described how the Commercial Office Supply division brings customer expectations into the design process: "Let's say, they're going to develop an improved tape of some kind. They don't wait until the product is finished…or nearly finished…to take it to their customers. They take the idea to customers right at the beginning of the process. They ask customers what they want from a particular tape. Very often, they'll hear things like: 'Don't make it too sticky.' 'I want to be able to pull it off the roll easily.' and, 'It's no good unless I can write on it.' Now, collecting opinions is the easy part of the process. The tough part is converting these soft expectations into technical requirements. This is done on a matrix *before* the development process gets very far. These soft expectations are converted into technical specifications for, say, adhesion, roughness and reflectance."
30. Robert Haavind, "Hewlett-Packard Unravels the Mysteries of Quality," *Electronic Business,* Oct. 16, 1989, pp. 101–105.
31. Ronald M. Fortuna, "The Quality Imperative," *Executive Excellence,* March 1990, p. 1.
32. Steve Kaufman, "Quest for Quality," *Business Month,* May 1989, pp. 60–65.
33. Jack Byrd, Jr. and Mark D. Carter, "A Just-in-Time Implementation Strategy at Work," *Industrial Management,* March/April 1988, pp. 8–10. See also Ira P. Krespchin, "What Do You Mean by Just-in-Time?" *Modern Materials Handling,* Aug. 1986, pp. 93–95.
34. John H. Sheridan, "World-Class Manufacturing: Lessons from the Gurus," *Industry Week,* Aug. 6, 1990, pp. 35–41. Shingo also believes that JIT extends to plant maintenance, and with a participative environment operators will protect their own equipment.
35. There are a number of references that point out the benefits as well as the pitfalls of JIT. These sources also contain suggestions for implementation. See the following: Bruce D. Henderson, "The Logic of Kanban," *Journal of Business Strategy,* Winter 1986, pp. 6–12. The author describes how the technique can provide a competitive advantage. The need to rethink traditional practices is discussed in Lynne Perry, "Simplified Manufacturing Is Best," *Industrial Management,* July/Aug. 1986, pp. 29–30. The way that small manufacturers can adapt the technique is described in Byron Finch, "Japanese Management Techniques in Small Manufacturing Companies," *Production & Inventory Management,* Vol. 27 Issue 3, 3rd Quarter 1986, pp. 30–38. The need for continued quality control and other requirements is outlined in Mark R. Jamrog, "Just-in-Time Manufacturing: Just in Time for U.S. Manufacturers," *Price Waterhouse*

Review, Vol. 32 Issue 1, 1988, pp. 17–29. The interface with other functions of the company is provided by R. Natarajan and Donald Weinrauch, "JIT and the Marketing Interface," *Production & Inventory Management Journal,* Vol. 31 Issue 3, 3rd Quarter 1990, pp. 42–46.

36. Joel E. Ross and Lawrence A. Klatt, "Quality: The Competitive Edge," *Management Decision (UK),* Vol. 24 Issue 5, 1986, pp. 12–16.
37. Peter Drucker, "The Emerging Theory of Manufacturing," *Harvard Business Review,* May/June 1990, p. 95.
38. James J. Webster, "Pulling — Not Pushing — For Higher Productivity," *Mechanical Engineering,* April 1988, pp. 42–44.
39. Gervase R. Bushe, "Cultural Contradictions of Statistical Process Control in American Manufacturing Corporations," *Journal of Management,* March 1988, pp. 19–31.

REFERENCES

Archer, N. P. and G. O. Wesolowsky, "Consumer Response Tactics in U.S. Manufacturing," *International Journal of Quality and Reliability Management,* April 1996, pp. 99–108.

Dellana, Scott and Mark Coffin, "Quality Management Tactics in U.S. Manufacturing," *Engineering Management Journal,* June 1996, pp. 27–34.

Pitman, Glen, "QFD Application in an Educational Setting," *International Journal of Quality and Reliability Management,* April 1996, pp. 99–108.

Reed, Richard, David Lemak, and Joseph Montgomery, "Beyond Process: TQM Content and Firm Performance," *Academy of Management Review,* Jan. 1996, pp. 173–202.

7

CUSTOMER FOCUS AND SATISFACTION

Quality begins and ends with the customer.

Joel Ross

Beginning in the 1980s, many organizations saw a growing number of its customers complaining loudly and litigiously in courtrooms, boardrooms, and waiting rooms. Some have referred to this as the era of "customer rebellion." There was a remarkable increase in the number of complaints from customers. Some businesses reacted by tuning out the voice of the customers, while others scrambled to establish formal mechanisms for tracking customer satisfaction. Today, customer satisfaction is not just a socially provocative concept; it has become the gold standard by which every organization is judged. It matters not whether a company is small, medium-sized, or large. It doesn't matter whether it makes widgets or provides a service; whether it is a for-profit or not-for-profit. Trying to compete solely on the basis of price is insufficient. Most banks, airlines, hospitals, hotels, and car rentals offer the same core services, with slight variations in price. What distinguishes them is customer service. Customer service has become a weapon of choice in the battle for market share.

Of all the Baldrige Award criteria, none is more important than customer focus and satisfaction. This category accounts for 300 points of the 1000-point value of the award.

> This category examines the company's relationships with customers and its knowledge of customer requirements and of the key quality factors that drive marketplace competition. Also examined are the company's methods to determine customer

> satisfaction, current trends and levels of customer satisfaction and retention, and these results relative to the competition.[1]

The principles discussed in this chapter and in the entire book apply equally to both service and manufacturing firms. Judging from what is known about U.S. manufacturing and service firms, not many companies would receive a grade of "A" for customer focus. A comprehensive study of 584 companies by the consulting firm Ernst & Young found that customer complaints were of "major or primary" importance in identifying new products and services among only 19% of banks and 26% of hospitals.[1a]

The widespread tendency to ignore complaints or fail to track them and identify the cause(s) can have very serious consequences. This is particularly true in services, where it is estimated that for every complaint a business receives, there are 26 other customers who feel the same way but do not air their feelings to the company.[2]

Failure to identify the root cause of complaints means that reduction of variation in the causative process is more difficult. A customer unable to get through to a sales representative is evidence of a malfunction in the telephone procedure (process) or the sales and marketing function. Thus, it becomes necessary to tie the customer to the process.

Evidence indicates that part of the cause of this failure to close the customer-process loop is inadequate support from top management for the total quality management (TQM) infrastructure and a continued focus on the techniques of TQM, particularly statistical process control (SPC).

The Ernst & Young study mentioned previously found that quality-performance measures such as defect rates and customer satisfaction levels play a key role in determining pay for senior managers in fewer than one in five companies. *Profitability* is still king. There is, of course, nothing wrong with a focus on cash flow and short-term profits, but long-term profit and market share require a base of satisfied customers that are retained by a focus on satisfaction. Some top executives may not like to believe the level or severity of customer complaints or may be offended by them. When Amtrak was criticized in the *Wall Street Journal* by a transportation analyst (Lind), the president of Amtrak responded (in the same issue):

> My own conclusion is that this [comment] is based on hopelessly incorrect assumptions about Amtrak and the railroad industry, and that Mr. Lind would be well advised to limit his comments and suggestions to the streetcar and transit business with which he is familiar and to avoid getting over his depth.[3]

While it may be true that the president of Amtrak is correct in this case, such an attitude expressed publicly could very well pervade the work force, who might perceive the message as justification for continuing the existing level of service.

Another reason for the lack of customer focus is the tendency of many firms to emphasize the techniques of TQM, such as SPC, and other outcome-oriented methods, such as productivity and cost reduction. Again, these are desirable and necessary, but a singular emphasis on these areas puts the cart before the horse.

The customer is not really interested in the sophistication of a company's process control, its training program, or its culture. The bottom line for the customer is whether he or she obtains the desired product. This truism is recognized by Deming, Juran, and Crosby.

PROCESS VS. CUSTOMER

Customer complaints are analogous to process variation. Both are undesirable and must be addressed. In both cases, the optimum output must be compared against an objective, a standard, or a benchmark. Both are integral parts of the quality improvement process. The integration of the customer and the process is shown conceptually in Figure 7-1.

From the company's point of view, customer satisfaction is the result of a three-part system: (1) company processes (operations), (2) company employees who deliver the product, and service that is consistent with (3) customer expectations. Thus, the effectiveness of the three-part system is a function of how well these three factors are integrated.

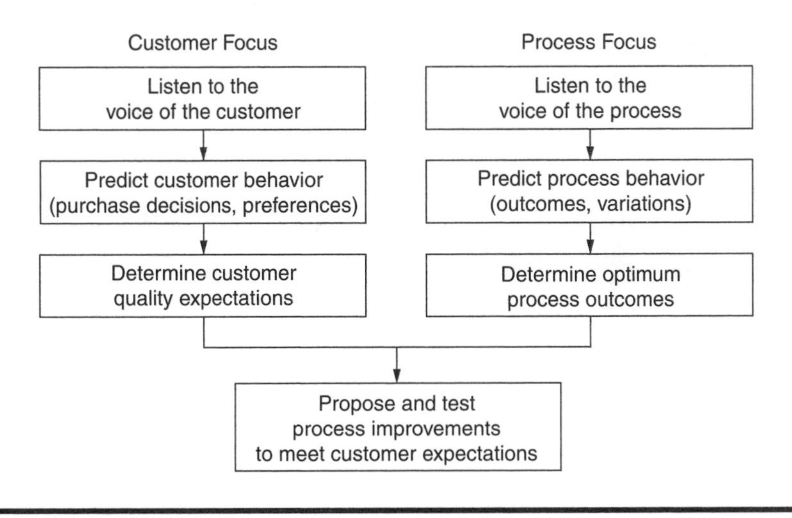

Figure 7-1 Integration of Customer and Process

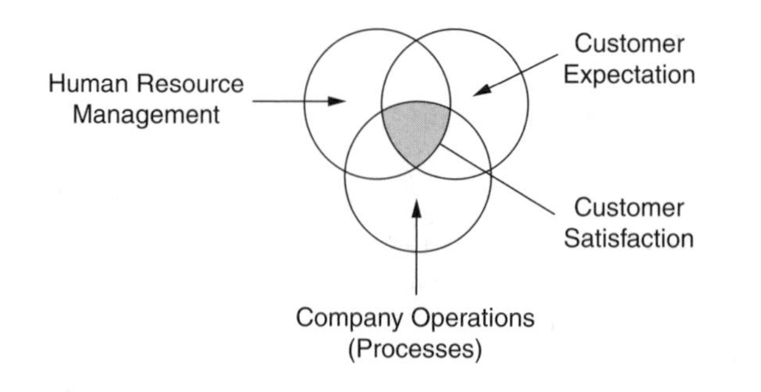

Figure 7-2 Customer Satisfaction: Three-Part System

This concept is shown in Figure 7-2. The overlap (shaded area) represents the extent to which customer satisfaction is achieved. The objective is to make this area as large as possible and ultimately to make all three circles converge into an integrated system. The extent to which this condition is achieved depends on the effectiveness of (1) the process, (2) employees, and (3) determination of what constitutes "satisfaction." Like any system, control is necessary. Thus, standards are set, performance is measured, and variation, if any, is corrected.

> Ritz-Carlton Hotel Company won the Baldrige Award in 1992. Many people thought that no hotel could do this because service in this industry is so difficult to measure and to deliver. The company meticulously gathers data on every aspect of the guest's stay to determine if the hotels are meeting customer expectations. Key to the research are the daily quality production reports that identify all problems and defects reported in each of 720 work areas. The data compiled range from the time it takes for housekeepers to clean a room to the number of guests who must wait in line to check in.[4]

INTERNAL CUSTOMER CONFLICT

Internal customers are also important in a TQM program. These are the people, the activities, and the functions within the company that are the customers of other people, activities, or functions. Hence, manufacturing is the customer of design, and several departments may be customers of data processing.

Conflict frequently arises between the needs of internal and external customers. In many cases, processes are designed to meet the needs of internal customers. Any "customer" who has been admitted to a hospital or outpatient service understands this. The registration process is designed to meet the needs of the admitting department, business office, or medical records. The result is a long wait to give information that will be provided again and again to personnel who represent admitting, the laboratory, finance, social work, and medicine. Who is the customer? Who is the beneficiary? Who is the recipient of the output? The patient gets the impression that he or she is a piece of raw material being moved along an amorphous assembly line known as healthcare.

It is not too difficult to identify other examples in both the private and public sectors. How about a university? It has been said that if you want to find out what kind of organization you are about to do business with, call on the phone!

A balance needs to be struck between the needs of these two customer groups. The solution is to determine the real needs of each and design the process to meet both.

DEFINING QUALITY

Supreme Court Justice Potter Stewart once said that while he could not define "obscenity," he knew it when he saw it. Wrestling with a definition of quality is almost as difficult, but necessary nevertheless. You cannot manage what you cannot measure.

There are eight dimensions of quality — performance, features, reliability, conformance, durability, serviceability, aesthetics, and perceived quality.[5] However, the shortfall regarding product quality is that the services connected with it are so frequently overlooked. Good packaging, timely and accurate shipping, and the ability to meet deadlines matter as much as the quality of the product itself. Customers define quality in terms of their total experience with the company. Many companies approach customer satisfaction in a narrow way by confining quality considerations to the product alone.[6]

A QUALITY FOCUS

It is impossible to avoid the constant bombardment of "quality" and "satisfaction" messages in advertising on television and radio and in print media. Much of this advertising, and the actions to deliver the product or service, is little more than vague rhetoric. Even the popular phrases "satisfaction guaranteed" or "low price guaranteed" do not state what the customer is supposed to get for his or her purchase.

Some companies have attempted to improve this rhetoric by supplementing the message with additional definitions of satisfaction. McDonald's guarantees customer satisfaction with the pledge: "If you are not satisfied, we'll make it right or the next meal is on us." What does the phrase *make it right* mean? The question is whether this guarantee relates to product quality and customer satisfaction or is merely a promotion. Perhaps the slogan should be changed to "enjoyment guaranteed."

Many firms back up a satisfaction guarantee with promises of a reward if they fail to meet their own standards or those of the customer. Hampton Inns refunds your money. At Pizza Hut you get it free if not served in 5 minutes. Some firms give you a $5 bill. Delta Dental Plans of Massachusetts sends you a check for $50 if you get transferred from phone to phone while seeking an answer to an insurance question. Automobile dealers and manufacturers are fond of promoting "quality service" without defining just what this is. Some back it up with such specifics as towing service, free rides to work, or loaner cars when the customer's car is kept overnight.

There are two advantages to backing up a guarantee with some penalty for failure to deliver. It can cure employee apathy and bring quality to the attention of employees on a personal basis. It also may leave the buyer with a perception of dedication and thereby serve to retain what otherwise may have been a lost customer. These customers may say to themselves and others: "Well, my pizza was 10 minutes late but they gave me a free one, so that proves they are serious about quality." Retaining this customer, who now has a better perception and higher expectation, may be worth the cost of the pizza and the foregone sale.

It should be remembered that any effort to tie the message of satisfaction to a failure-to-deliver penalty is ineffective if the variation or failure is not traced back to process improvement and the cause of the variation. Why was the pizza delivered late? Why was the customer shifted from phone to phone? Why did the dealer keep the car overnight? The variation-cause connection is identified by problem solving and the process improvement through process control.

Break Points

The need to improve customer satisfaction in *measurable* amounts is well known. But what is the measure and how much improvement is needed? If a customer is willing to stand in line for two minutes but finds five minutes unacceptable, anything between is merely satisfactory. Zero to one minute is outstanding. On-time delivery below 90% may be judged by customers as unacceptable, while over 98% is considered outstanding. Improvement programs should be geared toward reaching either a two-minute or five-minute range for standing in line and either 90% or 98% for delivery times. These are the market *break points,* where improving performance will change customer behavior, resulting in higher prices or sales volume. Forget the improvement program that targets one minute waiting in line or the delivery program that targets between 90% and 98%.

A Central Theme

Although individuals and teams may have targets that are directed at process improvement in their specific activities, a common theme or focus may integrate the many individual or group efforts that may have their own priority. At Motorola, the theme is *six sigma;* at Hewlett-Packard, it is a *tenfold reduction in warranty expense.* At General Electric, no part will be produced that cannot meet a *one-part-in-a-million* defect rate. At MBNA of America (a credit card company), the target is *customer retention.* In other companies it can be a reduction in defects or cycle time. Such a theme tends to be pervasive because so many individuals can relate their activities to it. It can serve to mobilize employees around an overall quality culture.

THE DRIVER OF CUSTOMER SATISFACTION

The benefits of having customers who are satisfied is well known and was outlined in Chapter 1. The issues in building customer satisfaction are to acquire satisfied customers, know when you have them, and keep them.[7]

The obvious way to determine what makes customers satisfied is simply to ask them. Before or concurrently with a customer survey, an audit of the company's TQM infrastructure needs to be made. IBM is one company that has identified the key excellence indicators for customer satisfaction. These key indicators are listed in Table 7-1.[8]

Despite the obvious need for customer input in determining new product/service offerings and improving existing ones, the widespread tendency is to determine perceived quality and perceived customer satisfaction based almost solely on in-house surveys.[9] Even when the company does attempt to get input, the survey may suffer from methodology shortcomings. Mailed questionnaires lose control over who responds, and respondents are less likely to reply if they are dissatisfied or if the name of the company or product is indicated. Just what is satisfaction? If the customer's expectation is low, satisfaction may be acceptable but perception will not improve. If perception is low but satisfaction is acceptable, how is this determined and what can be done? Suppose that 95% of respondents indicate satisfaction but do not perceive the product as one of the best. Survey results can be misconstrued and lead to complacency.[10]

Table 7-1 Key Excellence Indicators for Customer Satisfaction

- Service standards derived from customer requirements
- Understanding customer requirements
  - Thoroughness/objectivity
  - Customer types
  - Product/service features
- Front-line empowerment (resolution)
- Strategic infrastructure support for front-line employees
- Attention to hiring, training, attitude, morale for front-line employees
- High levels of satisfaction — customer awards
- Proactive customer service systems
- Proactive management of relationships with customers
- Use of all listening posts
  - Surveys
  - Product/service follow-ups
  - Complaints
  - Turnover of customers
  - Employees
- Quality requirements of market segments
  - Surveys go beyond current customers
  - Commitment to customers (trust/confidence//making good on word)

> The hotel chain Ritz-Carlton, a Baldrige winner, relies on technology to keep comprehensive computerized guest history profiles on the likes and dislikes of more than 240,000 repeat guests. Researchers survey more than 25,000 guests each year to find ways in which the chain can improve delivery of its service.[11]

GETTING EMPLOYEE INPUT

Employee input can be solicited concurrent with customer research. It could help identify barriers and solutions to service and product problems, as well as serving as a customer–company interface.[12] Such surveys can help identify changes that may be necessary for quality improvement. In addition to customer-related considerations, employee surveys can measure:[13]

- TQM effectiveness
- Skills and behaviors that need improvement
- The effectiveness of the team problem-solving process
- The outcomes of training programs
- The needs of internal customers

Corning Inc., a leader in the glassware industry, asked line and staff groups worldwide to assess themselves using the Baldrige criteria. Each group was to develop a few quality strategies that would address the most critical elements identified in the assessment. Certain measures, referred to as Key Result Indicators, that focused on evidence of customer deliverables and process outcomes were required.[14]

MEASUREMENT OF CUSTOMER SATISFACTION

The accelerating interest in the measurement of customer satisfaction is reflected in the over 170 consulting firms that specialize in this activity.[15] Some firms use the "squeaky wheel" or "if it ain't broke, don't fix it" approach and measure customer satisfaction based on the level of complaints. This has a number of disadvantages. First, it focuses on the negative aspects by measuring dissatisfaction rather than satisfaction. Second, the measure is based on the complaints of a vocal few and may cause costly or unneeded changes in a process. As indicated at the beginning of this chapter, for every complaint there are 26 others who feel the same way but do not air their feelings.

There are two basic steps in a measurement system: (1) develop key indicators that drive customer satisfaction and (2) collect data regarding the perceptions of quality received by customers.[16]

Key indicators of customer satisfaction are what the company has chosen to represent quality in its products and services and the way in which these are delivered. The building blocks that the system is designed to track are (1) expectations of the customer and (2) company perceptions of customer expectations.

In Chapter 1, a number of indicators for the physical product (e.g., reliability, aesthetics, adaptability, etc.) are identified. For service businesses or for services that accompany a product, the range of indicators depends on the nature of the service. One authority[17] has suggested that some important areas to consider are outcome, timeliness of the service, satisfaction, dependability, reputation of the provider, friendliness/courteousness of employees, safety/risk of the service, billing/invoicing procedures, responsiveness to requests, competence, appearance of the physical facilities, approachability of the service provider, location and access, respect for customer feelings/rights, willingness to listen to the customer, honesty, and an ability to communicate in clear language. These indicators, if appropriate and addressable, are converted to action items that reflect specific delivery systems where the product or service meets the customer. For example, in a bank, customer needs and systems would combine to deliver short teller lines, friendly and courteous staff, ATMs that work, and low fees on accounts.

Data collection is required in order to identify the needs of customers and the related problems of process delivery. The data-gathering process surveys both customers and employees. By including employees, customer needs and barriers to service can be identified, as well as recommendations for process improvement. Different orientations are emphasized for customers and employees: The former are asked for *their* expectations and the latter are asked what they think *customers* expect.

THE ROLE OF MARKETING AND SALES

Marketing and sales are the functions charged with gathering customer input, but in many firms the people in these functions are unfamiliar with quality improvement.[18] Shortcomings in marketing as identified by critics include:[19]

- Partnering with dealers and distribution channels
- Focusing on the physical characteristics of products and overlooking the related services
- Losing a sense of customer price sensitivity
- Not measuring or certifying suppliers such as advertisers
- Failing to perform cost/benefit analyses on promotion costs
- Losing markets to generics and house brands

> According to one source, Motorola is a world-class producer of products but is less than world class in marketing. Historically, the company has been oriented toward engineering and technology. Its six sigma quality is well known. The publisher of *Technologic Computer Letter* says, "With many product lines Motorola has an extremely compelling story to tell but it is used to hiding its light under a bushel and does not make its advantages heard."[20]

Quality and customer satisfaction have not played an important role in the sales function (*process*). Consider the stereotype of a salesperson: He or she is detail (rather than process) oriented and trained in technical product knowledge (rather than customer knowledge). Salespeople are feature oriented: "We've got six models, four colors, and it comes with a money-back guarantee." They are trained and rewarded for getting new customers, as opposed to retaining existing ones.

THE SALES PROCESS

According to Hiroshi Osada of the Union of Japanese Scientists and Engineers (JUSE), TQM needs to begin with the salespeople.[21] Yet TQM has migrated to the sales force in only a few companies.[22] Even fewer perceive sales as a *process* that lends itself to analysis and improvement for customer satisfaction. To repeat a previous caveat, "If you can't measure it, you can't manage it." Another can be added: "You can't measure it if it's not a process." Both of these cliches are as true for sales and marketing as for any other process. The objective is quality outcomes. In order for TQM to become a part of sales and marketing, managers and employees must move toward a deeper understanding of its processes — selling, advertising, promoting, innovating, distribution, pricing, and packaging — *as they relate to customer satisfaction.*

Marketing applications need not be confined to the marketing department. Other functions can borrow these techniques to improve the satisfaction of external and internal customers. A brokerage firm should care not only about sending accurate statements on time, but should also be concerned with whether the statement format fits the customer's needs. In an issue of *Marketing News,* Research Professor Eugene H. Fram of the Rochester Institute of Technology suggests the following types of non-traditional marketing extensions:[23]

- ***Adapted*** marketing refers to a non-marketing function that adapts traditional techniques. Relationship selling is an example. If a human resource department sends the same recruiter each year to a campus, this person can use principles of relationship selling to further company goals. This type of selling can also be used within the organization and between departments. The classic conflict between production (cost) and sales (delivery) can be reduced.
- ***Morale*** marketing can improve morale. Consider what the terms "Team Taurus" or "Team Xerox" did for morale in those companies.
- ***Sensitivity*** marketing borrows from the basic marketing principle that says that one must understand the customer's needs in order to fulfill them and to build long-term relationships. In a marketing sense, individuals, groups, and departments are better able to achieve quality and productivity if they are sensitive to the needs, concerns, and priorities of both internal and external customers.

SERVICE QUALITY AND CUSTOMER RETENTION

Customer defection is a problem and customer retention an opportunity in both manufacturing and service firms. Manufacturers have generally been good about measuring satisfaction with products, but now they are moving into service areas. The publicity surrounding the Baldrige Award accounts for much of this. Other reasons relate to the size and growth of service industries and the growing importance of service as a means of strategically competing in the marketplace.

Service industries are playing an increasingly important role in a nation's economy. Over 80% of the working population in the United States is employed in the service sector and the percentage is growing. When this employment is combined with service jobs in the manufacturing sector, it becomes evident that the importance of services is increasing. Many executives feel that the management of services is one of the most important problems they face today. Yet most of us know from personal experience that the quality of services is declining, despite the efforts of some companies to improve it.

Because so many services are intangible, the interaction between employees and customers is critical. Chase Manhattan Bank realizes that an employee's ability to meet or exceed customer expectations when conducting a routine transaction influences the customer's satisfaction with the organization. In fact, this interaction influences satisfaction more than the actual product or service obtained. The one-on-one or face-to-face contact between the customer and the deliverer of the service (nurse, flight attendant, retail clerk, restaurant server) is extremely important.

Manufacturers are careful to measure material yield, waste scrap, rework, returns, and other costs of poor quality processes. Service companies also have these costs, which are reflected in the cost of customers who will not come back because of poor service. These are customer defections and they have a substantial impact on cost and profits. Indeed, it is estimated that customer defections can have a greater impact than economies of scale, market share, or unit cost.[24] Despite this, many companies fail to *measure* defections, determine the *cause* of defections, and improve the *process* to reduce defections.

CUSTOMER RETENTION AND PROFITABILITY

What is the ultimate desired outcome of customer focus and satisfaction? Is it achieving profit in the private sector or productivity in the public or non-profit sectors? The answer must be yes. Oddly enough, however, an accurate cause-and-effect relationship has yet to be established between profit and customer satisfaction. This is due, in part, to the difficulty of measuring satisfaction and relating it to profit. However, there is a proven relationship between *customer retention and profit.*

One way to put a value on customer retention is to assign or estimate a "lifetime retention value," the additional sales that would result if the customer were retained. Taco Bell calculates the lifetime value of a retained customer as $11,000. An automobile dealer believes that the lifetime value of retaining a customer is $300,000 in sales.[25] Conversely, MBNA America (a credit card company) has found that a 5% improvement in customer *defections* increases its average customer value by 125%.

The system for improving *customer retention* and profit is illustrated in Figure 7-3. The drivers are *employee satisfaction* and employee retention. The system components are:

- ***Internal service quality,*** which establishes and reinforces a climate and organization culture directed toward quality.
- ***Employee retention,*** which is achieved through good human resource management practices and organization development methods such as teams, job development, and empowerment. Employee retention depends on employee satisfaction, which in turn can be related to external service and customer satisfaction.
- ***External service quality,*** which is delivered through the organization's quality infrastructure.
- ***Customer satisfaction*** and follow-up, in order to reduce customer defections and improve retention.

To reiterate, there is a proven relationship between customer retention and profit.

BUYER–SUPPLIER RELATIONSHIPS

Almost every company purchases products, supplies, or services in an amount that frequently equals around 50% of its sales. Traditionally many of these companies have followed the "lowest bidder" practice, where price is the critical criterion. The focus on price, even for commodity products, is changing as companies realize that careful concentration of purchases, together with long-term supplier–buyer relationships, will reduce costs and improve profits.[26] Deming realized this and suggested that a long-term relationship between purchaser and supplier is necessary for best economy.[27] If a buyer has to rework, repair, inspect, or otherwise expend time and cost on a supplier's product, the buyer is involved in a "value/quality-added" operation, which is not the purpose of having a reliable supplier. In that never-neverland of the perfect buyer–supplier relationship, no rework or inspection is necessary.

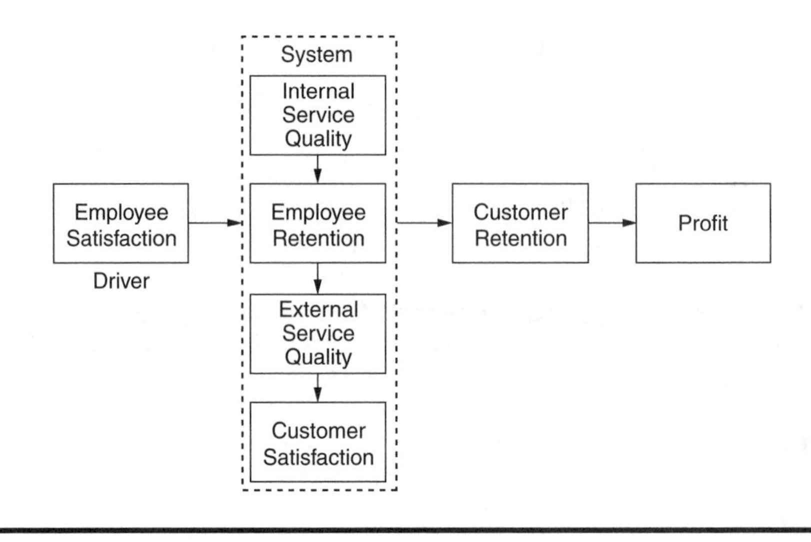

Figure 7-3 Profitability and Customer Retention

A partnership arrangement is emerging between a growing number of manufacturers and suppliers. At Eastman Kodak, the Quality Leadership Process (QLP) has improved the company's production processes, reduced overall manufacturing costs, and improved quality by transforming traditional manufacturer–supplier roles. Because one-half of all components used in manufacturing are supplied by outside vendors, realignment of the supplier base has become a central strategy of QLP.[28]

> Motorola has advanced the supplier–customer relationship further than most companies. The system is based on a basic economic principle: whenever someone buys from someone else, there is a mutually beneficial transaction and pleasing both sides is important. With this in mind, Motorola has begun to market itself as a customer. The company's director of materials and purchasing says, "If the sauce is good for the goose, it should be good for the gander, and we are genuinely trying to cooperate, collaborate and do some strategic things with our suppliers. Our goal is to become a world-class customer and that means that it is important for us to learn what the buyer needs to do in order for suppliers to see us as a world-class customer."[29]

Several guidelines will help both the supplier and customer benefit from a long-term partnering relationship:

- ***Implementation of TQM by both supplier and customer*** — Many customers (e.g., Motorola, Ford, Xerox) are requiring suppliers to operationalize the basic principles of TQM. Some have even required the supplier to apply for the Baldrige Award. This joint effort provides a common language and builds confidence between both parties.

- ***Long-term commitment to TQM and to the partnering relationship between the parties*** — This may mean a "life cycle" relationship that carries partnering through the life cycle of the product, from market research and design through production and service.
- ***Reduction in the supplier base*** — One or more automobile companies have reduced the number of suppliers from thousands to hundreds. Why have ten suppliers for a part when the top two will do a better job and avoid problems?
- ***Get suppliers involved in the early stages of research, development, and design*** — Such involvement generates additional ideas for cost and quality improvement and prevents problems at a later stage of the product life cycle.
- ***Benchmarking*** — Both customer and supplier can seek out and agree on the best-in-class products and processes.

How does one become a quality supplier? This of course depends on the criteria of the buyer, but it is reasonably safe to assume that if the following criteria are met, a company can reasonably expect to be classified in the quality category. The following criteria are required to be certified as *quality* in the automobile industry:

- *Management philosophy* of the CEO should support TQM.
- Techniques of *quality control* should be in place (SPC, etc.).
- Desire for a long-term *life cycle relationship.*
- Best-in-class *inventory and purchasing systems.*
- *Facilities* should be up to TQM standards.
- *Automation* level should meet quality standards.
- *R&D and design* should support customer expectations.
- Willingness to *share costs.*

EXERCISES

7-1 Describe how a program directed toward customer focus and satisfaction interacts with:

The information and analysis component of the TQM approach
Strategic quality planning
Human resource development and management
Management of process quality

7-2 Select a function or activity (e.g., design, order processing, accounting, data processing, engineering, market research) and identify a measure of quality that you would expect if you were an *internal* customer of that function or activity.

7-3 Choose a specific product or service in a particular industry and devise an action plan for obtaining customer input and feedback. How would the information generated by such a plan be used for process improvement?

7-4 Illustrate how a firm might focus on *internal* product or service specifications rather than customer expectations and desires.

7-5 Choose a product or service and list four or five characteristics that you as a customer would want and expect. Based on your experience, do you think that the firm will deliver?

7-6 How would you establish a system to measure customer satisfaction?

ILLUSTRATIVE CASES

Gulf Coast Health Systems is the primary hospital for a community of 20,000 in the Panhandle community of Florida. Stakeholders are defined as patients and their families, the community, the healthcare staff, employers and payors, and students (nursing and medical students from the university) and their sponsoring institutions. Healthcare staff includes physicians, midwife and practitioner nurses, physician assistants, and nurse anesthetists. The payor mix includes Medicare (35%), managed care/HMO (34%), indemnity (18%), Medicaid (6%), and self-pay (4%). Although the satisfaction of all stakeholders is important, particular attention is paid to patients and their families. Satisfaction is determined by an analysis of complaints as well as surveys of:

1. Overall satisfaction
2. Competitive comparison
3. Likelihood to return to Gulf Coast Health Systems
4. Reason for selection of Gulf Coast Health Systems
5. Evaluation of specific service attributes
6. Evaluation of billing practices

Questions

- Evaluate the process of stakeholder satisfaction determination.
- How would you manage the complaint process for feedback and improvement?
- Name four or five indicators of quality healthcare in this hospital.

Milliken & Company of Spartanburg, South Carolina, was an early (1989) winner of the Baldrige Award and later the European Quality Award. One of the several measures of customer satisfaction is on-time delivery measured by comparing the promised and the actual shipment dates. Employee satisfaction is measured by using annual surveys, a morale index, turnover and absenteeism, and the degree of employee involvement in company activities. The company believes that high ratings on employee satisfaction will result in high ratings on customer satisfaction.

Question

- Do you think that employee satisfaction affects customer satisfaction? In what way?

ENDNOTES

1. "Malcolm Baldrige National Quality Award Criteria — 1993," Washington, D.C.: U.S. Department of Commerce, National Institute of Standards and Technology, 1993, p. 29.

1a. Hays, Richard D., "The Strategic Power of Internal Service Excellence," *Business Horizons,* July/Aug. 1996, pp. 15–20.

2. "Satisfaction-Action," *Marketing News,* Feb. 4, 1991, p. 4.

3. "Management: Quality Programs Show Shoddy Results," *Wall Street Journal,* May 14, 1992, Section B, p. 1.

4. Edward Watkins, "How Ritz-Carlton Won the Baldrige Award," *Lodging Hospitality,* Nov. 1992, p. 23.

5. See David A. Garvin, *Managing Quality,* New York: The Free Press, 1988, pp. 49–59. Garvin has defined the eight dimensions of quality as performance, features, reliability, conformance, durability, serviceability, aesthetics, and perceived quality. Computer-maker NCR goes to great expense to define quality as appropriateness, reliability, aesthetics, and usability. Industrial designers have given the company a silver medal for design. For a comprehensive report on how product design enhances profits and market share, see a special report, "Hot Products: How Good Design Pays Off" (cover story), *Business Week,* June 7, 1993, pp. 54–78.

6. Oren Harari, "Quality Is a Good Bit-Box," *Management Review,* Dec. 1992, p. 8.

7. Gerald O. Cavallo and Joel Perelmuth, "Building Customer Satisfaction, Strategically," *Bottomline,* Jan. 1989, p. 29.

8. These indicators are taken from class material in an IBM in-house workshop, "MDQ (Market Driven Quality) Workshop." The company was kind enough to share this class material with the author (JER) and several other professors from the College of Business, Florida Atlantic University. For this, I thank them.

9. It was found, for example, that in the hospital industry fewer than 5% of referring physicians play a prominent role in identifying new service opportunities. Nearly 40% of U.S. hospitals indicate that senior management "always or almost always" takes the dominant role in identifying new services." U.S. hospitals seek minimal input from patients. *The International Quality Study, Healthcare Industry Report,* American Quality Foundation and Ernst & Young, 1992.
10. For some ideas on getting customer input, see Joel E. Ross and David Georgoff, "A Survey of Productivity and Quality Issues in Manufacturing: The State of the Industry," *Industrial Management,* Jan./Feb. 1991.
11. Edward Watkins, "How Ritz-Carlton Won the Baldrige Award," *Lodging Hospitality,* Nov. 1992, p. 24.
12. Luane Kohnke, "Designing a Customer Satisfaction Measurement Program," *Bank Marketing,* July 1990, p. 29.
13. Kate Ludeman, "Using Employee Surveys to Revitalize TQM," *Training,* Dec. 1992, pp. 51–57.
14. David Luther, "Advanced TQM: Measurements, Missteps, and Progress through Key Result Indicators at Corning," *National Productivity Review,* Winter 1992/93, pp. 23–36.
15. Lynn G. Coleman, "Learning What Customers Like," *Marketing News,* March 2, 1992, pp. CSM-1–CSM-11. This article contains a directory of 170 customer satisfaction measurement firms.
16. For a more detailed description of how to establish a measurement system, see J. Joseph Cronin, Jr. and Steven A. Taylor, "Measuring Service Quality: A Reexamination and Extension," *Journal of Marketing,* July 1992, pp. 55–68. See also Luane Kohnke, "Designing a Customer Measurement Program," *Bank Marketing,* July 1990, pp. 28–30; Gerald O. Cavallo and Joel Perelmuth, "Building Customer Satisfaction, Strategically," *Bottomline,* Jan. 1989, pp. 29–33.
17. Dean E. Headley and Bob Choi, "Achieving Service Quality through Gap Analysis and a Basic Statistical Approach," *Journal of Services Marketing,* Winter 1992, pp. 5–14. This is a good primer on gap analysis and the use of basic statistical techniques.
18. Joe M. Inguanzo, "Taking a Serious Look at Patient Expectations," *Hospitals,* Sep. 5, 1992, p. 68. This article points out that there is very little employee involvement in measuring satisfaction and practically none from patients.
19. Allan J. Magrath, "Marching to a Different Drummer," *Across the Board,* June 1992, pp. 53–54.
20. B. G. Yovovich, "Becoming a World-Class Customer," *Business Marketing,* Sep. 1991, p. 16.
21. Dick Schaaf, "Selling Quality," *Training,* June 1992, pp. 53–59.
22. John Franco, president of Learning International in Stamford, Connecticut, has conducted a series of round table discussions with sales executives. He reported: "When we ask participants how many of them are from companies that have a quality movement underway, we find about half do. But when we ask them whether that effort has migrated to the sales force, fewer than 10% say it has." Dick Schaaf, "Selling Quality," *Training,* June 1992, pp. 53–59.
23. Eugene H. Fram and Martin L. Presberg, "TQM Is a Catalyst for New Marketing Applications," *Marketing News,* Nov. 9, 1992.

24. Frederick F. Reicheld and W. Earl Sasser, Jr., "Zero Defections: Quality Comes to Services," *Harvard Business Review,* Sep./Oct. 1990, pp. 105–111. Reprint No. 90508.
25. Harvard Business School video series, "Achieving Breakthrough Service," Boston: Harvard Business School, 1992.
26. Robert D. Buzzell and Bradley Gale, *The PIMS Principles,* New York: The Free Press, 1987, p. 62. The data from over 3000 strategic business units show that concentrating purchases *improves* profitability, at least up to a point. "The positive net effect of a moderate degree of purchase concentration suggests that the efficiency gains that can be achieved via this approach to procurement are usually big enough to offset the disadvantages that might be expected as a result of an inferior bargaining position."
27. W. Edwards Deming, *Out of the Crisis,* Cambridge, Mass.: Massachusetts Institute of Technology, Center for Advanced Engineering Study, 1986, p. 35.
28. Joseph P. Aleo, Jr., "Redefining the Manufacturer–Supplier Relationship," *Journal of Business Strategy,* Sep./Oct. 1992, pp. 10–14.
29. B. G. Yovovich, "Becoming a World-Class Customer," *Business Marketing,* Sep. 1991, p. 29.

REFERENCES

Martinez, Erwin V., "You Call This Service?" *Technology Review,* April 1996, pp. 64–65.

Seawright, Kristie, "A Quality Definition Continuum," *Interfaces,* May/June 1996, pp. 107–113.

Shermach, Kelly, "Don't Grow Complacent When Your Customers Are 'Satisfied,'" *Marketing News,* May 20, 1996, p. 2.

8

BENCHMARKING

Benchmarking is a way to go backstage and watch another company's performance from the wings, where all the stage tricks and hurried realignments are visible.

Wall Street Journal

In Joseph Juran's 1964 book *Managerial Breakthrough,* he asked the question, "What is it that organizations do that gets results so much better than ours?" The answer to this question opens the door to *benchmarking,* an approach that is accelerating among U.S. firms that have adopted the total quality management (TQM) philosophy.

The essence of benchmarking is the continuous process of comparing a company's strategy, products, and processes with those of world leaders and best-in-class organizations in order to learn how they achieved excellence and then setting out to match and even surpass it. For many companies, benchmarking has become a key component of their TQM programs. The justification lies partly in the question: "Why re-invent the wheel if I can learn from someone who has already done it?" C. Jackson Grayson, Jr., chairman of the Houston-based American Productivity and Quality Center, which offers training in benchmarking and consulting services, reports an incredible amount of interest in benchmarking. Some of that interest may be explained by the criteria for the Malcolm Baldrige Award, which includes "competitive comparisons and benchmarks."[1]

THE EVOLUTION OF BENCHMARKING

The method may have evolved in the 1950s, when W. Edwards Deming taught the Japanese the idea of quality control. Other U.S. management innovations followed. However, the method was rarely used in the United

States until the early 1980s, when IBM, Motorola, and Xerox became the pioneers. Xerox became the best-known example of the use of benchmarking.

Xerox

The company invented the photocopier in 1959 and maintained a virtual monopoly for many years thereafter. Like "Coke" or "Kleenex," "Xerox" became a generic name for all photocopiers. By 1981, however, the company's market share shrank to 35% as IBM and Kodak developed high-end machines and Canon, Ricoh, and Savin dominated the low-end segment of the market. The Xerox vice president of copier manufacturing remarked: "We were horrified to find that the Japanese were selling their machines at what it cost us to make ours…we had been benchmarking against ourselves. We weren't looking outside." The company was suffering from the "not invented here" syndrome, as Xerox managers did not want to admit that they were not the best.

The company instituted the benchmarking process, but it met with resistance at first. People did not believe that someone else could do it better. When faced with the facts, reaction went from denial to dismay to frustration and finally to action. Once the process began, the company benchmarked virtually every function and task for productivity, cost, and quality. Comparisons were made for companies both in and outside the industry. For example, the distribution function was compared to L.L. Bean, the Freeport, Maine, catalog seller of outdoor equipment and clothing and everyone's model of distribution effectiveness.

By the company's own admission, it would probably not be in the copier business today if it were not for benchmarking. Results were dramatic:

- Suppliers were reduced from 5000 to 300.
- "Concurrent engineering" was practiced. Each product development group has input from design, manufacturing, and service from the initial stages of the project.
- Commonality of parts increased from about 20% to 60 to 70%.
- Hierarchical organization structure was reduced, and the use of cross-functional "Teams Xerox" was established.
- Results included:
  - Quality problems cut by two-thirds.
  - Manufacturing costs cut in half.
  - Development time cut by two-thirds.
  - Direct labor cut by 50% and corporate staff cut by 35% while increasing volume.

It should be noted that all of these improvements were not the direct result of benchmarking. What happened at Xerox (and what happens at most companies) is that in adopting the process, the climate for change and continuous improvement followed as a natural result. In other words, benchmarking can be a very good intervention technique for positive change.

Ford

The entire automobile industry may have undergone substantial change as a result of Ford's Taurus and Sable model cars. Operating performance and reliability were significantly improved, and the gains were recognized by U.S. car buyers as well as others in the industry. "Team Taurus," a cross-functional group of employees, was empowered to bring the car to market and was given considerable authority to act outside of the normal company bureaucracy.

The team defined 400 different areas that were considered important to the success of a mid-size car. A best-in-class competition was chosen for each area. Fifty different mid-sized car models were chosen. Few were Ford models. Based on the 400 benchmarks, specific teams were assigned responsibility to meet or beat the best-in-class for each area of performance, and 300 features were "copied" and incorporated into the car design. Target dates were set for beating the remaining features. "Quality Is Job One" became the fight song for Ford employees.

The Taurus was, and is, a resounding success. Some auto analysts credit the Taurus experience with the partial resurgence of quality in the U.S. automobile industry. The benchmarking process provided additional benefits. During the examination of competitors' features, valuable insights into the design process were gained. Cycle time was reduced. Buyer–supplier relationships were improved as supplier input was solicited for the design. All manufacturing processes were improved as a by-product of the benchmarking process.

Motorola

In the early 1980s, the company set a goal of improving a set of basic quality attributes *tenfold* in five years. Based on *internal* benchmarking, the goal was reached in three years. The company then began to look outside, sending teams to visit competitor plants in Japan. To their chagrin, the teams found that Motorola would have to improve its tenfold improvement level another two to three times just to match the competition.

Borrowing process benchmarks from companies as diverse as Wal-Mart, Benetton, and Domino's Pizza, the company now routinely fields benchmarking requests from those same Japanese companies it toured the first time around.[2]

THE ESSENCE OF BENCHMARKING

The process is more than a means of gathering data on how well a company performs against others both in and outside the industry. It is a method of identifying new ideas and new ways of improving processes and hence meeting customer expectations. Cycle time reduction and cost cutting are but two process improvements that can result. The traditional approach of measuring defect rates is not enough. The ultimate objective is *process improvement* that meets the attributes of customer expectation. This improvement, of course, should meet both strategic and operational needs.

A properly designed and implemented benchmarking program will take a total system approach by examining the company's role in the supply chain, looking upstream at the suppliers and downstream at distribution channels. How competitive are suppliers in the world market and how well are they integrated into the company's own core business processes — product design, demand forecasting, product planning, and order fulfillment.[3]

BENCHMARKING AND THE BOTTOM LINE

There are two basic points of view regarding how to get started in benchmarking. One minority view maintains that an *initial* action plan that tries to match the techniques used by world-class performance may actually make things worse by doing too much too soon. A three-year study of 580 global companies conducted by the management consulting firm Ernst & Young concluded that it may be best to start measuring existing financial performance measures. Two key measures are return on assets (which is simply after-tax income divided by total assets) and value added per employee. Value added is sales minus the costs of materials, supplies, and work done by outside contractors. Labor and administrative costs are not subtracted from sales to arrive at value added.[4]

The focus on financial results is not recommended by the majority of executives familiar with the benefits of benchmarking. Some believe that it is easy to be fooled by financial indicators that lull the company into thinking that it is doing well when what in reality occurs is a transitory financial phenomenon that may not hold up over the longer term. A more important payoff is quality processes that lead to a quality product.

Robert C. Camp headed up the now-famous study at Xerox in which the buzzword "benchmarking" was coined in 1980. When asked whether the best work practices necessarily improve the bottom line, he replied: "The full definition of benchmarking is finding and implementing best practices in our business, practices that meet customer requirements. So the flywheel on finding the very best is, 'Does this meet customer requirements?' There is a cost of quality that exceeds customer requirements. The basic objective is satisfying the customer, so that is the limiter."[5]

THE BENEFITS OF BENCHMARKING

Given the considerable effort and expense required for effective benchmarking, why would an organization embark on such an effort? The answer is justified by three sets of benefits.

Cultural Change

Benchmarking allows organizations to set realistic, rigorous new performance targets, and this process helps convince people of the credibility of these targets. This tends to overcome the "not invented here" syndrome and the "we're different" justification for the status quo. The emphasis on looking to other companies for ideas and solutions is antithetical to the traditional U.S. business culture of individualism. Robert Camp, the former Xerox guru quoted earlier, indicates that the most difficult part for a company that is starting the process is getting people to understand that there may be people out there who do things better than they do. According to Camp, overcoming that myopia is extremely important.

Performance Improvement

Benchmarking allows the organization to define specific gaps in performance and to select the processes to improve. It provides a vehicle whereby products and services are redesigned to achieve outcomes that meet or exceed customer expectations. The gaps in performance that are discovered can provide objectives and action plans for improvement at all levels of the organization and promote improved performance for individual and group participants.

Human Resources

Benchmarking provides a basis for training. Employees begin to see the gap between what they are doing and what best-in-class are doing. Closing the gap points out the need for personnel to be involved in techniques of problem solving and process improvement. Moreover, the synergy between organization activities is improved through cross-functional cooperation.

STRATEGIC BENCHMARKING

It is paradoxical that two AT&T divisions (AT&T Network Systems Group, Transmission Systems Business Unit, and AT&T Universal Card Services) were 1992 winners of the Baldrige Award. Like several other winners, the company has turned this win into an advantage and organized a separate operation to market this expertise. Training is the product offered by the AT&T Benchmarking Group of Warren, New Jersey.[6] The process is illustrated in Figure 8-1.

The paradox is that ten years earlier, in 1983, AT&T was convinced that it could be a major player in the computer industry. The company owned Bell Laboratories, the largest R&D facility in the world, and had extensive experience in the manufacture of telecommunications equipment, a related product.

Five years after entering the industry and after losing billions of dollars, the company was still trying to be a significant player in the market. The near disaster could be traced directly to the company's failure to (1) realize that the key success factors (KSFs) in the industry included sales, distribution, and service (functions that AT&T had very little experience in) and (2) conduct *strategic benchmarking* against such best-in-class competitors as IBM and Compaq. Moreover, the company apparently failed to define its market segment, the criteria used for customer purchasing decisions, and how the company's product could be differentiated in the chosen segment. If, for example, IBM, Compaq, or AT&T wanted to benchmark NCR, they would find that NCR has gone to great expense to define the criteria of product quality as "usability, aesthetics, reliability, functionality, innovation and appropriateness."[7]

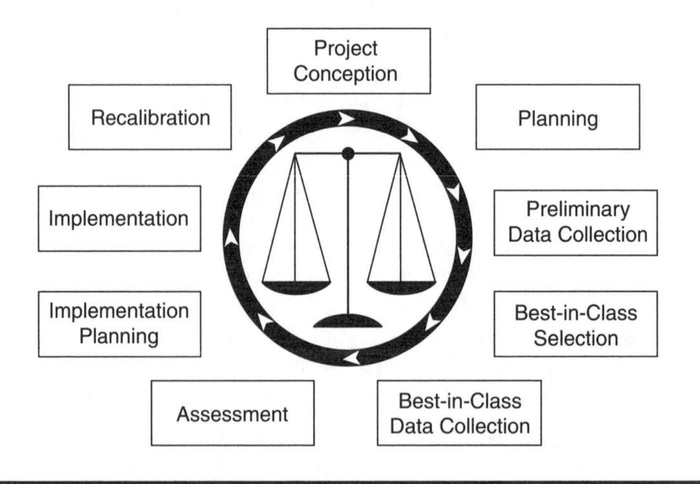

Figure 8-1 AT&T Benchmarking Process

One way to determine how well you are prepared to compete in a segment and to help define a best-in-class competitor is to construct a KSF matrix similar to the one shown in Figure 8-2. Following this determination, a matrix such as the hypothetical one shown in Figure 8-3 can be constructed to measure market differentiation criteria against competitors. Note that the criteria for comparison are based on the customer's purchase decision. This type of strategic analysis can be followed by one involving specific processes — operational benchmarking. Strategy drives performance and hence quality. Indeed, quality can and should become the central theme of strategy. Note that Figures 8-2 and 8-3 can be used to benchmark best-in-class *outside* the industry.[8]

Competitive Analysis
Computer Industry
_________ Segment

		Performance Rating		
Key Success Factors	Weight	Our Company	Competitor #1	Competitor #2
Sales Force				
Distribution				
Suppliers				
R&D				
Service				
Cost Structure				
etc.				
etc.				

Figure 8-2 Key Success Factor Matrix

Computer Industry
_________ Segment
Customer's Purchase Decision

		Performance Rating			
Criteria	Weight	Our Company	Competitor #1	Competitor #2	etc.
Reliability					
Performance					
Features					
Durability					
Service					
Software					
etc.					
etc.					

Figure 8-3 Measuring Market Differentiation Criteria against Competitors

OPERATIONAL BENCHMARKING

This category focuses on a particular activity within a company's functional operations and then identifies ways to emulate or improve on the practices of best-in-class. Whereas strategic benchmarking is largely concerned with the macro analysis of the environment, the industry, and the competitors, operational benchmarking is more detailed in terms of data gathering and the rigor of the analysis. Most of the focus is on cost and differentiation. Because the customer's purchasing decision (PD) is a function of price and differentiation, it is necessary to differentiate through *quality* [$PD = f(P \times Q)$] and improve price through *cost* reduction. Both lead to an analysis of the cost and activity chains of interconnected processes.

The scope of benchmarking extends to both strategic and operational processes. The scope of these two categories of benchmarking at Westinghouse (a Baldrige winner) is displayed in Table 8-1.

Table 8-1 How Westinghouse Uses Competitive Benchmarking Data

Process benchmarks	*Product benchmarks*
Categories	**Categories**
Assessment	Development
Performance	Features
Technology	Functionality
Financial	Architecture
Organizational	Availability
Development	**Marketing**
Goals	Target markets
Analysis	Market positioning
Countermeasures	Price strategies
Implementation	
Improvement	**Sales**
Gap analysis	Product positioning
Targets	Bid responses
Countermeasures	Customer talks
Comparisons and competitive analysis	**Comparisons and competitive analysis**
Scope	Features
Complexity	Functionality
Technology	Architecture
Performance	Availability
Cost	Market position
Strength/weakness	Price
Documentation	Strength/weakness
	Documentation

THE BENCHMARKING PROCESS

There is no standard or commonly accepted approach to the benchmarking process. Each consulting group[9] and each company[10] uses its own method. Whatever method is used, the major steps involve (1) measuring the performance of best-in-class relative to critical performance variables, (2) determining how the levels of performance are achieved, and (3) using the information to develop and implement a plan for improvement. These steps are discussed in further detail in the following sections.

Determine the Functions/Processes to Benchmark

Those functions or processes that will benefit the most should be targeted for benchmarking. It is wise to choose those that absorb the highest percentage of cost and contribute the greatest role in differentiation, always thinking in terms of process improvements that will have a positive impact on the customer's purchasing decision. Because no company can excel at everything, it is necessary to delineate targets. Benchmarking "manufacturing," for example, is much too broad and the subject is too ill-defined. If the elements to be benchmarked cannot be framed, data gathering is not focused and subsequent actions may be destructive.

Many companies focus their efforts on product comparisons. In manufacturing industries this may mean product tear-downs (e.g., Ford, Xerox) and re-engineering of design standards and assembly processes. This approach should take second place to improving time to market, first-time quality of design, and design for purchasing effectiveness, which are the primary drivers of both quality and cost. Of course, these actions should be undertaken after customer satisfaction has been defined with customer input.

> The healthcare industry provides an example of the potential for cost and quality improvement. For one procedure alone, coronary artery bypass grafts (CABGs, DRGs 106-7), Americans paid for more than 130,000 in 1991. Of the patients treated, 6033 died. Ancillary charges alone reached $2.67 billion. Baxter Healthcare Corporation of Deerfield, Illinois, which benchmarked CABGs in ten hospitals, calculated that $1.57 billion in ancillary charges alone could be saved if all hospitals benchmarked the processes of the benchmarked ten.[11]

Select Key Performance Variables

Functions, activities, and processes can be measured in terms of specific output measures of operations and performance. In general, these measures fall into four broad categories:

***Cost and productivity,* such as overhead costs and labor efficiency** — Total dollars per unit or per ton is a starting point in manufacturing. Other variables might include production yield of raw material, direct labor per unit produced, etc. Unless the project team begins with total costs before it breaks them down by process or activity, some very important overhead charges may be neglected when benchmarked against firms with different accounting systems. See Chapter 10 ("Productivity and Quality") for additional measures.

Comparing one company's financial statements and cost breakdowns against those of another would be a good method for a "me-too" strategy *if* access were available to the detailed statements of a competitor or the best-in-class and *if* they were based on similar accounting methodology. These are two big "ifs." A better way is to identify the underlying cost *drivers* of the many functions and activities that, when combined, make up total costs. For example, raw material costs may be driven by sales, purchase volume, source, or freight; direct labor by wage and benefit rates, skilled vs. unskilled, or union vs. non-union; indirect labor by the ratio of direct to indirect, salary levels, and so on.

> A team at Mercy Hospital in San Diego decided to benchmark medical records because the activity represented the largest portion of clinical support. The team left a benchmarking visit to a sister hospital empty-handed because they found that the two hospitals were quite different in this activity. A team member commented: "They weren't equivalent to us at all. It didn't do the functions we did, it wasn't open 24 hours a day like us, and it was more decentralized — a lot of what we do, they do in various other departments and clinics."

Timeliness — Often overlooked, timeliness is a major factor in internal processes as well as customer satisfaction. The measure is frequently expressed in cycle time or turnaround time, such as time to fill an order or time to answer the phone. Some manufacturing executives have been known to visit automobile races to measure pit stops as benchmarks for setup time or line changeover time.

Differentiation and quality — Measures of differentiation and quality are needed for both processes and product. Quality measures should capture the errors, defects, and waste attributable to an entire process and express them relative to the total output achieved. Defects tend to cascade down a chain of processes, becoming increasingly expensive to correct.

Differentiation and quality of product are essentially the same, because quality is what differentiates a product. The variables should include any factors that affect a customer's purchasing decision (see, for example, Figure 8-3).

Business processes — These are the processes not directly related to product design, production, sales, and service. They include the many staff and internal service activities that are costed under general and administrative (G&A) expense. One has only to look at the organizational chart to identify areas for cost reduction and for improvement of productivity and quality. Human resources, data processing, accounts receivable, marketing services, maintenance, security, data center, warehousing, public relations...the list goes on. Many companies have had severe cash flow and profit problems due to a failure to control the cost and output of these business or support processes. Whereas direct labor and material costs may make up the largest segment of total costs in a manufacturing firm and can be benchmarked more easily, G&A costs are more elusive and more difficult to measure; however, they represent fertile ground for improvement. Another area is internal quality and internal customers. A good place to start may be to use the techniques of activity analysis and activity-based costing.

IDENTIFY THE BEST-IN-CLASS

This is a major step in the benchmark analysis. The objective is to identify companies whose operations are superior, the so-called best-in-class, so that the company's own operations can be targeted.

The quickest way to identify excellent performers is simply to visit some companies that have won the Baldrige Award. A lot could be learned in a hurry, but these companies may not have the time or may not have similar processes. Other sources include (1) available databases, (2) sharing agreements between companies, and (3) out-of-industry companies.

Databases are an expanding source of comparison information. The most current and most comprehensive of these is maintained by the Houston-based American Productivity and Quality Center (AP&QC). Some of the chief difficulties that organizations encounter are identifying top-performing companies in specific functions and finding companies that have already conducted studies in specific areas. Helping others overcome these difficulties is the role of the AP&QC. It serves as a central networking source and has the support of top benchmarkers.

The cost of membership in the AP&QC ranges from $6,000 to $12,500, depending on the number of employees. Dissemination of benchmarking information is through face-to-face networking meetings, electronic bulletin boards, and on-line access to abstracts of company benchmark studies.[12]

As the popularity of benchmarking accelerates, so does the number of consortium efforts among industry peer groups. For example, a number of hospitals have formed the MECON-PEER database to provide information and analysis software for examining individual operations and compare them with similar operations nationwide. Some of the participants have discovered an additional use for the database: putting muscle into a budget squeeze and justifying additional resources based on benchmarking activities of peers.

Even universities are emerging as benchmarkers. Oregon State University pioneered the process in the academic world, and its success led to the creation of NACUBO, a database of the National Association of College and University Business Officers.

A number of companies are also developing *in-house* databases. This is particularly effective in large multidivision companies, where economies of scale in data sharing can be achieved. One such company is AT&T. The extent of the competitive benchmarking data maintained by the Network Systems Group for use by all company divisions is shown in Table 8-1 (see earlier).[13]

Cooperative sharing agreements between companies are another source of best-in-class identification. Members of the agreement may or may not be competitors and may or may not be in the same industry. DEC, Xerox, Motorola, and Boeing joined forces to standardize benchmarking procedures in training.

Out-of-industry companies may be the best source of information for many firms in the early or intermediate stages of project implementation. A benchmark planner at Johnson & Johnson suggests that 90% of all opportunities for breakthrough improvement lie in studying practices outside the industry. Perishable food companies often teach other manufacturers about supply–demand balancing, demand forecasting, production scheduling, and distribution management. Pharmaceutical companies are quite knowledgeable in production record keeping, quality assurance, and batch traceability.

Although many companies are mistakenly paranoid about sharing strategic and operating information, many others are not. Most Baldrige winners and applicants and people from many best-in-class companies are just regular people and are proud of what they have accomplished.

> When Mid-Columbia Medical Center of The Dalles, Oregon got serious about TQM and benchmarking, it formed the "MCMC University." The director, dubbed "Professor," decided to benchmark the training function and spent five days taking notes at

> Disney University, and then went on to attend Ritz-Carlton's training session for a week. "They were flattered," the "Professor" said. "We were the only people who had ever asked them if we could attend." Training videos were supplied by Northwest Tool & Die Company, Disney, Harley-Davidson, and Johnson Sausage.

Table 8-2 contains a selected list of companies noted for their best practices in the functions shown.

Table 8-2 Selected Best-Practice Companies

Company	*Function*
American Airlines	Information systems (long line)
American Express (Travel Services)	Billing
AMP	Supplier management
Benetton	Advertising
Disney World	Optimum customer experience
Domino's Pizza	Cycle time (order and delivery)
Dow Chemical	Safety
Emerson Electric	Asset management
Federal Express	Delivery time
General Electric	Management processes
GTE	Fleet management
Herman Miller	Compensation and benefits
Hewlett-Packard	Order fulfillment
Honda	New product development
IBP	Productivity
L.L. Bean	Distribution
3M	Technology transfer
Marion Merrell Dow	Sales management
Marriott	Admissions
MBNA America	Customer retention
Merck	Employee training
Milliken	Cross-functional processes
Motorola	Flexible manufacturing
NEXT	Manufacturing excellence
Ritz-Carlton	Training
Travelers	Healthcare management
US Sprint	Customer relations
Wal-Mart	Information systems
Xerox	Benchmarking

MEASURE YOUR OWN PERFORMANCE

At this step in the process, your own performance should have been premeasured; otherwise, there is nothing to compare against the benchmarking data. Moreover, data analysis of best-in-class may proceed aimlessly unless the benchmarker understands what information is being sought.

Having determined with some degree of accuracy the performance of the target firm and the extent of your own performance, it follows that an analysis of the *gap* between the two is necessary. The trickiest part of the process is to compare internal and external data on an equivalent basis. This does not mean that both sets of data must be comparable in the same exact form.

Performing a "gap analysis" of the variation with the benchmarked process involves the problem-solving process treated in Chapter 6. This analysis will reveal:

- The extent, the size, and the frequency of the gap
- Causes of the gap; why it exists
- Available methods for closing the gap and reaching the performance level of the benchmarked process

ACTIONS TO CLOSE THE GAP

Once the *cause(s)* of the gap is determined through problem analysis, alternative courses of action to close the gap become evident. Selecting the right alternative course of action is a matter of rational decision making. Among the criteria for weighing the courses of action are time, cost, technical specifications, and, of course, quality. It should be added here that the best source of information on closing the gap may be the best-in-class, because that company has already experienced what the benchmarking organization is going through.

The action plan lists each action step, the time of completion, the person responsible, and the cost, if appropriate. The results expected from each action step should also be listed in order to provide a measure of whether the objective or output of each step is achieved.

The action plan itself represents a process and lends itself to the basics of process control. Hence monitoring, feedback, and recalibration are required.

PITFALLS OF BENCHMARKING

Curt W. Reimann, who heads the Baldrige Award program at the National Institute of Standards and Technology, finds that a lot of people think benchmarking is "instant pudding." It will not improve performance if the proper infrastructure of a total quality program is not in place. Indeed, there is significant evidence that it can be harmful. Unless a corporate culture of quality and the basic components of TQM (such as information systems, process control, and human resource programs) are in place, trying to imitate the best-in-class may very well disrupt operations.

Other potential pitfalls include the failure to:

- Involve the employees who will ultimately use the information and improve the process. Participation can lead to enthusiasm.
- Relate process improvement to strategy and competitive positioning. Design to factors that affect the customer's purchasing decision.
- Define your own process before gathering data or you will be overwhelmed and will not have the data to compare your own process.
- Perceive benchmarking as an ongoing process. It is not a one-time project with a finite start and complete date.
- Expand the scope of the companies studied. Confining the benchmarking firms to your own area, industry, or to competitors is probably too narrow an approach in identifying excellent performers that are appropriate for your processes.
- Perceive benchmarking as a means to process improvement, rather than an end in itself.
- Set goals for closing the gap between what is (existing performance) and what can be (benchmark).
- Empower employees to achieve improvements that they identify and for which they solve problems and develop action plans.
- Maintain momentum by avoiding the temptation to put study results and action plans on the back burner. Credibility is achieved by quick and enthusiastic action.

EXERCISES

8-1 What benefits can be gained from benchmarking?

8-2 Identify two or three functions or activities, other than product characteristics, that could be benchmarked by:

- A manufacturer
- A service company

8-3 How can benchmarking become an intervention technique for organizational change?

8-4 Summarize some actions taken by Xerox, Ford, and Motorola while implementing their benchmarking programs.

8-5 What are the pros and cons of benchmarking based on financial performance?

8-6 Select an industry and list three or four key success factors (e.g., advertising, distribution, engineering, sales) for that industry. Which firm(s), in your opinion, would be appropriate to benchmark?

ILLUSTRATIVE CASE

Midstate University[14] is a comprehensive, publicly supported, four-year doctoral-granting institution located in Indianapolis, Indiana. The area is populated by numerous businesses and industries with national and international markets. Transfer students, largely from community colleges, account for 19% of undergraduates and minorities 11%. Approximately 70% of undergraduates are single and reside on campus. Approximately 75% of graduate students are commuters. Other stakeholders have been identified as businesses that employ graduates, parents, Indiana taxpayers, the board of trustees, society at large, graduate schools, public education, government funding agencies, and alumni.

Midstate's councils, departments, units, and teams identify benchmarking needs within the university. Any group proposing a benchmark must answer the following questions: (1) Is the issue critical to successful progress at Midstate? (2) Will the issue result in improving educational outcomes or operating performance? (3) Will addressing the issue result in adding value to students and/or stakeholders? (4) Will the issue challenge the university to be innovative and stretch itself in its results? Among the overall processes or issues that are benchmarked are student retention and minority student retention, using other Indiana public universities as the benchmarking source. Telephone registration, room-and-board costs, marketing, and faculty and staff salaries are benchmarked against six peer institutions.

Questions

- What other benchmarks would you propose for a university (for example, costs of operation, quality of instruction, stakeholder satisfaction)?
- Are peer institutions the appropriate source of benchmarks or standards? What other source would you recommend?

ENDNOTES

1. Rick Whiting, "Benchmarking: Lessons from the Best-in-Class," *Electronic Business,* Oct. 7, 1991, pp. 128–134. This article provides a good justification for benchmarking and the principles behind it.
2. Bob Gift and Doug Mosel, "Benchmarking: Tales from the Front," *Healthcare Forum,* Jan./Feb. 1993, pp. 37–51.
3. A. Steven Walleck, "Manager's Journal: A Backstage View of World-Class Performers," *Wall Street Journal,* Aug. 26, 1991, Section A, p. 10. This article contains good examples of benchmarking applications in several companies.
4. See "Quality," a special report in *Business Week,* Nov. 30, 1992, p. 66. This report suggests various benchmarking measures for three types of firms: the novice, the journeyman, and the master.
5. Adrienne Linsenmeyer, "Fad or Fundamental?" *Financial World,* Sep. 17, 1991, p. 34.
6. The address of the group is 10 Independence Blvd., Warren, NJ 07059. Florida Power & Light Company, the only U.S. winner of the Japanese Deming Prize, formed Qualtec, a consulting group offering services in quality management.
7. *Wall Street Journal,* May 26, 1992, Section C, p. 15.
8. Perhaps the largest *strategic* database is the PIMS (Profit Impact of Marketing Strategy) collection maintained at the Strategic Planning Institute in Cambridge, Massachusetts. The database contains the strategic and financial results of over 3000 strategic business units. A member firm can search for strategic "look-alike" firms and benchmark the determinants of good or not so good performance. See Robert D. Buzzell and Bradley T. Gale, *The PIMS Principles,* New York: The Free Press, 1987. See also Bradley T. Gale and Robert D. Buzzell, "Market Perceived Quality: Key Strategic Concept," *Planning Review,* March/April 1989, pp. 6–48.
9. For example, Kaiser Associates, Inc. has a seven-step process that is outlined in a company publication, *Beating the Competition: A Practical Guide to Benchmarking,* Vienna, Va.: Kaiser Associates, 1988.
10. For example, Alcoa's steps include (1) deciding what to benchmark, (2) planning the benchmarking project, (3) understanding your own performance, (4) studying others, (5) learning from the data, and (6) using the findings. See Alexandra Biesada, "Benchmarking," *Financial World,* Sep. 17, 1991, p. 31.
11. Bob Gift and Doug Mosel, "Benchmarking: Tales from the Front," *Healthcare Forum,* Jan./Feb. 1993, p. 38.
12. David Altany, "Benchmarkers Unite," *Industry Week,* Feb. 3, 1992, p. 25.
13. Taken from a company brochure entitled *A Summary of AT&T Transmission Systems: Malcolm Baldrige National Quality Award Application.* AT&T's database contains data from over 100 companies and over 250 benchmarking activities for key processes such as hardware and software development, manufacturing, financial planning and budgeting, international billing, and service delivery. Over 20,000 entries describe benchmarking trips or visits with internal and external customers. Sources of competitive benchmarking information include customers, visits to other companies, trade shows and journals, professional societies, standards committees, product brochures, outside consultants, and installation data.

14. This case is based on the 1995 Midstate University Case Study used in the Malcolm Baldrige National Quality Award training course.

REFERENCES

Callahan, Terrence W., "Improving Performance Through Benchmarking," *Business Credit,* Jan. 1996, p. 42.

Choperena, Alfredo, "Fast Cycle Time — Driver of Innovation and Quality," *Research-Technology Management,* May/June 1996, pp. 36–40.

Hiebler, Robert J., "Benchmarking: Knowledge Management," *Strategy and Leadership,* March/April 1996, pp. 22–29.

Junkins, Jerry, "Confessions of a Baldrige Winner," *Management Review,* July 1994, p. 58.

9

ORGANIZING FOR TOTAL QUALITY MANAGEMENT

If you still believe in hierarchy, job descriptions and functional boundaries, and are not experimenting with new approaches to boundaryless/networked/virtual organizations engaged in ever-changing partners, you are already in deep yogurt.

Tom Peters
Forbes

The process of quality management requires both time and structure. Organizations that embark upon the quality journey must prepare for the long haul. One of the early discoveries for organizations that pursue the goals of continuous quality improvement is the lack of structure necessary for success. Synthesizing quality values and policies into every person's job and every operation is a complex task that must be supported by an appropriate organizational infrastructure. Management texts universally define organizing as a variation of a statement such as "the process of creating a structure for the organization that will enable its people to work together effectively toward its objectives."[1] Thus, the process recognizes a structural as well as a behavioral or "people" dimension.

This chapter is concerned with the macro dimension of organization: the overall approach the company might take to establish a quality infrastructure. The micro dimension (organizing the "quality department" or the duties of the "top quality manager") is technical in nature and beyond the scope of this book. Both Deming[2] and Crosby[3] treat this in some detail.

Historically, organizations have tended to focus on the classical principles of specialization of labor, delegation of authority, span of control (a limited number of subordinates), and unity of command (no one works for two bosses). The result in many cases was the traditional pyramidal organization chart, cast in stone and accompanied by budgets, rules, procedures, and the chain of command hierarchy. Task specialization was extreme in some cases. The classic bureaucracy thus emerged.

Prior to the current emergence of total quality management (TQM) in the early 1980s, responsibility for quality was vague and confusing. Executive management grew detached from the idea of managing to achieve quality. The general work force had no stake in increasing the quality of its products and services. Quality had become the business of specialists — product specification engineers and process control statisticians who determined acceptable levels of product variability and performed quality control inspection on the factory floor.

Today, it is generally recognized that there are two prerequisites for a TQM organization. The first is a quality attitude that pervades the entire organization. Quality is not just a special activity supervised by a high-ranking quality director.[4] This attitude (culture, vision) was examined in Chapter 2 and is largely a challenge for top management. The second prerequisite is an organizational infrastructure to support the pervasive attitude. Companies must have the means and the structure to set goals, assign them to appropriate people, and convert them to action plans. People must be aware of the importance of quality and trained to accomplish the necessary tasks.

ORGANIZING FOR TQM: THE SYSTEMS APPROACH

A system can be defined as an entity composed of interdependent components that are integrated for achievement of an objective. The organization is a social system comprising a number of components such as marketing, production, finance, research, and so on. These organizational components are activities that may or may not be integrated, and they do not necessarily have objectives or operate toward achievement of an objective. Thus, synergism, a necessary attribute of a well-organized system, may be lacking as each activity takes a parochial view or operates independently of the others. This lack of synergism cannot continue under the TQM approach to strategic management because interdependency across functions and departments is a necessary precondition.

The concept of an organizational system is shown in Figure 9-1. Inputs to the system are converted by organization activities into an output. Indeed, the sole reason for the existence of the organization and each activity within it is to add value to inputs and produce an output with greater value. A measure of this conversion of inputs into outputs is known

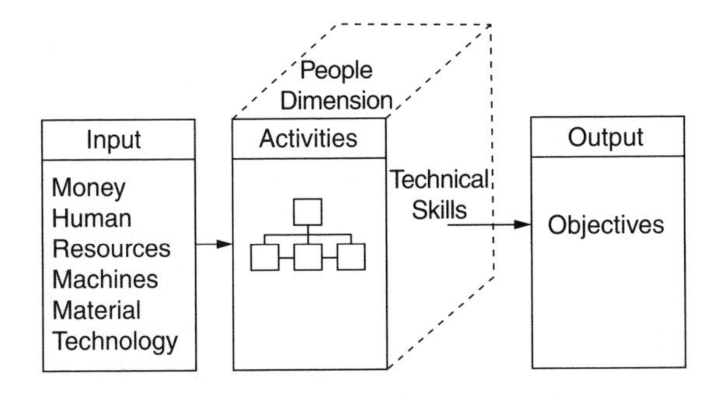

Figure 9-1 The Organizational System

as productivity, and the ratio of output to input must be a positive number if the system is to survive in the long run.

The activities of the organization are subsystems of the whole, but are also individual systems with inputs and outputs that provide input to other systems such as customers and other internal activities. This *chain* of input/output operations is depicted in Figure 9-2.

Despite the simplicity of the concept, it most often fails in practice. Activity supervisors and individuals within activities do not understand the objective or results of their "subsystem," nor can they define their output in measurable terms. When asked to define the output of their jobs, they will answer: "I am responsible for maintenance," or "I work in finance," or "my job is to ship the product." In each case these are statements of activity and not output, objective, or results expected. *Quality* output is stated in such vague terms as "do a good job" or "keep the customer happy." People can describe what they do (activity) but not what they are supposed to get done (objective or result). They may be very efficient at doing things right but ineffective in doing the right things. This failure is critical to organization output as well as structure.

Michael Porter, in his excellent book *Competitive Advantage,*[5] has taken the systems theory a major practical step forward with his concept of *the value chain*. He suggests that "competitive advantage (in this case quality) cannot be understood by looking at a firm as a whole. It stems from the many discrete activities a firm performs in designing, producing, marketing, delivering, and supporting its product." While Porter's concept is expanded to include any of the many sources of competitive advantage, the value chain concept will be used here to focus on the organizational structure for TQM.

The discrete activities of an organization can be represented using the generic value chain shown in Figure 9-3. Note that the activities

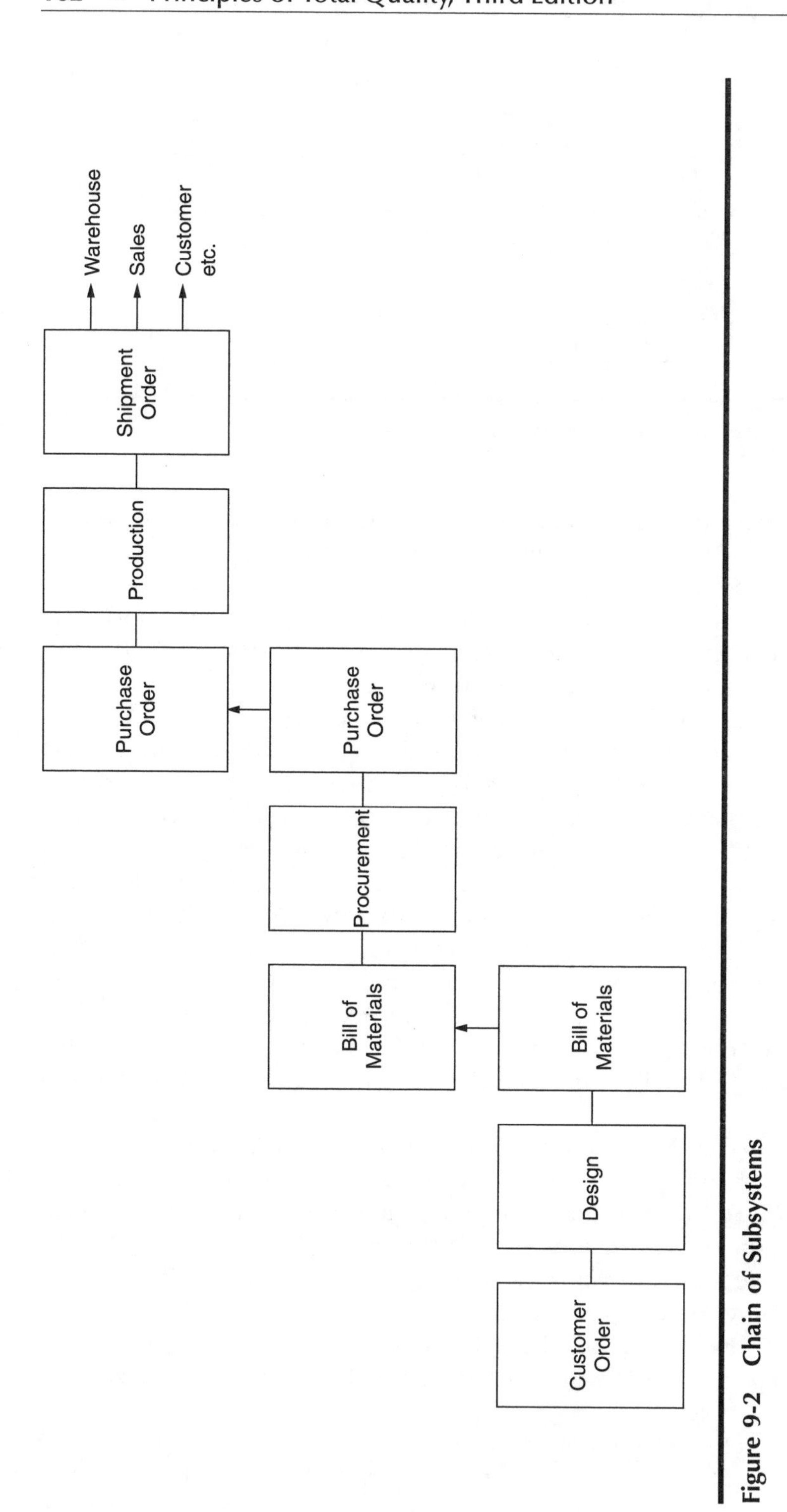

Figure 9-2 Chain of Subsystems

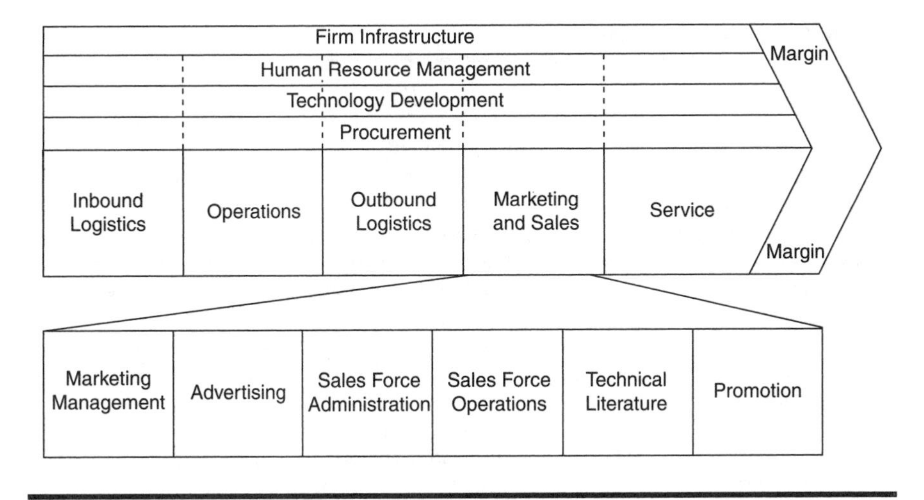

Figure 9-3 Subdividing a Generic Value Chain

or organizational functions are comprised of primary and support activities, which may or may not be changed from those listed in Figure 9-3, depending on the firm's industry and its particular strategy. Selected examples of chain activities from Porter's book are summarized in Table 9-1.

Customers, channels, and suppliers also have value chains, and the firm's output of product or service becomes an input to the customer's value chain. The firm's differentiation and its competitive advantage depend on how the activities in its value chain relate to the needs of the customer, channel, or supplier. If quality has been chosen as a competitive advantage, it now remains to determine the customer's value chain and how the product or service can add value to the customer's system. Following this determination, the value chain should be organized into the required discrete activities, each one of which can improve the quality of the output for the purpose of meeting the customer's expectations. Before asking what you can do for the customer, ask what the customer expects to accomplish. The answer forms the basis for a quality organization. In this regard, it should be kept in mind that there are linkages between a firm's value chain and those of its customers, as well as downstream linkages with channels and suppliers. An excellent example of this is Wal-Mart, where a key competitive advantage was achieved through the value chain activity of technology development; in Wal-Mart's case, it was the sophisticated computer-based information system that improved the output of many other activities such as distribution, purchasing, and warehousing.

Table 9-1 Chain Activities

Primary activities	*Support activities*
Inbound logistics	Materials handling
Warehousing	**Procurement**
Inventory control	Dispersion of the procurement function throughout the firm
Vehicle scheduling	**Technology development**
Returns to suppliers	Efforts to improve products and processes
Operations	**Human resource management**
Machining	Recruiting
Management	Hiring
Packaging	Training
Assembly	Development
Maintenance	Compensation
Testing	**Firm infrastructure**
Outbound logistics	Supports entire chain
Material handling	General management
Order processing	Planning
Scheduling	Finance, accounting
Finished goods warehouse	Quality management
Marketing and sales	
Advertising	
Promotion	
Sales force	
Pricing	
Service	
Installation	
Repair	
Training	
Parts supply	

A spokesperson for Winnebago Industries, manufacturer of motor homes, concludes, "You must pick the right distribution network. In our case, it is our dealers. We believe we are only as strong as our dealer network. They are our first, last, primary, and most critical link to our end customers."[6]

Globe Metallurgical of Cleveland, the first small company to win the Baldrige Award, realized the importance of suppliers in their own value chain. Globe's management determined that the most effective method of assuring compliance with statistical process control and quality approaches in the suppliers' facilities would be to visit each supplier location with a quality improvement team and to train the hourly employees at each location. The program is a vital aspect of Globe's quality system.[7]

> BSQ Group, an architectural firm in Tulsa, Oklahoma, designs and constructs stores for Wal-Mart. Although the firm's immediate customer is Wal-Mart, they organize their value chain to go downstream with linkages to Wal-Mart's customer: "Many people believe that quality is generally in the eyes of the beholder. Well, in the case of Wal-Mart, that beholder is the store's customer. They are the ones that are helping us define the quality standards that we currently strive to present and it's with them in mind that we begin our study."[8]

ORGANIZING FOR QUALITY IMPLEMENTATION

The traditional approach to organization sees the process as a mechanical assemblage of functions and activities without a great deal of attention to strategy and desired results. The process takes the product as given and groups the necessary skills and activities into homogeneous functions and departments. This approach to building an organization structure has been criticized by Peter Drucker: "What we need to know are not all the activities that might conceivably have to be housed in the organization structure. What we need to know are the load-bearing parts of the structure, the *key activities*."[9]

Key activities will differ depending on the nature of the organization, its products, and its strategy. What is a key activity in one may not be in another. Advertising may be a key activity in the value chain of Coca-Cola, but not in Boeing Aircraft, where design is the key activity. Back-office activity may be a key activity in Merrill Lynch, but not in McDonald's. Firms frequently fail to prioritize or identify key activities in the value chain because of a tendency to organize around the chart of accounts. Some firms focus on those activities where cost, rather than quality or other source of differentiation, is the major consideration.

The value chain concept provides a systematic way to identify the key activities necessary for quality differentiation and a way to group them into homogeneous departments and functions. Indeed, an organization structure that corresponds to the value chain is the most economic and effective way to deliver quality and therefore achieve a competitive advantage.

It should be noted that the quality assurance department is generally not the load-bearing key activity when organizing for TQM. Quality assurance activities can be found in nearly every function of the company if these functions are viewed as links in the value chain. Any activity or function is a potential source of quality differentiation. The ill-defined or elusive word "quality" may be too narrow if it focuses on product or service alone. Moreover, such limited focus may exclude the many other activities that impact the customer's value chain. Not only those functions normally classified as "line" but a variety of "staff" functions as well can be the source of quality in the organization structure. Consider the following sample activities:

Activity	*Value to customer*
Purchasing	Improved cost and quality of product
Engineering and design characteristics	Unique product
Manufacturing	Product reliability
Order processing	Response time
Service	Customer installation
Scheduling	Response time
Inspection	Defect-free product
Spare parts	Maintenance
Human resources	Customer training

By listing the activities of the organization and comparing them to a value chain such as Figure 9-3, one can see the many potential ways that quality differentiation can be achieved. It should also be noted that these activities can lower customer costs as well.

Production of quality does not stop when the product leaves the factory. Distribution and service are part of the production process. Careful identification of customer value will reveal a number of other opportunities for quality differentiation. For example, buyers and potential customers frequently perceive value in ways they do not understand or because of incomplete knowledge. Scanning a daily newspaper or magazine quickly reveals the many ways that both manufacturers and service firms signal subjective, qualitative measures of quality. Do you buy Pepsi Cola for taste or brand image? Do you contemplate the purchase of a Volvo for performance or long life and safety? Consulting and accounting firms signal quality by the appearance and presumed professionalism of employees. Banks are known to build impressive facilities to indicate quality. Charles Revson, formerly of Revlon, once said, "I'm not selling cosmetics, I'm selling hope." The several criteria that the buyer may use to make a buying decision means that there may be an equal number of activities that become *key* activities in the creation of customer value. Porter provides several illustrative signaling criteria,[10] to which firm examples and organization activities that become key in delivery of the criteria have been added here:

Criteria	*Firm example*	*Activity involved*
Reputation	Appliances	Advertising
Appearance	Apparel	Design
Label	Athletic shoes	Graphics
Facilities	Bank	Maintenance
Time in business	Whiskey	Distribution
Customer list	Magazine publisher	Marketing
Visibility of top management	Consumer products	Hot line

Of course, having signaled a particular criterion to buyers and potential buyers, it is necessary to deliver as promised, measure the effectiveness of the criterion, and keep customer feedback communication lines open to ensure satisfaction.

Delivery of quality products or services depends on how well the many activities of the company are organized and integrated. The measurement of effectiveness is fundamental to the TQM process (see Chapter 7). It now remains to organize for customer feedback, another *key activity* that impacts other functions and activities throughout the organization.

Measuring customer satisfaction, or dissatisfaction, is an essential but often overlooked activity. What happens when a customer chooses a bank's trust department based on the criterion of experienced personnel, only to be shunted off to a recent college graduate or ignored by a "customer representative?" Research indicates that customers who are satisfied with a bank's quality will tell, on the average, three other people, while those who are dissatisfied will tell eight or nine others about poor quality.[11] How does a customer feel when returning an item under warranty, only to be patronized by a retail clerk? One survey found that for every problem incident reported to corporate headquarters, there are at least 19 other similar incidents that simply were not reported or that were handled by the retailer or the front line without being recorded. Most companies spend 95% of their resources handling complaints and less than 5% analyzing them.

There is a strong correlation between consumer satisfaction with response to problems or questions and the likelihood of purchasing another product from the same company.[12] Yet few customers bother to complain, and of those who do, only a small fraction reach top management. What is needed is the institutionalization of customer service throughout the organization as a key activity to be performed by everyone. Despite this evident need, many companies have neither the activities nor the supporting policies. For many that do, there is a conflict between organization and policies that may have an opposite effect. Covertly measuring quality by using mystery shoppers, holding motivational meetings that employees perceive as paternalistic and patronizing, and paying for sales rather than service are among those policies that may conflict with the need to provide quality products and service.[13] It may be difficult for employees to be quality conscious in the face of policies that discourage this attitude.

THE PEOPLE DIMENSION: MAKING THE TRANSITION FROM A TRADITIONAL TO A TQM ORGANIZATION

The typical company (Figure 9-4a) operates with a vertical, functional organizational structure based on reporting relationships, budgeting procedures, and specific and detailed job classifications.[14] Departmentation is by function, and communication, rewards, and loyalties are functionally oriented. Processes are forced to flow vertically from the top down, creating costly barriers to process flow.

The systems approach to organizing suggests three significant changes, one conceptual and two requiring organizational realignment:

- The concept of the inverted organizational chart
- A system of intracompany internal quality
- Horizontal and vertical integration of functions and activities

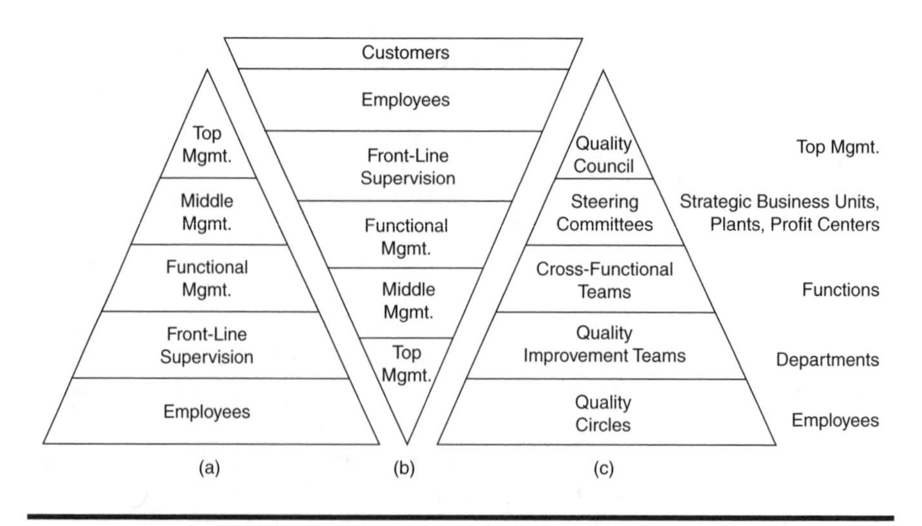

Figure 9-4 Transition from Traditional to TQM Organization

The Inverted Organizational Chart

If you've seen one organizational chart, you've seen them all: the symmetrical pyramid with the chairman at the top and the cascading of authority to successive levels (14 at General Motors) until the functions are shown near the bottom of the chart. Front-line supervisors are rarely shown and non-supervisory personnel almost never appear.

Where are the front-line supervisor and the employees? These are the people who deliver quality to the customer. In the eyes of the customer, they *are* the company. The sports fan cares not for the owner or the manager. The players deliver the quality. And so it is with the flight attendant, the bank teller, the auto mechanic, the salesperson explaining a product, the person answering the telephone…even the college professor.

Perhaps it is time to put first things first. To make the transition from traditional to TQM management, it may be desirable to *conceptualize* a new organizational chart. Invert the existing one (Figure 9-4b) and put the customer at the top, followed by the employees and front-line supervisors. These are the deliverers of quality. This concept does not change the hierarchy and flow of authority, but the boss is no longer the boss in the old-fashioned sense. He or she is now a facilitator, a coach, and an integrator, whose job is to remove barriers that prevent subordinates from doing their jobs. The same role now falls on middle and top management. Quality is now the responsibility of everyone and not just the quality assurance department.

Internal Quality

The Juran Institute of Wilton, Connecticut, delivers a program called "Managing Business Process Quality," which is a technique for executing cross-functional quality improvement among intracompany functions and activities.[15] A key factor in this approach is an organization-wide focus on the customer, including both *internal* and *external* customers. An enlarged definition of quality should be used to embrace all business processes, rather than just manufacturing.

The systems approach, by definition, requires the integration of organizational activities for achievement of a common goal. This goal, under the TQM form of organization, remains the satisfaction of customer requirements, but customers are now considered to be both outside as well as within the organization.[16] The process applies whether relating to a final customer or an internal customer; it is a participative process involving supplier and customer in an active dialogue. Examples include:

> Metropolitan Life Insurance Company has made a major commitment to improve quality by implementing a *horizontal management* approach that is built on management commitment, employee involvement, and knowledge of internal suppliers.[17]

> Campbell USA has aimed its latest quality emphasis, its "Quality Proud" program, at the administrative and marketing activities of the company. Job descriptions, promotions, pay, and bonuses for all employees are linked to the results of the new program.[18]

As a major step in its transformation to a total quality organization, the organization formerly known as DEC (acquired by Compaq in 1998) asked each of its 125,000 employees to answer in writing the following questions:

1. What business process are you involved in?
2. Who are your customers (that is, the next step in the processes you are involved in)?
3. Who are your suppliers (that is, the preceding step in the processes you are involved in)?
4. Are you meeting the expectations of your customers?
5. Are your suppliers meeting your expectations?
6. How can the processes be simplified and waste eliminated?[19]

> DEC reported that this simple survey had a massive impact. In the short run, countless redundant activities were discovered and eliminated. In the long run, DEC employees now think in terms of meeting both internal and external customer expectations. (This concept is also illustrated in Figure 9-2.)

Aside from the obvious benefits of improvements in quality, productivity, and cost, a system of internal customer quality is important for a number of other reasons:

- External customer satisfaction cannot increase unless internal customer satisfaction does.
- No quality improvement effort can succeed without employee buy-in and proactive participation.
- Focus on internal quality promotes a quality and entrepreneurial culture.
- An understanding of internal quality policy is an aid in communication and decision making.
- It is a significant criterion in the Malcolm Baldrige National Quality Award (Section 5.6).

ROLES IN ORGANIZATIONAL TRANSITION TO TQM

Members of a successful organization need a sound understanding of their roles during the transition to a TQM program. People at all levels require orientation as to how they will be impacted under the new philosophy of employee involvement. The improvement process involves a group of complementary activities that provide an environment conducive to improvement of performance for both employees and managers. Each level has a role to play.

The role of ***top management*** is critical. Many of the most successful companies launched their programs by creating a quality council or steering committee (Figure 9-4c) whose members comprise the top management team. Some multidivision companies encourage a council in each division or strategic business unit. The council provides a good vehicle for management to demonstrate its leadership in the quality initiative. At Motorola, the CEO, who is also the chief quality officer of the corporation, chairs the Operating and Policy Committee in all-day meetings twice each quarter.[20]

Opinions differ as to who should lead or coordinate the TQM effort. One source suggests a new role similar to that of a financial controller, a role that is justified on the basis that quality is now a strategic business planning and management function.[21] Others disagree and suggest that the company should avoid setting up a quality bureaucracy headed by a high-profile quality director. There is general agreement that it should not be headed by a staff department such as personnel or quality assurance. The process should be line led and given back to the business managers who implement it on a daily basis. To reiterate, quality should not be led by a non-line manager.

The major changes are strategic and organizational and have been outlined in this and previous chapters. It now remains for top management to manage the transition.[22]

The role of ***middle managers*** has traditionally been an integrative one. They are the drivers of quality and the information funnel for change both vertically and horizontally — the go-between for top management and front-line employees. They implement the strategy devised by top management by linking unit goals to strategic objectives. They develop personnel, make continuous improvement possible, and accept responsibility for performance deficiencies.[23]

Front-line supervision has been called the missing link in TQM.[24] At Federal Express, a Baldrige winner, the communication effort is focused on the front-line supervisors because most employees report directly to them. The company realizes that the real purveyors of quality are the employees, and a basic quality concept is candid, open, two-way communication.

Supervisors can make or break a quality improvement effort. They are called upon to provide support to employee involvement teams and create a climate that builds high levels of commitment in groups and individuals.

Quality assurance and the quality professional are faced with good news and bad news as TQM emerges as the load-bearing concern of company strategy. On the one hand, the accelerating emphasis on quality has given them more visibility, and in some cases the reporting relationships have moved to higher levels in the organization. On the other hand, they may now be perceived as a staff support function as quality becomes more widespread and led by line managers.

Philip Crosby indicates that the quality professional must become more knowledgeable about the process of management.[25] The limited tools of inspection techniques and statistical process control have become less important as the more sophisticated approaches of TQM begin to pervade all functions and activities, rather than just manufacturing.

SMALL GROUPS AND EMPLOYEE INVOLVEMENT

In a *Harvard Business Review* article, David Gumpert described a small "microbrewery" where the head of the company attributed their success to a loyal, small, and involved work force. He found that keeping the operation small strengthened employee cohesiveness and gave them a feeling of responsibility and pride.[26]

This anecdote tells a lot about small groups (hereafter called teams) and how they can impact motivation, productivity, and quality. If quality is the objective, employee involvement in small groups and teams will greatly facilitate the result because of two reasons: motivation and productivity.

The theory of motivation, but not necessarily its practice, is fairly mature, and there is substantial proof that it can work. By oversimplifying a complex theory, it can be shown why team membership is an effective motivational device that can lead to improved quality.[27]

A team-based structure often leads to improved productivity as a result of greater motivation (Table 9-2), reduced overlap, and better communication, unlike a functionally based classical structure characterized by territorial battles and parochial outlooks. There is always the danger that functional specialists may pursue their own interests with little regard for the overall company mission. Membership in a team, particularly a cross-functional team, reduces many of these barriers and encourages an integrative systems approach to achievement of common objectives, those that are common to both the company and the team. There are many success stories. To cite a few:

Table 9-2 Team Membership and Motivation

Motivating factors	*Team membership*
Job development (the work)	
Vertical loading	Provides responsibility
Job closure	Team members see results
Feedback	Self-established goals
Achievement	Targets set by teams
Growth/self-development	Training, more responsibility
Recognition	By peers and supervisors
Communication	Team is vehicle for communication (see Chapter 3)

Globe Metallurgical, Inc., the first small company to win the Baldrige Award, had a 380% increase in productivity, which was attributed primarily to self-managed work teams.[28]

The partnering concept requires a new corporate culture of participative management and teamwork throughout the entire organization. Ford increased productivity 28% by using the team concept with the same workers and equipment.[29]

Harleysville Insurance Company's Discovery program provides synergism resulting from the team approach. The program produced a cost savings of $3.5 million, along with enthusiasm and involvement among employees.[30]

At Decision Data Computer Corporation, middle management is trained to support "Pride Team."[31]

Martin Marietta Electronics and Missiles Group has achieved success with performance measurement teams (PMTs).[32]

Publishers Press has achieved significant productivity improvements and attitude change from the company's process improvement teams (PITs).[33]

Florida Power & Light Company, the utility that was the first recipient of the Deming Prize, has long had quality improvement teams as a fundamental component of their quality improvement program.[34]

TEAMS FOR TQM

The several subsystems or components of a TQM approach were examined in previous chapters. The most critical of these components is employee involvement, and it is the one around which the management system of TQM should be based. It is the most important of the components of TQM and also the most complex. Consider the analogy of an iceberg. Approximately 10% of an iceberg is visible, while 90% is hidden from view. Imagine that the organizational chart is an iceberg. The visible 10% is top management and functional management. The 90%, where the true potential for quality exists, is comprised of front-line supervision and non-management employees. Does it not make good sense to tap into the 90% which represents a reservoir of ideas for quality and productivity improvements? The vehicle for doing this is some form of *team*.

A 1989 General Accounting Office study found that over 80% of all companies had implemented some form of employee involvement.[35] However, the statistic is misleading because responding companies considered a suggestion system as an employee involvement program, which is hardly a systems approach or a linking vehicle. Moreover, the methods most likely to have enduring effects are those that covered the smallest percentage of employees.

Quality Circles

The most widespread form of an employee involvement team is the quality circle, defined as "a small group of employees doing similar or related work who meet regularly to identify, analyze, and solve product-quality and production problems and to improve general operations."[36] Although the concept has had some success in white-collar operations, the major impact has been among "direct labor" employees in manufacturing, where concerns are primarily with quality, cost, specifications, productivity, and schedules. By their very nature, quality circles were limited to concerns of the small group of members and few cross-functional problems were considered.

The major growth of the circles occurred in the late 1970s and early 1980s, as thousands of companies adopted the concept. Like so many previous movements (e.g., management by objectives, value analysis, zero-based budgeting), however, the concept never met expectations and widespread abandonment resulted. As many as 50% of Fortune 500 companies disbanded their circles in the 1980s.[37] The major reason for failure was a general lack of commitment to the concept of participation and the lack of interest and participation by management.[38] From a TQM perspective, quality circles lack the prerequisites of integration with strategy, company goals, and management systems. Organizations can go beyond using circles by creating task forces, work teams, and cross-functional teams.[39]

Task teams are a modification of the quality circle. The major differences are that task teams can exist at any level and the goal or topic for discussion is given, whereas in quality circles members are generally free to choose the problems they will solve. Task teams with the best chance for success are those that represent an extension of a pre-existing, successful quality circle program.[40]

Self-managing work teams are an extension of quality circles but differ in one major respect: members are empowered to exercise control over their jobs and optimize the efficiency and effectiveness of the total process rather than the individual steps within it. Team members perform all the necessary tasks to complete an entire job, setting up work schedules and making assignments to individual team members. Peer evaluation is another characteristic.[41]

Cross-Functional Teams

> Several organizations have integrated a range of proven TQM techniques into their programs, including cross-functional process improvement teams. Some of the techniques involve a sophisticated, closed-loop suggestion system designed to discover and address problems. Under such techniques, only an employee who makes a suggestion can dispose of it. He or she also has the responsibility of working with other employees to implement or reshape the suggestion in order to determine whether it is feasible. The techniques empower employees and promote team building, two essential elements of quality management.

The centuries-old hierarchical form of organization with a vertical chain of command was the norm until recently, when organizational complexity demanded horizontal as well as vertical coordination in order to plan and control processes that flowed laterally. If no lateral coordination is achieved, the organization becomes a collection of islands of specialization without integration, a requirement of the systems approach. Linking business process improvement (billing, procurement, recruiting, record keeping, design, sales, etc.) to the key business objectives of the organization is necessary if quality is to become real and relevant. There is widespread agreement that cross-functional teams provide the best vehicle for linking these activities and processes. The concept of linkages is shown in Figure 9-5. Note that a cross-functional approach achieves the objectives of:

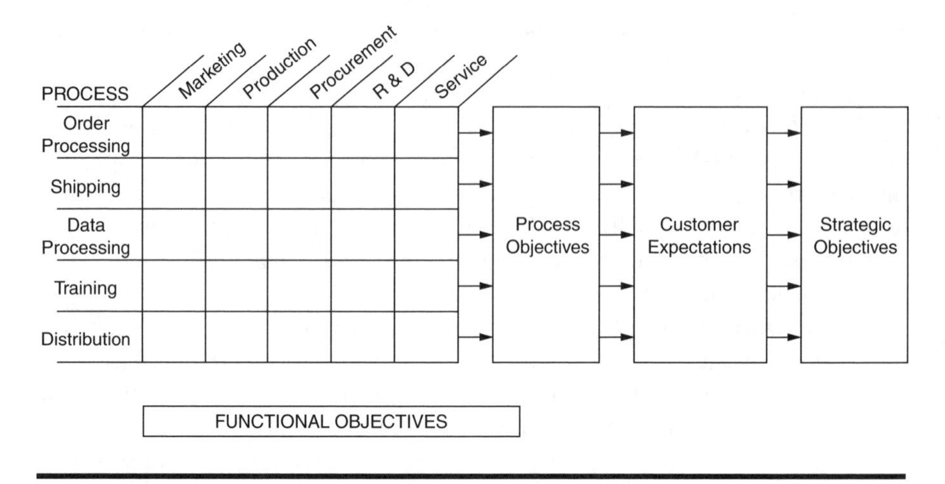

Figure 9-5 Cross-Functional Linkages

- Customers
- Functions
- Processes
- The organization

Team expert Michael Donovan summarizes a number of trends that will shape the structure and process of employee involvement efforts in the future:[42]

From	*To*
Perception of employee involvement as a program	Perception as an ongoing process
Voluntary participation	Participation by all members as a natural work team
Quality circles	Several types of teams at many levels
Project focus	Goal focus
Limited management involvement	Active management involvement
Functional management skills	Building participative leadership and facilitation skills into management roles
Employee participation in operating problems	Employee participation in broader issues

EXERCISES

9-1 How does an organizational structure that is focused on classical principles (specialization of labor, unity of command, span of management, delegation of authority) tend to inhibit the implementation of TQM?

9-2 Define the concept of *synergism*. How does organizing around the principles of TQM tend to integrate the organization and achieve synergism?

9-3 What is the concept of the *value chain*? How can it be useful in building an organizational structure?

9-4 In organizing for customer satisfaction, what would be a key activity for:
- A brokerage firm
- An aircraft manufacturer
- A retail store

9-5 Explain the concept of the inverted organizational chart.

9-6 Explain how membership in a small group might lead to improved motivation and hence improved quality.

ILLUSTRATIVE CASES

Existing management literature acknowledges that there are fundamental operational differences between small and large firms, but the question remains: Do these differences hinder the implementation of total quality management? One study[43] found that there are no operational differences in TQM implementation attributable to firm size and that small and large firms that produce high-quality products implement TQM equally effectively. A second study[44] found that two of the most serious problems small companies may face when implementing TQM are the owner-manager's lack of business experience and knowledge and the shortage of financial and human resources required.

Question

- What is your opinion regarding the differences between large and small businesses and their ability to implement a program of quality improvement?

HR Focus reported the case of an industrial equipment manufacturer seeking the help of a consulting firm because the team it formed to improve the process of engineering, manufacturing, and assembling one of its product lines was not performing as expected. Production problems were mounting and customers were becoming unhappy. The consulting firm conducted surveys and interviews and through observation found out that team members felt frustrated, angry, and burned out. Several factors undermined team performance, including a lack of management support, a weak measurement system, and a lack of incentives. The consultant recommended that the team be disbanded and that its scope and membership be revamped. It was also recommended that the company be restructured around product lines.

Questions

- Do you agree with the consultant? Why or why not?
- Describe the scope and membership of the revamped team.
- Why do you think the consultant recommended restructuring around product lines?

ENDNOTES

1. For example, Michael H. Mescon, Michael Albert, and Franklin Khedouri, *Management,* New York: Harper & Row, 1988, p. 323.
2. W. Edwards Deming, *Out of the Crisis,* Cambridge, Mass.: Massachusetts Institute of Technology, Center for Advanced Engineering Study, 1982, pp. 465–474.
3. Philip B. Crosby, *Quality Is Free,* New York: McGraw-Hill, 1979, pp. 69–70.
4. The Conference Board, *Global Perspectives on Total Quality,* New York: The Conference Board, 1991, p. 9.
5. Michael Porter, *Competitive Advantage: Creating and Sustaining Superior Performance,* New York: The Free Press, 1985.
6. Presentation at the Total Quality Service Management Conference, Dallas, May 21–23, 1990.
7. Kenneth Leach, Vice-President, Administration of Globe Metallurgical, Inc. at the Third Annual Quality Conference, June 22, 1990.
8. Presentation at the Total Quality Service Management Conference, Dallas, May 21–23, 1990.
9. Peter Drucker, *Management: Tasks, Responsibilities, Practices,* New York: Harper & Row, 1974, p. 530.
10. Michael Porter, *Competitive Advantage: Creating and Sustaining Superior Performance,* New York: The Free Press, 1985, p. 144. Signals of value are those factors that buyers use to infer the values a firm creates.
11. Keith Brinksman, "Banking and the Baldrige Award," *Bank Marketing,* April 1991, pp. 30–32.
12. American Society for Quality Control, *'88 Gallup Survey of Consumers' Perceptions Concerning the Quality of American Products and Services,* Milwaukee: ASQC, 1988.
13. Mark Graham Brown, "How to Guarantee Poor Quality Service," *Journal for Quality and Participation,* Dec. 1990, pp. 6–11.

14. In 1981, Cleveland Twist Drill, a Cleveland-based manufacturer of cutting tools with $400 million in sales, had over 500 job classifications in a direct labor force that numbered fewer than indirect labor. Joseph L. Bower et al. *Business Policy,* Homewood, Ill.: Irwin, 1991, p. 588.
15. "How to Profit from Managing Business Process Quality," presentation at the Total Quality Service Management Conference, Dallas, May 21–23, 1990.
16. David Mercer, "Key Quality Issues," in *Global Perspectives on Total Quality,* New York: The Conference Board, 1991, p. 11. Mercer is the project director of the European Council on Quality of The Conference Board Europe.
17. Keith D. Denton, "Horizontal Management," *SAM Advanced Management Journal,* Winter 1991, pp. 35–41.
18. Herbert M. Baum, "White-Collar Quality Comes of Age," *Journal of Business Strategy,* March/April 1990, pp. 34–37.
19. U.S. General Accounting Office, "Quality Management Scoping Study," Washington, D.C.: General Accounting Office, 1991, p. 23.
20. A company handout entitled "The Motorola Story," written by Bill Smith, Senior Quality Assurance Manager, Communications Sector. The committee's meetings are described: "The Chief Quality Officer of the corporation opens the meetings with an update on key initiatives of the Quality Program. This includes results of management visits to customers, results of Quality System Reviews (QSR's) of major parts of the company, cost of poor quality reports, supplier–Motorola activity, and a review of quality breakthroughs and shortfalls. This is followed by a report by a major business manager on the current status of his/her particular quality initiative. This covers progress against plans, successes, failures, and what he projects to do to close the gap on deficient results, all pointed at achieving Six Sigma capability by 1992." Discussion follows among the leaders concerning all of these agenda items.
21. Al P. Staneas, "The Metamorphosis of the Quality Function," *Quality Progress,* Nov. 1987, pp. 30–33.
22. There are a number of good sources that provide suggestions for managing change. See, for example, Tom Peters, "Making It Happen," *Journal for Quality and Participation,* March 1989, pp. 6–11; Nina Fishman, "Playing the Transition Game Successfully," *Journal for Quality and Participation,* June 1990, pp. 52–56; John Herzog, "People: The Critical Factor in Managing Change," *Journal of Systems Management,* March 1991, pp. 6–11; Ronald Elliott, "The Challenge of Managing Change," *Personnel Journal,* March 1990, pp. 40–49; Edmund Metz, "Managing Change: Implementing Productivity and Quality Improvements," *National Productivity Review,* Summer 1984, pp. 303–314; and Richard Sparks and James Dorris, "Organizational Transformation," *Advanced Management Journal,* Summer 1990, pp. 13–18.
23. G. Harlan Carothers, Jr., "Future Organizations of Change," *Survey of Business,* Spring 1986, pp. 16–17.
24. Nina Fishman and Lee Kavanaugh, "Searching for Your Missing Quality Link," *Journal for Quality and Participation,* Dec. 1989, pp. 28–32.
25. Nancy Karabatsos, "Quality in Transition: Part One," *Quality Progress,* Dec. 1989, pp. 22–26.
26. David E. Gumpert, "The Joys of Keeping the Company Small," *Harvard Business Review,* July/Aug. 1986, pp. 6–14.

27. With apologies to Maslow and Herzberg, who have provided what is probably the most practical approach to motivation. See Abraham Maslow, "A Theory of Human Motivation," *Psychological Review,* No. 50, 1943, pp. 370–396 and Frederick Herzberg, "One More Time: How Do You Motivate Employees?" *Harvard Business Review,* Jan./Feb. 1968, pp. 56–57. A complete review and summary of the writings of both of these theorists can be found in almost any principles of management textbook. For example, see Michael Mescon, Michael Albert, and Franklin Khedouri, *Management,* New York: Harper & Row, 1988.
28. James H. Harrington, "Worklife in the Year 2000," *Journal for Quality and Participation,* March 1990, pp. 56–57.
29. John Simmons, "Partnering Pulls Everything Together," *Journal for Quality and Participation,* June 1989, pp. 12–16.
30. Rick L. Lansing, "The Power of Teams," *Supervisory Management,* Feb. 1989, pp. 39–43.
31. Larry Gerhard and Walter T. Sparrow, "Pride Teams, A Quality Circle that Works," *Journal for Quality and Participation,* June 1988, pp. 32–36.
32. Vladimir J. Mandl, "Team Up for Performance," *Manufacturing Systems,* June 1990, pp. 34–41.
33. Gary Ferguson, "Printer Incorporates Deming — Reduces Errors, Increases Productivity," *Industrial Engineering,* Aug. 1990, pp. 32–34.
34. In company presentation at the Miami headquarters.
35. As reported in Brian Usilaner and John Leitch, "Miles to Go...Or Unity at Last," *Journal for Quality and Participation,* June 1989, pp. 60–67.
36. Joel E. Ross and William C. Ross, *Japanese Quality Circles and Productivity,* Reston, Va.: Reston Publishing, 1982, p. 6. For those contemplating the establishment of quality circles or other quality improvement teams, this book provides an action plan for the process.
37. James H. Harrington and Wayne S. Rieker, "The End of Slavery: Quality Control Circles," *Journal for Quality and Participation,* March 1988, pp. 16–20. For an example of how the Avco division of Textron revitalized its quality circles with management support, see Peggy S. Tollison, "Managers Are People Too: A Case Study on Developing Middle Management Support," *Quality Circles Journal,* March 1987, pp. 12–15.
38. Rick Lansing, "The Power of Teams," *Supervisory Management,* Feb. 1989, pp. 39–43.
39. Edward E. Lawler and Susan A. Mohrman, "Quality Circles: After the Honeymoon," *Organizational Dynamics,* Spring 1987, pp. 42–54.
40. Carol Gabor, "Special Project Task Teams: An Extension of a Successful Quality Circle Program," *Quality Circles Journal,* Sep. 1986, pp. 40–43.
41. Michael J. Donovan, "Self-Managing Work Teams — Extending the Quality Circle Concept," *Quality Circles Journal,* Sep. 1986, pp. 15–20.
42. Michael Donovan, "The Future of Excellence and Quality," *Journal for Quality and Participation,* March 1988, pp. 22–24.
43. Sanjay L. Ahire and Damodar Golhar, "Quality Management in Large vs. Small Firms," *Journal of Small Business Management,* April 1996, pp. 1–13.
44. See Cengiz Haksever, "Total Quality Management in the Small Business Environment," *Business Horizons,* March/April 1996, pp. 33–40.

REFERENCES

Masters, Robert J., "Overcoming the Barriers to TQM's Success," *Quality Progress,* May 1996, pp. 53–55.

Milas, Gene H., "Guidelines for Organizing Employee TQM Teams," *IIE Solutions,* Feb. 1996, pp. 36–39.

Ross, Timothy L., "Self-Management and Gainsharing: A Winning Duo," *National Productivity Review,* Summer 1996, pp. 55–63.

Trotman, Alex, "Preparing for Change," *Executive Excellence,* Jan. 1996, pp. 18–19.

10

QUALITY AND PRODUCTIVITY

In Japan, we are keeping very strong interest to improve quality by use of methods which you started. When we improve quality we also improve productivity.

Dr. Yoshikasu Tsuda
University of Tokyo

During the mid-1980s, the President's Council for Management Improvement wrestled with the productivity process mandated by Ronald Reagan. However, corporate chief executives encouraged the president to get away from processes that stressed productivity and instead to focus on quality. These events led to the creation of the Malcolm Baldrige Award and the subsequent popularity of total quality management (TQM) in U.S. industry.

The relationship among quality, market share, and profitability was examined in Chapter 1, and it was shown that higher quality leads to both increased profits and greater market share. The following questions now arise: Are productivity and quality related? Are they two sides of the same coin? Can you have both? The answer, of course, is *yes*.

Despite a growing body of evidence that indicates a positive correlation, the misconception exists that productivity and cost must be sacrificed if quality is to be improved. In an annual survey of its members in 1990, the Institute of Industrial Engineers found the general opinion to be that only when productivity and quality are considered together can competitiveness be enhanced.[1]

There may be some justification for the belief that increased quality means decreased productivity, but it seems to be the view of those who rank production ahead of quality as the top priority. It is argued that a program to improve quality causes disruptions and delays that result in reduced output. While this may be the case in the short run, it generally is not true over a longer time period. As will be discussed in Chapter 11 ("The Cost of Quality"), such an argument usually fails when the costs associated with poor quality are considered.

The argument for a positive relationship was made by Deming, who based it on the reduced productivity that is caused by quality defects, rework, and scrap. He concluded, "Improvement of quality transfers waste of man-hours and of machine-time into the manufacture of good products and better service."[2] Feigenbaum maintains that a certain "hidden" and non-productive plant exists to rework and repair defects and returns, and if quality is improved, this hidden plant would be available for increased productivity.[3] These arguments are straightforward; any quality improvement that reduces defects is, by definition, an improvement in productivity. The same can be said, of course, for services and for those firms in service businesses. The cost of quality improvement rarely exceeds the savings from increased productivity.

To build a case for or against quality improvement based on output or defect reduction alone is to oversimplify. A more convincing case can be built around the proven benefits of TQM. When the broader picture is considered, it can be shown that increasing quality also increases productivity, and the two are mutually reinforcing.[4] Productivity has come to mean more output for the same or less cost. TQM embraces a broader concept and can be perceived as *including* the benefits of productivity when properly implemented. Productivity has become a tactical short-term approach associated with cost reduction, greater efficiency, better use of resources, and organizational restructuring. TQM is longer term and more comprehensive, and as such is concerned with cultural change and creating visions, mission, and values.

Examples of productivity improvements resulting from TQM abound:

> Under Joseph Juran's guidance, the Internal Revenue Service's processing center in Ogden, Utah, adopted quality as a core value, but also achieved productivity increases of $11.3 million from team and management initiatives.
>
> NASA's Productivity Improvement and Quality Enhancement (PIQE) program has evolved into a multiprogram approach incorporating TQM in the agency and in the contractor work force, which comprises about 60% of NASA's total.[5]
>
> The introduction of computer-integrated manufacturing (CIM), combined with TQM and self-directed work teams, resulted in a 50% increase in productivity at Monsanto Chemical's Fibers Division.[6]

THE LEVERAGE OF PRODUCTIVITY AND QUALITY

If quality has a leverage effect on market share and profitability (as pointed out here and in Chapter 1), what are the bottom-line consequences of productivity improvement?

Confining the illustration to the question of profitability leverage, three hypothetical income statements will demonstrate how small (10%) increases in productivity will yield much greater results than a similar increase in sales:

	I	*II*	*III*
	Before	*Sales up 10%*	*Productivity improved 10%*
Sales	$100	$110	$100
Variable costs	70	77	63
Fixed costs	20	20	20
Profit	$10	$13 (+30%)	$17 (+70%)

In situation I, sales are $100, variable costs $70, and fixed costs $20, yielding a profit of $10. In situation II, a sales increase of 10% yields a 30% profit increase, while situation III shows a 70% profit increase with *no increase in sales.* The leverage is even more dramatic if a smaller and more realistic return on sales is used. There are also potential additional companion benefits that can be achieved in quality. Again, the answer lies in TQM and the continuous improvement of all processes.

MANAGEMENT SYSTEMS VS. TECHNOLOGY

Since the time of Adam Smith's historic 18th-century book *The Wealth of Nations,* we have been taught to believe that labor specialization accompanied by mechanization was the answer to economic growth and productivity. The Industrial Revolution proved this to be so. Even today, the conventional wisdom of economists tells us that the rate of productivity growth is largely a function of changes in real capital relative to labor.

There is a continuing debate in Washington regarding the "reindustrialization of U.S. industry," or "supply-side economics," as it is came to be known in the administrations of Ronald Reagan and George H. W. Bush. The primary domestic objective of these administrations was the improvement of the productivity of U.S. industry by encouraging greater savings and thus investment in capital stock. Competitiveness, it was said, required an overhaul of U.S. technology. It was generally believed that Japan's quality and productivity advantage came from advanced technology.

It would be a mistake to attribute Japan's success to technology alone and a bigger mistake to consider technology to be the only answer to improved U.S. quality and productivity. It is not labor replacement that is needed but rather improved processes. Why, for example, would a company invest in advanced computer equipment to improve an information system that is flawed or a manufacturing process that is antiquated? In the first case, the technology will provide bad information more quickly so that poor decisions can be made faster. In the second case, process labor may be replaced only to find an increase in lead time, inventory turn, or cost of quality.

Many people think of technology as automation and mechanization, machines and computers, and semiconductors and new inventions, but the term has a much broader meaning. It is a means of transforming inputs into outputs. Thus, technology includes methods, procedures, and techniques which enable this transformation. It includes both machines and methods. This is worth repeating: technology includes methods that improve processes to improve the output/input ratio. Company after company has achieved remarkable increases in both quality and productivity with little or no investment in the hardware side of technology.

No one can argue convincingly against the use of the hardware side of technology to improve both quality and productivity. The problem is that automation and machines require time and money, both of which are in short supply. Management systems take little of either and may be equally or more effective. The solution is to improve the system — the process — before introducing technology. General Motors has spent more on automation than the gross national product of many countries, yet the excessive cycle time from market research to manufacture resulted in the production of cars that were not competitive. While GM was taking eight years to produce a Saturn, Honda took half as long to market a more competitive car. Honda accomplished this by controlling cycle time and processes.

The general tendency is to focus on technology to reduce labor cost and to overlook the improved quality that can be achieved through improvement of related processes and tapping the potential of the work force. Good companies buy technology to improve processes, reduce lead times, boost quality, and increase flexibility.

Capital spending in service industries has exploded, but there has been very little increase in productivity or quality. Jonathan M. Tisch, president and CEO of Loews Hotels, remarked: "Productivity in manufacturing is advancing five times as fast as in the service sector. In the late 1950s we needed roughly one employee for every four occupied rooms and that was the average across the industry. Today's average, nationwide, is one employee for every two rooms. In other words, productivity is half what it used to be. Despite the advent of the computer and the introduction of many so called labor-saving devices."[7] The focus in both manufacturing and service industries has been on labor productivity, but for most businesses capital intensity does not improve labor productivity enough to keep return-on-investment above the cost of capital. For those businesses that become more capital intensive relative to sales, a decline in return on investment is the result, even if a normal increase in productivity is achieved.[8]

PRODUCTIVITY IN THE UNITED STATES

The productivity record in the United States is not good. Our capital-intensive industries — home of industrial engineering and the assembly line, production planning, and the computer — have been beaten by Japan and the leading nations of Western Europe as U.S. labor productivity continues to compare unfavorably with the rest of the industrialized world. It is a critical issue for the nation and for individual firms.

Reasons for Slow Growth[9]

When it comes to identifying causes for what has been called the "productivity crisis," every economist, industrialist, and government official seems to have a favorite culprit. Among the most popular explanations are the following issues.

Management inattention — U.S. Secretary of Commerce Malcolm Baldrige (who died in 1987 and for whom the Baldrige Award is named) stated: "Between our own complacency and the rise of management expertise around the world, we now too often do a second-rate job of management, compared to our foreign competitors." One survey by A.T. Kearney, Inc. (management consultants) concluded that the key to productivity is better management and not continued efforts to produce more pounds of automobile per worker. The decade of the 1980s is noted for top management's diversion from productivity, quality, and growth to leveraged buyouts, restructuring, downsizing, and in many case executive perks and golden parachutes.[10]

Short-term gain — The trend has been to focus on short-term financial ratios while failing to take action to ensure long-term growth and productivity.[11] While no one would recommend overlooking financial data, this type of information suffers from the shortcomings of all accounting data. Moreover, financial figures tend to favor the productivity of capital while overlooking the other inputs of labor, material, and energy.

Direct labor — Focus on direct labor has historically been the one variable cost around which financial control systems are designed. Today, the direct labor share of total production costs is down to 8 to 12% on average.[12] Some firms fold these costs into overhead or general and administrative expenses, categories that are frequently overlooked when searching for ideas to improve productivity and quality.

Capital — Capital stock formation is largely dependent on savings. Yet U.S. workers appear to be spending more and saving less, leaving fewer dollars for capital formation. The net savings ratios for the major industrialized nations of the world[13] are as follows:

	Average 1980–89	*1990*	*1991*
United States	6.0	4.6	4.3
Japan	16.0	14.3	14.5
Germany	12.5	13.4	12.8
Three other major European countries (France, Italy, U.K.)	14.1	12.0	12.2

Although these ratios may have shifted somewhat since 1991, there has been no substantial difference in relationships.

Research and development — Expenditures for research and development in the United States surpass every other nation, yet overseas rivals are outpacing the United States in spending growth. Opinion is mixed as to the impact of R&D on productivity. Some evidence indicates that spending is directed toward product improvement rather than productivity improvement. This is good and is expected. However, as previously suggested, many quality investments also improve productivity. An R&D "peace dividend" may be expected from political events in the former Soviet Union and Eastern Europe as R&D dollars move from defense to programs in industry.

Inflation — Is inflation the cause of productivity decline or is inflation the result of the decline?[14] It is almost certain that lower productivity combined with higher wages does result in inflation. To the extent that inflation results in increased relative cost of plant and equipment as compared to labor and the relative cost of operating capital, there can be little doubt that these are investment *disincentives.*

Government regulation, the shift to a service economy, and the lack of goals and programs are among other reasons that have been advanced for the poor record of U.S. productivity. The cumulative effect, although significant, is difficult to estimate.

MEASURING PRODUCTIVITY

Measuring productivity is somewhat easier than measuring quality because the latter is determined by the customer and may be fragmented and elusive. On the other hand, productivity can also be difficult to measure because it is measured by the output of many functions or activities, many of which are also difficult to define.[15] What is the *measurable* output of design, market research, training, or quality assurance?

Despite these difficulties, measures are needed for each activity and in most cases for each individual front-line supervisor. Standards are needed for comparison against past performance, the experience of competitors, and as a basis for action plans to improve.

Carl G. Thor, president of the American Productivity and Quality Center in Houston, is a pioneer in the productivity measurement process and has worked for many years on the development of a measurement system. His principles of measurement for both productivity and quality include:[16]

- Meet the customer's need — that person who plans to use it. The customer may be external or internal.
- Emphasize feedback directly to the workers in the process that is being measured.
- The main performance measure should measure what is important. This may not be the case with the traditional cost control report.
- Measures should be controllable and understandable by those being measured. This principle may be enhanced by the participation of those being measured.
- Base measures on available data. If not available, apply cost–benefit analysis before generating new data. Information is rarely worth more than the cost of obtaining it.

BASIC MEASURES OF PRODUCTIVITY: RATIO OF OUTPUT TO INPUT

Total factor is the broadest measure of output to input and can be expressed as:

$$\frac{\text{Total output}}{\text{Labor} + \text{Materials} + \text{Energy} + \text{Capital}}$$

This measure is not only concerned with how many units are produced or how many letters are typed, but also considers all aspects of producing goods and services. Hence, this measure is concerned with the efficiency of the entire plant or company.

Partial factor measures are established by developing ratios of total output (e.g., number of automobiles, patients, depositors, students, widgets, etc.) to one or more input categories and are expressed as follows for the partial factor of labor:

$$\frac{\text{Total output}}{\text{Labor input}}$$

The same applies to material, capital, and energy. All measures are ratios of quantities. Although some ratios can be expressed in quantitative terms such as units produced per man-hour, others must combine unlike quantities of inputs, such as tons and gallons of products, employee-hours, pounds, kilowatt-hours, etc. To solve this problem, a set of weights representative of the relative importance of the various items can be used to combine unlike quantities. Base period prices are the recommended weights to be used for calculating total productivity, although other weighting systems such as "man-hour equivalents" can be used.

Total Productivity Measurement Model (TPM)

Total Productivity is defined as the ratio of total output to total input, where total input consists of labor, material, equipment, energy, and capital.[17]

Functional and departmental measures are more likely to benefit the company than an effort to apply comprehensive, company-wide coverage. Most firms rely largely on budgetary dollar accounting data to analyze their operations, even though these data include the effects of inflation, taxes, depreciation, and the arbitrary accounting cost allocations previously mentioned. Because these accounting figures are frequently not significantly related to the activity or process under study, it is desirable to develop measures that reflect output and input in more realistic terms. Where financial measures are used, it is appropriate to deflate them to a base benchmark.

It is important to establish function and activity measures because these organizational entities are where productivity and quality are delivered and where processes are improved. It is here where process design and control happen. A sampling of illustrative measures is provided in Table 10-1.

Individual measures provide the individual supervisor and worker with the basic target for improvement of both quality and productivity through individual action planning. Improvement can only occur if measured against some benchmark (target, yardstick, standard, objective, or result expected).

Table 10-1 Function and Activity Measures

Function/activity	*Measure*
Customer support	Cost per field technician, cost per warranty callback
Data processing	Operations employees per systems design employees
Quality assurance	Units returned for warranty repair as percentage of units shipped
Order processing	Orders processed per employee, sales per order-processing employee
Production control	Order cycle time, inventory turn, machine utilization, total production to production schedule
Shipping	Orders shipped on time, packing expense to total shipping expense
Testing	Man-hours per run-hour, test expense to rework expense

The simplest and most effective way to set a standard is to list the responsibility of the job on a piece of paper and then list the measures (results expected) that would indicate that the job is being performed satisfactorily. This provides a benchmark from which improvement can proceed. For example:

Responsibility	*Measure*
Maintenance	Maintain an uptime machine rate of 95%
Assembly	Assemble 32 units per man-hour of direct labor
Accounts receivable	Maintain an accounts receivable level of 42 days

Having established these measures, or standards, the individual can then write a *productivity or quality improvement objective* (results expected). Taking the preceding examples, these improvement objectives could be written:

My productivity (or quality) improvement objective is

Action verb	*Results expected*	*Time*	*Cost*
Improve	Machine uptime from 94%	By June 30	At no increase in man-hours or preventive maintenance costs
Increase	Actual production from 90% of schedule to 95%	Commencing this quarter	At the same cost of manufacture

Industry and competitive measures are important for benchmarking against the competition, best-in-class, and others in the industry. These are examined in Chapter 8, "Benchmarking."

Many companies set measures of total factor productivity such as output per labor hour, material usage rates, ratio of direct to indirect labor, etc., but such macro measures provide little in the way of functional or departmental measures from which an improvement plan can be developed. Unlike return on investment (a measure of capital productivity), which can be broken down into each of its determinants, broad macro measures mean little to those lower in the hierarchy who need specific objectives in order to develop an action plan.

WHITE-COLLAR PRODUCTIVITY

Productivity of white-collar workers is no less important than that of direct labor or manufacturing employees. Indeed, in terms of numbers and expense, staff and non-production employees outnumber production employees by a wide margin. Yet the problem of measurement of output is more elusive. Measuring the units assembled per man-hour is not too difficult, but how many reports should an accountant prepare, not to mention the most difficult of all measures — managerial productivity. Peter Drucker tells us that it is "usually the least known, least analyzed, least managed of all factors of productivity."[18] Research has shown that white-collar employees are productive only about 50% of the time. The remainder is non-productive time and can be traced to personal delays (15%) and improper management (35%). Causes of wasted time include:

- Poor scheduling
- Slack start and quit times
- Lack of communication between functions
- Information overload
- Poor staffing
- Inadequate communication of assignments
- Unproductive meetings and telephone conversations

Measuring the Service Activity

Although the manufacturing worker (one who physically alters the product) has been measured for decades by time standards, time studies, and work sampling, it is not as easy to set standards for the non-manufacturing employee or the service activity. It is unlikely that measurement can be achieved in the same way as is done for the manufacturing worker. Nevertheless, a system can be devised to describe the productivity of an

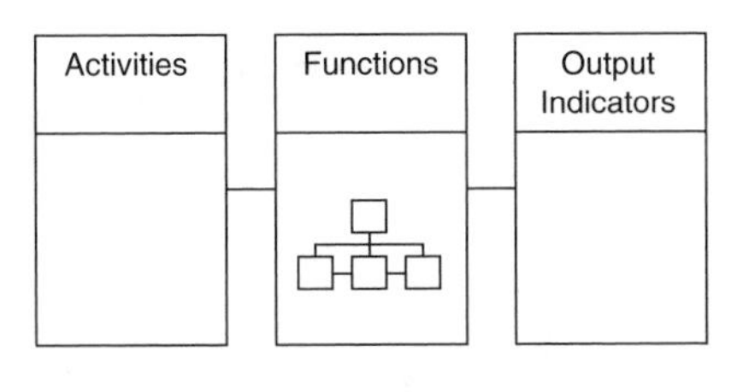

Figure 10-1 Measuring White-Collar (Indirect) Activity

activity at a point in time and then provide a baseline for judging continuous improvement over time. The system is particularly appropriate for multiplant or multidivisional companies with similar products or services and for individual companies within an industry.

The basis for a system of measurement starts with the existing functions and activities of the organization. Each activity is a subset of a particular function. For example, the *activity* of recruiting is a part of the human resource *function,* accounts receivable is a part of the accounting function, and so on. The typical organization may identify a hundred or more activities that can be grouped into ten or more functions. This concept is shown in Figure 10-1.

The next step is to identify the *output indicators* that "drive" the activities or cause work in the activities. In other words, if it were not for the work caused by or resulting from the *indicators,* there would be little need for the *activities.* If, for example, there were no personnel employed, there would be no need for employee relations. If there were no purchasing, there would be no need for vendor invoicing. The resources utilized in the activity of vendor invoicing are therefore a dependent variable of the purchasing function. In other words, if activities are the "input" in the productivity ratio of output to input, then the indicators are the "output."

IMPROVING PRODUCTIVITY (AND QUALITY)

Improvement means increasing the ratio of the output of goods and services produced divided by the input used to produce them. Hence, the ratio can be increased by either increasing the output, reducing the input, or both. This concept is illustrated in Figure 10-2, along with a sampling of actions and techniques for improving the productivity ratio. This might be called the productivity wheel.

Historically, productivity improvement has focused on technology and capital equipment to reduce the input of labor cost. Improved output was

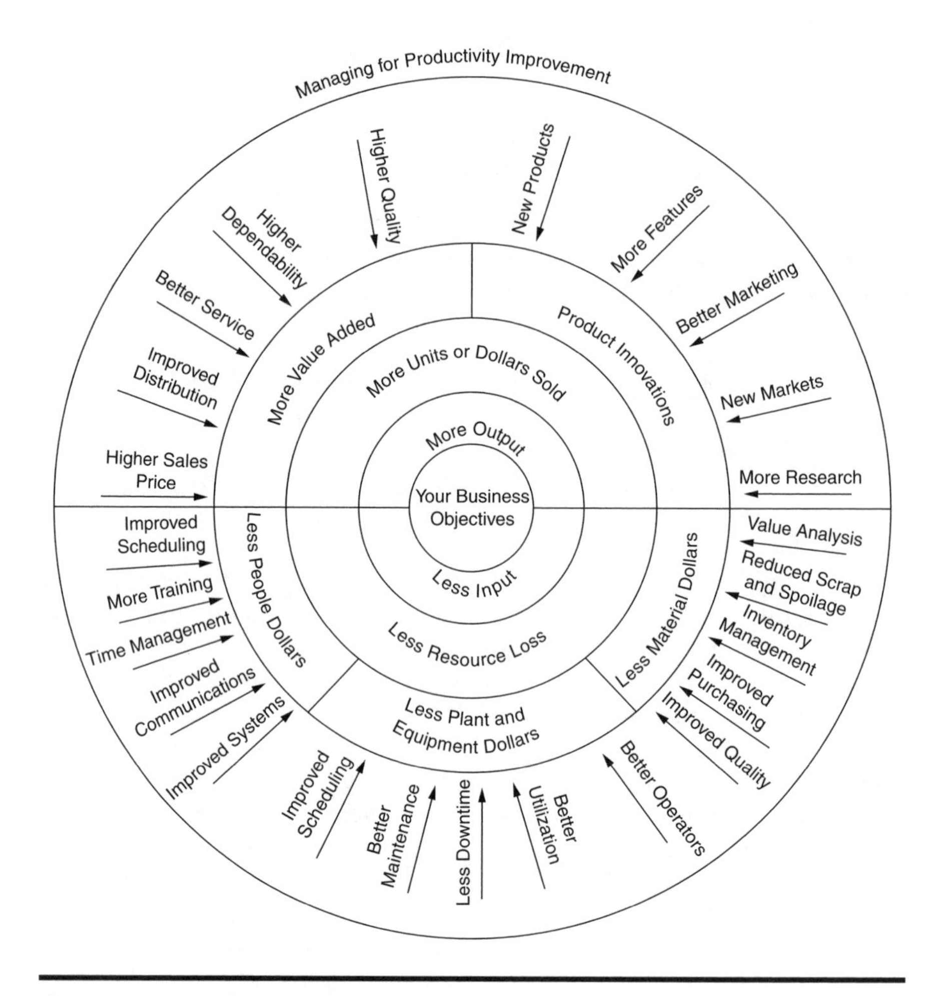

Figure 10-2 Productivity Wheel

generally thought to be subject to obtaining more production by applying industrial engineering techniques such as methods analysis, work flow, etc. Both of these approaches are still appropriate, but the current trend is toward better use of the potential available through human resources. Each worker can be his or her own industrial engineer — a mini-manager, so to speak. This potential can be tapped by allowing and encouraging people to innovate in one or more of the five ways described in the next section. Employee ideas can improve productivity, and in most cases this is accompanied by an improvement in quality as well.

Five Ways to Improve Productivity (and Quality)

Cost reduction is traditionally the most widely used approach to productivity improvement (see Figure 10-3) and is an appropriate route if implemented correctly. However, many companies maintain a somewhat outdated "across-the-board" mentality that directs each department to "cut costs by 10%." Staff services are slashed and training reduced, and the result is an inefficient sales force, reduced advertising, and diminished R&D. Maintenance is delayed and machine downtime is increased. The results may be a non-competitive product and loss of market share.

Under this "management-by-drive" approach, people are perceived as a direct expense, and the immediate route to cost reduction is seen as cutting this expense as much as possible. This policy usually leads to employee resentment and is frequently counterproductive. It may result in trading today's headache for tomorrow's upset stomach.

Managing growth is a more positive approach, but growth without productivity improvement is *fat*. The improvement may suggest an investment or cost addition, but the investment must return more than the cost, thus increasing the ratio. Capital and technological improvements, systems design, training, organization design, and development are among the many ways to manage growth while improving productivity and quality. The approach does not necessarily mean additional investment in capital improvement. It can also mean reducing the amount of input per unit of output during the growth period. This may be termed *cost avoidance*.

Working smarter means more output from the same input, thus allowing increases in sales or production with the same gross input and lower unit cost. Many companies think that working smarter means putting a "freeze" on budgets while expecting a higher level of output. Although this may be necessary as a stopgap measure, it is hardly a rational course of action to improve productivity over the longer term. Better ways of improving this ratio might be getting more output by reducing manufacturing cost through product design, improving processes, or getting more production from the same level of raw materials by increasing inventory turnover.

Paring down is similar to cost reduction, except that as sales or production is off, input should be reduced by a proportionately larger amount, thus increasing the ratio. This productivity improvement can frequently be achieved through "sloughing off." In many organizations, there are many more opportunities than are generally realized to reduce marginal or unproductive facilities, employees, customers, products, or activities. Peter Drucker puts it this way: "Most plans concern themselves only with the new and additional things that have to be done — new products, new processes, new markets, and so on. But the key to doing

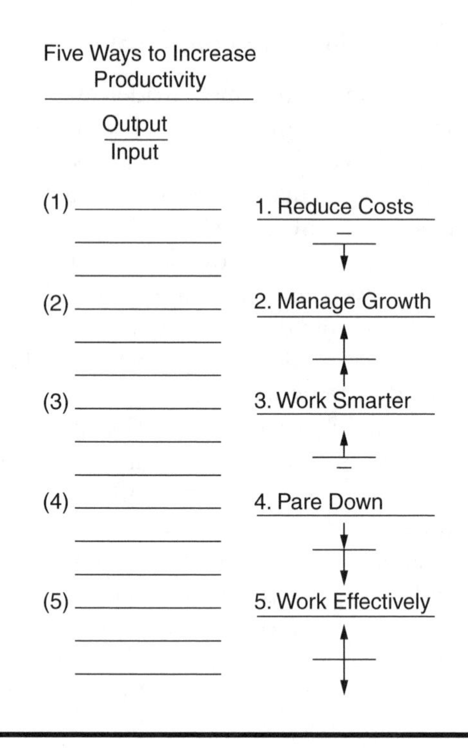

Figure 10-3 Productivity Improvement

something different tomorrow is getting rid of the no-longer-productive, the obsolescent, the obsolete." This "sloughing off" could apply to customers as well. Remember the 80/20 rule.

Working effectively is the best route to productivity and quality improvement; simply stated, you can get more for less. Some ways in which this can be accomplished are suggested in Figure 10-2.

Examples of Increasing Productivity While Improving Quality

Experience has shown that front-line supervisors and employees have a wealth of innovative ideas for productivity and quality improvement. They have only to be asked. In workshops and seminars conducted for hundreds of participants, there has been a high degree of enthusiasm for setting improvement objectives, defining problems, and organizing action plans for improvement. A few that were converted to action plans and resulted in substantial cost reduction as well as improved productivity and quality are presented here as illustrative examples. Each improvement objective will improve the output/input ratio in one or more of the five ways outlined earlier.

- Improve assembly output by 30% by reducing the excessive number and types of fasteners
- Reduce repetitive machine downtime by problem solving
- Set material standards and reduce rework by 10%
- Decrease work in process from 45 to 30 days by improved scheduling and shop floor layout
- Improve clerical costs by 30% by avoiding duplication with adequate work procedures
- Set standards for setup and improve setup time by 10%
- Improve tool revision cost by 50% by decreasing lead time from design
- Improve process flow and get 30% increased output of presses
- Improve flow of finished goods by improving warehouse layout
- Reduce labor cost by training technicians to replace engineers
- Get more output with less input by cross training and reduction of specialization
- Get more output with same input by better production planning
- Improve bill of materials by reducing custom parts
- Reduce assembly hours by using modular assembly
- Improve reliability by simplified design and design for customer maintainability

CAPITAL EQUIPMENT VS. MANAGEMENT SYSTEMS

Improvements in both productivity and quality have been slowed by two traditional management systems. The first has been the tendency to look to capital equipment as a solution to the problem of labor productivity. In the age of "high-tech," additions to capital have been viewed as the answer to boosting output. There is nothing wrong with this approach. Indeed, as pointed out previously, remarkable gains have been made in mechanization and automation since the Industrial Revolution. However, there are a number of arguments against depending on technology alone. It costs money and takes time, neither of which is an abundant resource.[19] Moreover, direct labor, the focus of capital equipment, is in the range of 8 to 12% of total cost of manufacturing. Technology has yet to make significant inroads in the productivity of indirect labor and service industries. Finally, high-tech must be accompanied by low-tech — the way workers, supervisors, and managers interact in adapting to new systems.

A basic principle of Economics 101 is illustrated in Figure 10-4. As additional increments of capital are used, productivity increases up to the point where benefits and cost are equal. This is classical economics at its best and reflects Washington thinking about U.S. industrial policy. Figure 10-4 also demonstrates how the productivity curve can be shifted

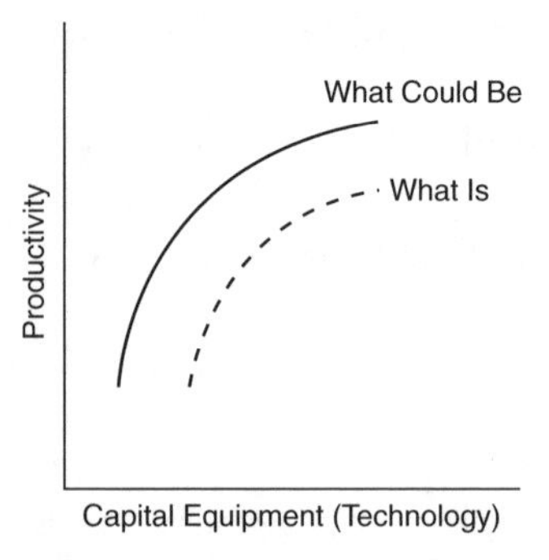

Figure 10-4 Productivity Curve

upward by means of improved management systems. This approach costs little and is available immediately. As discussed in earlier chapters, process control and related methods can improve both quality and productivity.

Another shortcoming of the capital investment argument as the primary or sole source of productivity and quality improvement relates to the historical focus on cost reduction. As discussed in Chapter 11, the traditional cost accounting methods of the past provide inadequate information for decision making in the 1990s. Today, decisions on capital expenditures must be based on overall productivity, improving quality, cutting cycle time, reducing inventory, and adding flexibility. Activity analysis is a first step and it is fundamental to improving *management systems*.

ACTIVITY ANALYSIS

Measurement of an activity output is not sufficient. Questions still remain: (1) Is the output/input ratio a positive number? (2) Can this ratio be improved? Most importantly, (3) does the *value added* by the activity contribute to the goal of the organization and the external or internal customer? The overwhelming majority of people in an organization cannot answer either of these questions, except in general and non-measurable terms. They define their activities in terms of what they are doing, not what they want to get done or whether the output is worth more than the input.

People characterized as *input* supervisors or employees are recognized by their dedication to collecting voluminous data for variance reports or closely examining the details of an expense account. The emphasis is on paperwork and the maintenance of records. They are the guardians of company rules and procedures, but are unconcerned about the value of their service to external or internal customers. The means become the end. Emphasis is on form and administration (doing things right) rather than process and results (doing the right things). They confuse *efficiency* with *effectiveness*. The design department is efficient at making repeated modifications to the product without regard for the impact on production. The sales force is efficient at calling on the wrong customers with the wrong product. Staff departments are efficient at providing services to internal customers who place no value on the service because they do not have to pay for it. The focus is on the budget rather than results.

Activity-focused supervisors and employees are intent on what they are doing, as opposed to what should be done. The accountant focuses on preparing the cost report rather than reducing overhead costs. The engineer is concerned only with the technical specifications of design without regard to cost, value analysis, or competitive considerations. When asked to define the results of their jobs, these people will reply with such platitudes as "improve the operations," "keep maintenance costs down," or "stay within the budget." It can be said of bureaucracy that focus on activity rather than results seems perfectly logical to those who are trapped within it. The activity may seem logical to the individual performing it, but to an outsider or a customer it is obviously wasteful.

The historical attention that is paid to budgets and cost control has encouraged a focus on activity rather than non-financial measures that plan and monitor sources of competitive value and *strategic* cost information. For most white-collar and service activities, the purpose of the output is to provide input to another downstream activity that can be viewed as the *internal* customer. A good starting point, therefore, is to determine whether the internal customer's expectation is met by the value provided by the upstream activity. The analysis of these activities begins by charting the flow throughout the organization and identifying sources of customer value in each. The central questions to be asked are, "What is the value added by the activity?" and "What is the output worth to the supplier and receiver?"

The major steps in conducting an activity analysis program include:

- Each unit, function, or activity develops a baseline budget that includes a breakdown of one year's costs.
- Set a cost, productivity, or quality target.
- Develop a mission statement for each unit that answers the question: "Why does it exist?"

- Identify each activity that supports the mission and the end products or services that result from that activity.
- Allocate end-product cost that equals the baseline budget.
- Identify receivers (customers) of the end product or service.
- Develop and implement ideas for improvement.[20]

EXERCISES

10-1 Give an example of how improving quality can also increase productivity.

10-2 Illustrate how productivity improvement may be more effective than increased sales in improving profitability.

10-3 How can improved management be as effective as technology and capital equipment in improving productivity?

10-4 Why has the rate of productivity increase been low in the United States?

10-5 Choose four or five functions or activities in staff or white-collar jobs and indicate a measure of productivity for each.

10-6 List three of the five ways to improve the productivity rate of input to output, and identify a specific action that could be taken to achieve the improvement.

ILLUSTRATIVE CASE

Agencies of the federal government have supported TQM in varying degrees since the Baldrige Award was established in 1987 as a result of Public Law 100-107. In 1993, Vice President Gore became the leader of a White House "total quality management" effort to examine each federal agency for ways to cut spending and improve services. The Federal Productivity Measurement System (FPMS), administered through the U.S. Department of Labor, reports the results of approximately 2500 output indicators. These data are analyzed to identify relationships between TQM implementation and federal productivity.

Private sector productivity is rising in most parts of the U.S. economy, but the public sector TQM movement has been less successful. The President's Award, the public sector equivalent of the Baldrige Award, has been in existence since 1989, but in 1990 and 1993 no agency could be found that was worthy of it. One significant study was undertaken at the IRS. An analysis of input and output indicators shows no statistically significant difference before and after TQM implementation. However, the IRS reports having saved millions of dollars as a result of improvement process teams and other TQM situations.

Question

- What are the differences, if any, between TQM and productivity in the private vs. the public sector?

ENDNOTES

1. Institute of Industrial Engineers, "Productivity and Quality in the USA Today," *Management Services (UK),* Jan. 1990, pp. 27–31.
2. W. Edwards Deming, *Quality, Productivity, and Competitive Position,* Cambridge, Mass.: Center for Advanced Engineering Study, Massachusetts Institute of Technology, 1982, pp. 1–2.
3. A. V. Feigenbaum, "Quality and Productivity," *Quality Progress,* Nov. 1977, p. 21.
4. This conclusion is suggested by the Profit Impact of Market Strategies (PIMS) database referred to in Chapter 1. The studies suggest that higher conformance and total quality costs are inversely related and better manufacturing-based quality results in higher output without a corresponding increase in costs. See K. E. Maani, "Productivity and Profitability through Quality: Myth and Reality," *International Journal of Quality & Reliability (UK),* Vol. 6 Issue 3, 1989, pp. 11–23. One empirical study concludes that improvements in quality level may be related to productivity increases. See Daniel G. Hotard, "Quality and Productivity: An Examination of Some Relationships," *Engineering Management International (Netherlands),* Jan. 1988, pp. 259–266. In one Conference Board research study of 62 firms that attempted to measure the results of quality on profitability, 47 indicated that profits have increased noticeably because of lower costs and/or increased market share. See Francis J. Walsh, Jr., *Current Practices in Measuring Quality,* New York: The Conference Board, 1989, p. 3. See also Colin Scurr, "Total Quality Management and Productivity," *Management Services,* Oct. 1991, pp. 28–30.
5. Joyce R. Jarrett, "Long Term Strategy...A Commitment to Excellence," *Journal for Quality and Participation,* July/Aug. 1990, pp. 28–33.
6. Raymond C. Cole and Lee H. Hales, "How Monsanto Justified Automation," *Management Accounting,* Jan. 1992, pp. 39–43.
7. At the Third National Productivity Conference in Dallas on May 21, 1990.
8. Robert D. Buzzell and Bradley T. Gale, *The PIMS Principles,* New York: The Free Press, 1987, pp. 10–11.
9. For a more detailed examination of the reasons for slow productivity growth in the United States, see Joel E. Ross *Productivity, People & Profits,* Englewood Cliffs, N.J.: Prentice-Hall, 1981 and Joel E. Ross and William C. Ross, *Japanese Quality Circles & Productivity,* Englewood Cliffs, N.J.: Prentice-Hall, 1982.
10. For one popular view of the unwillingness of managers to manage, see Robert H. Hayes and William J. Abernathy, "Managing Our Way to Decline," *Harvard Business Review,* July–Aug. 1989, pp. 67–77.

11. "Productivity and Quality in the 90's," *Management Services (UK),* June 1990, pp. 28–33. This article reports on a survey of British managers, the majority of whom believe that managers are more interested in short-term financial gain than in long-term productivity. Similar surveys in the United States have had similar results.
12. "The Productivity Paradox," *Business Week,* June 6, 1988, p. 103.
13. OECD, *Economic Outlook,* July 1991, p. 3.
14. Nobel laureate economist Milton Friedman stated that higher wages and the price–wage spiral are an *effect* of inflation, not a *cause.* Milton Friedman and Rose Friedman, *Free to Choose: A Personal Statement,* New York: Harcourt Brace Jovanovich, 1980.
15. Coopers & Lybrand conducted a survey to determine what federal executives know and think about quality management. About half of the respondents said that the lack of dependable ways to measure quality is a major obstacle. The same could be said of productivity measures. David Carr and Ian Littman, "Quality in the Federal Government," *Quality Progress,* Sep. 1990, pp. 49–52.
16. See Carl G. Thor, "How to Measure Organizational Productivity," *CMA Magazine,* March 1991, pp. 17–19. A company-wide system for measuring productivity is quite complex. The American Productivity and Quality Center conducts a three-day seminar on the topic. See also Brain Maskell, "Performance Measurement for World Class Manufacturing," *Management Accounting (UK),* July/Aug. 1989, pp. 48–50. This article identifies seven common characteristics used by world-class manufacturing firms: (1) performance measures are directly related to the manufacturing strategy, (2) primarily non-financial measures are used, (3) the measures vary among locations, (4) the measures change over time as needs change, (5) the measures are simple and easy to use, (6) the measures provide rapid feedback to operators and managers, and (7) the measures are meant to foster improvement instead of only monitoring.
17. Sumanth, D.J. Total Productivity Engineering and Management, McGraw-Hill, New York, 1984.
18. Peter Drucker, *Management,* New York: Harper & Row, 1974, p. 70.
19. Carl Thor, president of the American Productivity and Quality Center in Houston, favors management systems. Regarding high-tech additions, he says: "You need a decade's worth of that kind of investment to have an effect." See a special report entitled "The Productivity Paradox," *Business Week,* June 6, 1988, pp. 100–112.
20. For additional ideas on activity analysis, see Thomas H. Johnson, "Activity-Based Information: A Blueprint for World-Class Management Accounting," *Management Accounting,* June 1988, pp. 23–30. See also Philip Janson and Murray E. Bovarnick, "How to Conduct a Diagnostic Activity Analysis: Five Steps to a More Effective Organization," *National Productivity Review,* Spring 1988, pp. 152–160; Paul L. Brown, "Quality Improvement through Activity Analysis," *Journal of Organizational Behavior Management,* Vol. 10, Issue 1, 1989, pp. 169–179. For a more detailed program for implementing an organization-wide productivity improvement program, see Joel E. Ross, *Productivity, People & Profits,* Englewood Cliffs, N.J.: Prentice-Hall, 1981.

REFERENCES

Masters, Robert J., "Overcoming the Barriers to TQM's Success," *Quality Progress,* May 1996, pp. 53–55.

Schonberger, Richard, "Backing Off from the Bottom Line," *Executive Excellence,* May 1996, pp. 16–17.

Shelley, G. C., "The Search for the Universal Management Elixir," *Business Quarterly,* Summer 1996, pp. 11–13.

Woodruff, Davis M., "Ten Essentials for Being the Low-Cost, High-Quality Producer," *Management Quarterly,* Winter 1995–96, pp. 2–7.

11

THE COST OF QUALITY

> Quality is measured by the cost of quality which is the expense of non conformance — the cost of doing things wrong.
>
> **Philip Crosby**
> *Quality Is Free*

What will it cost to improve quality? What will it cost to not improve quality? These are basic questions that managers need to ask as they focus on the bottom line and company strategic decisions. These questions about the cost of quality have served to draw attention to the quality movement. No one will deny the importance of quality, but it is the confusion surrounding the payoff and the trade-off between cost and quality that is unclear to many decision makers.

It is becoming increasingly clear that whereas the answer to the cost of poor quality may be difficult to obtain, the potential payoff from improvement is extraordinary. Hewlett-Packard estimated that the cost of not doing things right the first time was 25 to 30% of revenues. Travelers Insurance Company found that the figure was $1 million per hour. On a positive note, Motorola has reduced the cost of poor quality by about 5% of total sales, or about $480 million per year.

COST OF QUALITY DEFINED

The cost of quality has been defined in a number of ways, some of which include:

- At 3M quality cost equals actual cost minus no failure cost. That is, the cost of quality is the difference between the actual cost of making and selling products and services and the cost if there were no failures during manufacture or use and no possibility of failure.[1]

- Quality costs usually are defined as costs incurred because poor quality may or does exist.[2]
- The cost of not meeting the customer's requirements — the cost of doing things wrong.[3]
- All activities that are carried out that are not needed directly to support departmental [quality] objectives are considered the cost of quality.[4]

These definitions leave unanswered the question: "How much quality is enough?" In theory, the answer is analogous to a principle of economics: basic marginal cost equals marginal revenue (MC = MR). That is, spend on quality improvement until the added profit equals the cost of achieving it. This is not so easy in practice. In economics, the MC and MR curves are difficult to define and more difficult to compute. The same is true of the cost/benefit curves of quality costs. What are the costs of added quality and the "hidden" costs of non-quality? What are the bottom-line benefits? Neither of these questions is easy to answer, particularly in view of the long-run strategic implications. The answer lies at the very essence of what a company is about.

THE COST OF QUALITY

The cost of quality or, more specifically, "non-quality" is a major concern to both national policymakers as well as individual firms. Because much of our national concern with competitiveness seems to be focused on Japan, it is interesting to note that some estimates of quality costs in U.S. firms indicate 25% of revenues, while in Japan the figure is less than 5%.[5] Estimates of potential savings are as high as $300 billion by nationwide application of total quality management (TQM).[6] Feigenbaum puts the estimate at 7% of the gross national product and suggests that this figure can be one of the tools used by policymakers in considering the quality potential of the U.S. economy in relation to the country's major competitors.[7]

The cost of poor quality in individual firms and the potential for improvement can be staggering. In *Thriving on Chaos,* Tom Peters reports that experts agree that poor quality can cost about 25% of the personnel and assets in a manufacturing firm and up to 40% in a service firm. There appears to be general agreement that the costs range between 20 and 30% of sales.[8]

The potential for profit improvement is very substantial. One has only to visualize a profit-and-loss statement with a net profit of 6% before tax and then compute what the profit would be if 20 to 30% of the operating budget were reduced. Add to this the additional strategic benefits and the potential is great indeed.

THREE VIEWS OF QUALITY COSTS

Historically, business managers have assumed that increased quality is accompanied by increased cost; higher quality meant higher cost. This view was questioned by the quality pioneers. Juran examined the economics of quality and concluded that benefits outweighed costs.[9] Feigenbaum introduced "total quality control" and developed the principle that quality is everyone's job, thus expanding the notion of quality cost beyond the manufacturing function.[10] In 1979, Crosby introduced the now popular concept that "quality is free."[11] Today, the view among practitioners seems to fall into one of three categories:[12]

1. ***Higher quality means higher cost*** — Quality attributes such as performance and features cost more in terms of labor, material, design, and other costly resources. The additional benefits from improved quality do not compensate for the additional expense.
2. ***The cost of improving quality is less than the resulting savings*** — This view was originally promoted by Deming and is widely held among Japanese manufacturers. The savings result from less rework, scrap, and other *direct* expenses related to defects. This is said to account for the focus on continuous improvement of processes in Japanese firms.
3. ***Quality costs are those incurred in excess of those that would have been incurred if the product were built or the service performed exactly right the first time*** — This view is held by adherents of the TQM philosophy. Costs include not only those that are direct, but also those resulting from lost customers, lost market share, and the many hidden costs and foregone opportunities not identified by modern cost accounting systems.

The attention now being given to the more comprehensive view of the cost of poor quality is a fairly recent development. Even today, many companies tend to ignore or downplay this opportunity because of a continuing focus on production volume or frustration with the problem of computing the trade-off between volume and quality. This computational difficulty is compounded by accounting systems that do not recognize the expenses as manageable. More on this will be provided later in this chapter.

One survey of 94 corporate controllers found that only 31% of the firms regularly measured costs of quality, and even among those firms productivity was ranked higher than quality as a factor contributing to profit. Not surprisingly, the major reason for failure to measure these costs was lack of top management commitment.[13]

Philip Crosby, of "quality is free" fame, is of the firm opinion that zero defects is the absolute performance standard, and the cost of quality is the price of non-conformance against that standard. His concept is catching on as more companies set goals such as parts per million, six sigma, and even zero defects. On the other hand, a goal of zero defects may be more costly than the payoff that might accrue. As one approaches zero defects, costs may begin to increase geometrically.

Another of Crosby's principles, which he calls "absolutes," is *measurement of quality:*

> The measurement of quality is the Price of Nonconformance, not indexes....Measuring quality by calculating the price of waste — wasted time, effort, material — produces a monetary figure that can be used to direct efforts to improve and measure the improvement.[14]

This monetary figure, according to Crosby, is a percentage of sales, and he suggests that the standard should be reduced to about *2 to 3%*. This measure has been generally accepted, and many firms use it as a target and measure of progress.

QUALITY COSTS

The costs of quality are generally classified into four categories: (1) prevention, (2) appraisal, (3) internal failure, and (4) external failure.[15] *Prevention* costs include those activities that remove and prevent defects from occurring in the production process. Included are such activities as quality planning, production reviews, training, and engineering analysis, which are incurred to ensure that poor quality is not produced. *Appraisal* costs are those costs incurred to identify poor quality products after they occur but before shipment to customers. Inspection activity is an example.

Failure costs are those incurred either during the production process (*internal*) or after the product is shipped (*external*). Internal failure costs include such items as machine downtime, poor quality materials, scrap, and rework. External failure costs include returns and allowances, warranty costs, and the hidden costs of customer dissatisfaction and lost market share. Recognition of the relative importance of external failure costs has caused many companies to broaden their perspective from product quality to total consumer satisfaction as the key quality measure.

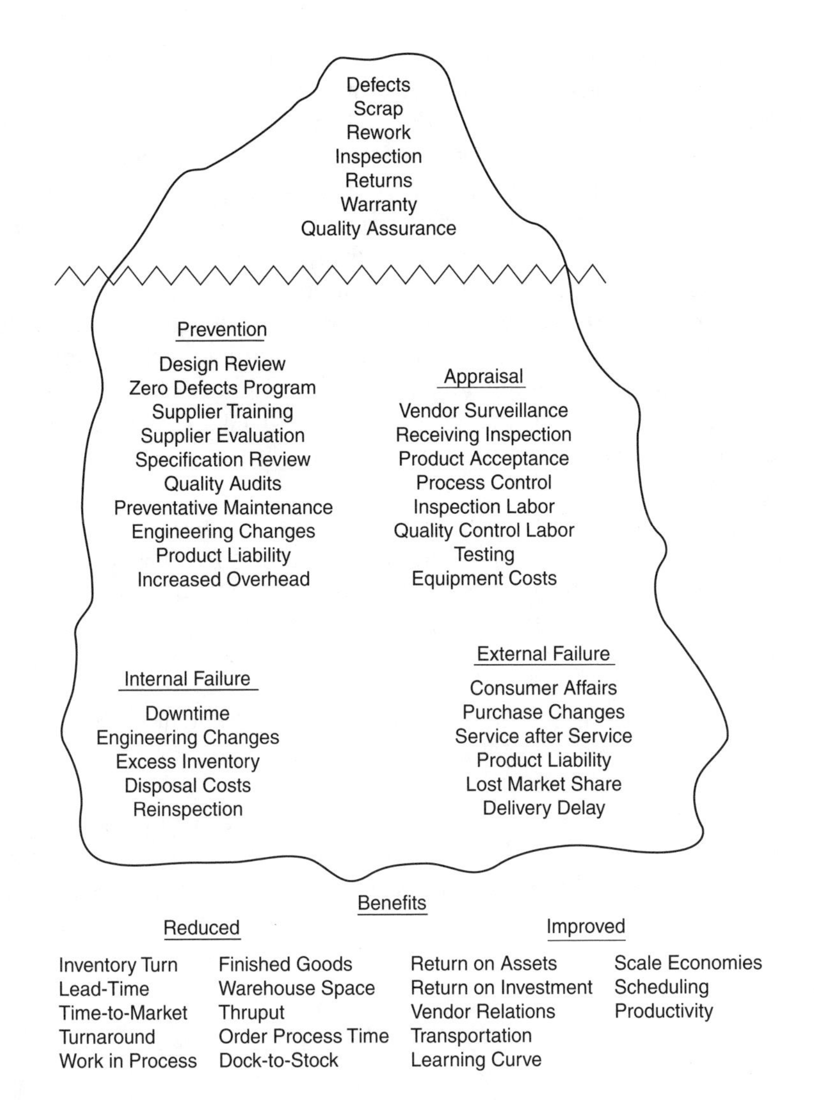

Table 11-1 Benefits of Costs-of-Quality Control

In Figure 11-1, the many costs of non-quality are classified into the four categories outlined earlier: (1) prevention, (2) appraisal, (3) internal failure, and (4) external failure. The figure is an attempt to convey the idea of an iceberg, where only 10% is visible and 90% is hidden from view. The analogy is a good one because the *visible* 10% comprises such items as scrap, rework, inspection, returns under warranty, and quality assurance costs; for many companies these comprise what they believe to be the total costs. When the *hidden* costs of quality are computed, controlled, and reduced, a firm can achieve the benefits shown at the bottom of Figure 11-1.

Of these types of costs, prevention costs should probably take priority because it is much less costly to prevent a defect than to correct one. The principle is not unlike the traditional medical axiom: "An ounce of prevention is worth a pound of cure." The relationship between these costs is reflected in the 1-10-100 rule depicted in Figure 11-2. One dollar spent on prevention will save $10 on appraisal and $100 on failure costs. As one moves along the stream of events from design to delivery or "dock-to-stock," the cost of errors escalates as failure costs become higher and the payoff from an investment in prevention becomes greater. Computer systems analysts are aware of this and understand that an hour spent on better programming or design can save up to ten hours of system retrofit and redesign. One general manager of Hewlett-Packard's computer systems division observed:

> The earlier you detect and prevent a defect the more you can save. If you catch a two cent resistor before you use it and throw it away, you lose two cents. If you don't find it until it has been soldered into a computer component, it may cost $10 to repair the part. If you don't catch the component until it is in the computer user's hands, the repair will cost hundreds of dollars. Indeed, if a $5000 computer has to be repaired in the field, the expense may exceed the manufacturing cost.[16]

When total customer satisfaction becomes the definition of a quality product or service, it creates a need to develop measures that integrate the customer perspective into a measurement system. This need moves

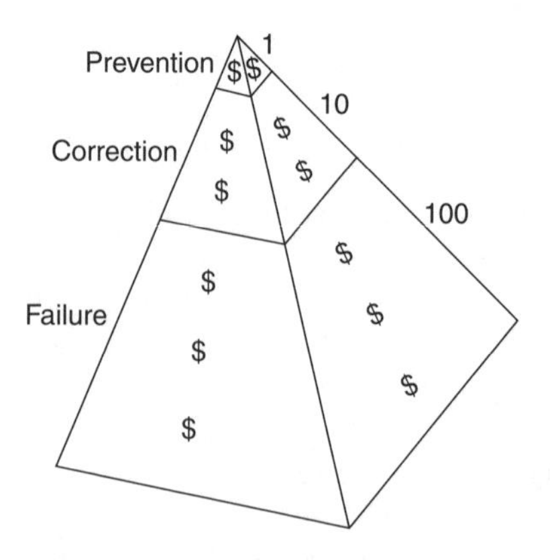

Figure 11-1 1-10-100 Rule

beyond the shop floor and into the many non-product features such as delivery time, responsiveness, billing accuracy, etc. This need also leads to a search for quality, and hence quality costs, in activities not usually recognized as incurring these costs. This will change as more companies realize that all activities can contribute to total customer satisfaction. Thus, quality costs include those factors that lie behind the obvious production processes. Moreover, it becomes necessary to identify the hidden quality costs associated with foregone opportunities.

What is frequently overlooked is the unrealized potential for improved productivity and quality to be achieved by identifying and measuring the difference between no failure (parts per million, six sigma, zero defects, etc.) cost and actual cost. What, for example, would be the payoff from just-in-time, better process control, improved inventory turn, and reduced cycle time in the many cross-functional processes and cost interrelationships in the stream of activities during the life cycle of a product or a service? Each of these actions would improve quality, use fewer resources, and improve return on investment (ROI). How these same actions could also increase market share and profitability was examined in Chapter 1. To quote Feigenbaum: "Quality and cost are a sum, not a difference — complementary, not conflicting objectives."[17]

MEASURING QUALITY COSTS

In a 1989 Conference Board survey of 149 large U.S. companies (96 manufacturing), it was found that 111 had a quality process or program. Of the 111 that had a program, 83 attempted to measure quality. The majority of the companies that attempted to measure quality costs compiled the information outside of the accounting system. The breakdown of cost categories reflected a major focus on the direct labor costs of scrap, rework, returns, and costs related to inventory including past-due receivables. There was little evidence to indicate that these costs, once collected, were used to manage processes leading to customer satisfaction quality. Rather, the systems appeared to resemble the traditional cost reduction syndrome discussed in Chapter 10.

An effective cost-of-quality planning and control system should be directed toward the basic reason for quality improvement; that is, support of a differentiation strategy. Of course, if a company has not developed a strategy, it becomes difficult to identify those costs of quality that support differentiation of satisfaction in the minds of the customers. For a multidivision or multiproduct firm, this strategy may be different for each market segment or strategic business unit. There is little advantage to investing in equipment, overhead, or process improvements that do not add customer value. What is good for Neiman Marcus may not be good for K-Mart.

The cost of differentiation reflects the *cost drivers* of the value activities on which uniqueness is based.[18] Differentiation can also result from the coordination of linked value activities that may not add much cost but nevertheless provide a cost savings and a competitive edge when integrated.

The measurement and reporting of quality costs to facilitate these strategic demands need to be provided to users of the information in a form that aids in decision making. Thus, the measurement and reporting of costs of quality should meet the three-part need to (1) report quality costs, (2) identify activities where involvement is suggested, and (3) indicate interlinking activities.

Activities and functions are not independent. They form a system of interdependencies that are connected by linkages and relationships. For example, purchasing from a low-quality supplier may lead to redesign, rework, scrap, increased field service, and direct labor variance. These linkages are difficult to recognize and are often overlooked. Nor is the conventional accounting system equipped to separate the cost of quality in these linked activities. Virtually all accounting classifications group activities along functional lines and force the reporting of quality costs into several general expense categories such as salaries, depreciation, training, etc. Analyzing the accounts can produce limited estimates of quality costs, but unless the costs are designed into the system, they will be elusive for decisions and action planning.

As one of the steps in the design of a planning and control system, it is useful to identify those activities and linkages between activities where costs occur. Some form of linear or matrix organizational chart or table is useful for this purpose. Departments or activities are listed across the top and costs of quality down the left-hand side. A number (e.g., 1 for primary responsibility or 2 for coordinating responsibility) can be entered at the intersection of the cost-of-quality category and the activity or function involved. The chart will show overlap among activities and will therefore indicate the need for cooperation, interfunctional teams, and the like. A similar chart can be devised to present cost of quality by activity. Thus, quality costs can be presented based on both cost and activity responsibility, and this form of presentation is more likely to get the attention of top management.

A similar chart can be constructed for reporting the dollar costs of quality. The same format could be used for both budgeting and reporting. Costs can be tabulated by organization unit, by time, by cost-of-quality categories, or by product. Quality costs can also be normalized for volume by using one or more of the following measures: per direct labor hour, per direct labor cost, per dollar of standard manufacturing cost, per dollar of sales, or per equivalent unit of product.[19]

The most elusive category for reporting is the cost of lost opportunities, which is an external failure cost. This represents the impact on profit from lost revenues resulting from purchase of competitive products and services or from order cancellations due to customer requirements not being met. An additional problem is assigning these *estimated* costs to a quality project or action plan that may prevent recurrence. It is also difficult to compile the elusive relationships among two or more costs that affect quality costs (i.e., prevention plus appraisal).[20]

The constant theme throughout a cost-of-quality system is that *costs are not incurred or allocated, but rather are caused.* Cost information does not solve quality problems, nor does it suggest specific solutions. Problems are solved by tracing the *cause* of a quality deficiency.

THE USE OF QUALITY COST INFORMATION

Quality cost information can be used in a number of ways:[21]

- To identify profit opportunities (every dollar saved goes to the bottom line)
- To make capital budgeting and other investment decisions (quality, as opposed to payback, is the driver of decisions to purchase new equipment or dispose of unneeded equipment; equipment for rework is not needed if the rework is eliminated or reduced)
- To improve purchasing and supplier-related costs
- To identify waste in overhead caused by activities not required by the customer
- To identify redundant systems
- To determine whether quality costs are properly distributed
- To establish goals for budgets and profit planning
- To identify quality problems
- As a management tool for comparative measures of input–output relationships (e.g., the cost of a reliability effort vs. warranty costs)
- As a tool of Pareto analysis to distinguish between the "vital few" and the "trivial many"
- As a strategic management tool to allocate resources for strategy formulation and implementation
- As an objective performance appraisal measure

General Electric's cost-of-quality system increasingly emphasizes non-product features such as inquiry responsiveness, delivery times, and billing accuracy. The emphasis is on root cause analysis and process improvement: simplifying procedures and reducing cycle time and driving down quality costs while improving customer satisfaction. Internal and external systems measure performance vs. customer expectations; these systems also track opportunities that have been lost by non-conformance to customer expectations.[22]

ACCOUNTING SYSTEMS AND QUALITY MANAGEMENT[23]

Accounting information by itself provides little help for reducing costs and improving quality and productivity. The tendency is to *allocate* rather than manage costs. Moreover, the allocation is normally a function of direct labor, an item that has shrunk to 15% or less of manufacturing costs. Overhead, at about 55%, is spread across all products using the same formula. Accounting also cannot identify or account for the many non-dollar hidden costs of quality and productivity.

Critics claim that management accounting systems should be designed to support the operations and strategy of the company, two dimensions in which quality plays a dominant role. This is increasingly evident in the "new" manufacturing environment, sometimes known as advanced manufacturing technology, which is characterized by a number of emerging trends. These trends and their implications for quality management were summarized in Chapter 6. Some of the decision-making needs and how traditional accounting practices may fall short in meeting them are listed here:

Decision needs	*Traditional accounting*
Activity management	Financial accounting
Investment management	Payback or ROI
Non-dollar measures	Dollar accounting
Process control	Cost allocation
Just-in-time	Inventory turn
Feedforward control	Historical control

ACTIVITY-BASED COSTING

The majority of companies that attempt to measure quality costs compile the information and statistics outside of the accounting system. These data are aggregated and do not reflect the true cost of quality or the activity in the process that is causing it. It is worth repeating that costs are not incurred or allocated; *they are caused.* The mere collection of data is of little use unless the data can help identify the drivers of quality costs so that problem identification leads to problem solution.

Activity-based costing (ABC), called "A Bean-Counter's Best Friend" by *Business Week,*[24] can be the system that promises to fill this gap.[25] ABC is a collection of financial and operation performance information that traces the significant activities of a firm to process, product, and quality costs. It is well suited to TQM because it encourages management to analyze activities and determine their value to the customer.

Imagine the case of a firm with excessive warranty costs. The following questions might arise:

- What is the cost of the returns?
- What is the cause of the returns and can the cause be traced to a specific activity? Is it the supplier, the design, or one of the many activities in production?
- How can the process(es) be improved to reduce the cost of returns?
- What is the trade-off between cost of process revision and reduction of warranty costs?
- What are the strategic implications? The concepts of ABC may lead to some answers.

The concepts of process control and activity analysis were described in Chapter 6 ("Management of Process Quality") and Chapter 10 ("Quality and Productivity"). ABC brings these interlinking concepts together through cross-functional analysis:

- ***Process control*** documents the process flow, identifies requirements of internal and external customers, defines outputs of each process step, and determines process input requirements.
- ***Activity analysis*** defines each activity within each process and identifies activities as value added or non-value added based on customer requirements.

Activity analysis applies to internal as well as external customers. When Rear Admiral John Kirkpatrick assumed command of the six U.S. Naval Aviation Depots, he inaugurated the use of TQM. One element of the system was that wherever possible, the internal customer was allowed to demand only those internal products or services desired.[26] Could this be a logical extension of customer satisfaction? If it can be applied to external customers, why not internal customers as well?

The third step is to develop cause-and-effect relationships by identifying *drivers* of cost or quality. In the case of cost, the drivers are the conditions that create or "drive" the need for an activity and hence the resources consumed. If the cost driver relates to a non-value activity, it can be eliminated or reduced. It is estimated that 50% or more of the activities in most businesses are cost added rather than value added.[27]

ABC recognizes that activities, not products, consume resources, and process value analysis is needed to assign costs to the activities that use them. The system recognizes that costs are driven by factors other than volume or direct labor. In the case of product costing, the costs are assigned based on their consumption of activities such as order preparation, storage time, wait time, internal product movement, field maintenance, and design. The focus on the process, not the product, suggests a transition to breaking down the floor into smaller cost centers and identifying the cost drivers of each.

Cost drivers are agents that cause activity to happen. Consider an engineering change order (ECO) that causes many activities to occur, such as documentation, production schedule changes, purchase of a new machine, or change in a process. If the ECO is issued to correct excessive field maintenance costs, manufacturing will absorb additional charges, marketing's distribution costs will increase, and customer satisfaction may erode because of delays and field repairs. By using the ABC concept, the true cost of the affected product can be determined as well as its cross-functional impact on budgets and performance.

This ECO example illustrates the impact of engineering and design on product life cycle costs. Roughly 80 to 85% of a product's lifetime costs, including maintenance and repair expenses, are locked in at this stage. ABC might provide guidelines to help engineers design a product that meets customer expectations and can be produced and supported at a competitive cost.

The Multiproduct Problem

> At Rockwell International Corporation, a capital budgeting request for an $80,000 laser was denied because at $4,000 per year in labor savings the payback would take 20 years. Further analysis showed that the process would be reduced from 2 weeks to 10 minutes, moving shipments out faster and saving $200,000 a year in inventory holding costs.[28]
>
> Tektronix, Inc. adopted ABC in a printed circuit board plant and found that one high-volume product drew on so many resources that it generated a negative margin of 46% and sapped profits from other products. These examples illustrate how "across the board" accounting *allocation* of costs, rather than *management* of costs, distorts the information required for good decision making.

There is great potential for inaccurate costing and control of multiproduct lines in a firm with a single overhead center, and inaccuracies in costing increase dramatically when allocation is achieved by direct labor, machine time, processing time, or some other "assignment" method. A major soft drink producer found that the costs of its array of brands varied as much as 400% from what traditional cost accounting methods reported.

In summary, ABC decomposes activities, identifies the drivers of the activities, and provides measures so that costs can be traced to the activities that cause the cost.

Strategic Planning and Activity-Based Costing

> At a meeting of IBM's board of directors in November 1991, various restructuring proposals were considered. One option was to unburden the lines of business from general overhead expenses. For example, the company may remove from its personal computer business the burden of helping pay for research on mainframe computers. (This action was subsequently taken in 1992.)

There is a cost dimension to most strategic decisions. Product lines, channels, locations, brands, segmentation, and differentiation need to be identified, and each decision establishes a linkage between demands and spending on resources. If costs are forecast on the arbitrary basis of some unit directly related to production, the real cost of a product or capital project may be made arbitrarily.[29] ABC can help reveal data for strategic decisions about which product lines to develop or abandon and which prices to increase or decrease. Tracing overhead to activities and then to products may also identify costs that do not contribute to quality and hence to differentiation.

ABC has leapfrogged traditional cost accounting, but it is a new and complicated system. For these reasons, the great majority of companies have not achieved a significant level of sophistication in its use. The basic concept of ABC is that costs of products and quality can be traced to the drivers of activities that consume the resources which *cause* these costs. Research reveals that there is widespread failure to compile the many prevention, appraisal, internal failure, and external failure costs that are "hidden" until identified by a cost-of-quality management system. If the costs are not identified, there is little chance of tracing them to the process or activity that is causing them. Only the "visible" rework, scrap, and repair/service costs are compiled by more than half of the respondents.

Summary

Is a cost-of-quality program essential to a quality improvement effort? The answer may be no, but a firm cannot spend unlimited resources without regard for both strategic issues and the cost/benefit equation. Moreover, a cost-of-quality effort is but one of a system of interlinking efforts that comprise a management philosophy of TQM.

EXERCISES

11-1 Select a firm (restaurant, hotel, airline, manufacturer) and list several costs related to quality failure. Estimate these costs.

11-2 What is the estimated cost of poor quality in U.S. industry?

11-3 What is the justification for Philip Crosby's claim that "Quality Is Free?"

11-4 Illustrate each of the four types of costs of quality.

11-5 Why should prevention costs take precedence over the other three classifications?

11-6 What are the benefits of a cost-of-quality measuring system?

11-7 Consider the following "defects" (or others with which you are familiar)

- A product (e.g., appliance, auto, clothing) requires rework while under warranty.
- A restaurant customer sends an unsatisfactory meal back.
- A house painter is required to return and redo an unsatisfactory job.
- A bank or retail firm makes a mistake in billing.

11-8 Referring to Figure 11-1, how many of the prevention, appraisal, and failure costs are incurred by the preceding "defects?" What is your estimate of the cost of these quality defects?

ENDNOTES

1. Doug Anderson, "How to Use Cost of Quality Data," in *Global Perspectives on Total Quality,* New York: The Conference Board, 1991, p. 37.
2. John F. Towey, "Information Please: What Are Quality Costs?" *Management Accounting,* March 1988, p. 40. Apparently this is a quasi-official definition adopted by the National Association of Accountants (NAA).
3. Roger G. Schroeder, *Operations Management,* New York: McGraw-Hill, 1989, p. 586.
4. J. M. Asher, "Cost of Quality in Service Industries," *International Journal of Quality & Reliability Management (UK),* Vol. 5 Issue 5, 1988, pp. 38–46.
5. William Band, "Marketers Need to Understand the High Cost of Poor Quality," *Sales & Marketing Management in Canada,* Nov. 1989, pp. 56–59.

6. Ned Hamson, "TQM Can Save Nearly $300 Billion for Nation," *Journal for Quality and Participation,* Dec. 1990, pp. 54–56. This potential is reflected in the quality improvement potential (QIP) index.
7. Armand V. Feigenbaum, "The Criticality of Quality and the Need to Measure It," *Financier,* Oct. 1990, pp. 33–36. This estimate reflects a national (QIP) index for the gross national product. Feigenbaum is president and chief executive officer of General Systems Company, Inc., which installs company-wide quality systems in manufacturing and service organizations.
8. Financial managers estimate the cost at 25 to 30% of sales. See Garrett DeYoung, "Does Quality Pay?" *CFO: The Magazine for Chief Financial Officers,* Sep. 1990, pp. 24–34. See also Lester Ravitz, "The Cost of Quality: A Different Approach to Noninterest Expenses," *Financial Manager's Statement,* March/April 1991, pp. 8–13. A 1990 study of quality in North American banks found that non-quality cost related to unnecessary rework and related factors represented 20 to 25% of a bank's operating budget. In Britain, the United Kingdom Institute of Management Services estimates that the cost of quality non-conformance amounts to 25 to 30% of sales. See John Heap and Lord Chilver, "Total Quality Management," *Management Services (UK),* June 1990, pp. 6–10.
9. J. M. Juran, Ed., *Quality Control Handbook,* New York: McGraw-Hill, 1951.
10. Armand V. Feigenbaum, *Total Quality Control,* New York: McGraw-Hill, 1961.
11. Philip Crosby, *Quality Is Free,* New York: McGraw-Hill, 1979.
12. An excellent discussion of these categories is contained in David A. Garvin, *Managing Quality,* New York: The Free Press, 1988, pp. 78–80.
13. Thomas N. Tyson, "Quality & Profitability: Have Controllers Made the Connection?" *Management Accounting,* Nov. 1987, pp. 38–42.
14. Taken from a promotional brochure by Philip Crosby Associates, Inc. of Winter Park, Florida.
15. The British Science and Engineering Research Council funded a study on quality-related costs as part of a two-year study. A literature review showed the domination of the prevention-appraisal-failure classification, a preoccupation with in-house costs, and little regard for supplier- and customer-related costs. See J. J. Plunkett and B. G. Dale, "A Review of the Literature on Quality-Related Costs," *International Journal of Quality & Reliability Management (UK),* Vol. 4 Issue 1, 1987, pp. 40–52. This classification can also apply to non-manufacturing areas. Xerox is one firm that has implemented a well-defined quality program aimed at achieving quality in non-manufacturing services. A model was developed to illustrate the costs of quality based on the prevention-appraisal-failure classification. See Michael Desjardins, "Managing for Quality," *Business Quarterly (Canada),* Autumn 1989, pp. 103–107.
16. As quoted in David A. Garvin, *Managing Quality,* New York: The Free Press, 1988, p. 79.
17. Armand V. Feigenbaum, "Linking Quality Processes to International Leadership," in *Making Total Quality Happen,* New York: The Conference Board, 1990, p. 6.
18. Michael E. Porter, *Competitive Advantage,* New York: The Free Press, 1985, pp. 127–130. Although Porter does not address the specifics of cost of quality, his discussion of differentiation costs provides an excellent dimension to the topic.

19. For more detailed information on methods of compiling quality cost information, see Wayne J. Morse, Harold P. Roth, and Kay M. Poston, *Measuring, Planning, and Controlling Quality Costs,* Montvale, N.J.: National Association of Accountants, 1987. This NAA research study provides a number of actual reporting formats used by responding companies in the survey. For the collection and reporting formats used by ITT and Xerox, see Francis J. Walsh, Jr., *Current Practices in Measuring Quality,* New York: The Conference Board, 1989 (Conference Board Research Bulletin).
20. James T. Godfrey and William R. Pasewark, "Controlling Quality Costs," *Management Accounting,* March 1988, pp. 48–51.
21. For sources of information regarding the use of cost-of-quality information, see the following: John F. Towey, "Why Quality Costs Are Important," *Management Accounting,* March 1988, p. 40. See also James M. Reeve, "TQM and Cost Management: New Definitions for Cost Accounting," *Survey of Business,* Summer 1989, pp. 26–30; J. J. Plunkett and B. G. Dale, "A Review of the Literature on Quality-Related Costs," *International Journal of Quality & Reliability Management (UK),* Vol. 4 Issue 1, 1987, pp. 40–52; John J. Heldt, "Quality Pays," *Quality,* Nov. 1988, pp. 26–28.
22. Elyse Allan, "Measuring Quality Costs: A Shifting Perspective," in *Global Perspectives on Total Quality,* New York: The Conference Board, 1991, p. 35 (Conference Board Report Number 958).
23. H. Thomas Johnson and Robert Kaplan, *Relevance Lost: The Rise and Fall of Management Accounting,* Boston: Harvard Business School Press, 1991. Peat Marwick, one of the Big Six accounting firms, scored a major coup by signing Kaplan to an exclusive contract in the field of activity-based cost accounting. See "A Bean-Counter's Best Friend," *Business Week/Quality 1991* (special bonus issue dated October 25, 1991, entitled "The Quality Imperative: What It Takes to Win in the Global Economy").
24. "A Bean-Counter's Best Friend," *Business Week,* October 25, 1991, pp. 42–43 (special bonus issue entitled "The Quality Imperative: What It Takes to Win in the Global Economy").
25. Because ABC has not been widely used, there is no widespread treatment of it in the literature. Perhaps the best source of information, at least from the accountant's point of view, is *Management Accounting.* See, for example, Thomas E. Steimer, "Activity-Based Accounting for Total Quality, *Management Accounting,* Oct. 1990, pp. 39–42. See also Michael R. Ostrenga, "Activities: The Focal Point of Total Cost Management," *Management Accounting,* Feb. 1990, pp. 42–49 and Norm Raffish, "How Much Does that Product Really Cost?" *Management Accounting,* March 1991, pp. 36–39. For a managerial perspective, it is suggested that a literature search be conducted for recent writings of Robert Kaplan.
26. Michael D. Woods, "How We Changed Our Accounting," *Management Accounting,* Feb. 1989, pp. 42–45.
27. Michael J. Stickler, "Going for the Globe. Part II: Eliminating Waste," *Production & Inventory Management Review and APICS News,* Nov. 1989, pp. 32–34.
28. "The Productivity Paradox," *Business Week,* June 6, 1988, p. 104.

29. Bernard C. Reimann, "Robert S. Kaplan: The ABCs of Accounting for Value Creation," *Planning Review,* July/Aug. 1990, pp. 33–34. The author reports Kaplan's contention that the "essence of strategy" is to regard all overhead expenses as variable and driven by something other than the number of units. Also, financial reporting is fine for reporting bottom-line financial performance but inadequate for strategic decisions.

REFERENCES

Bonilla, Nelson E., "Total Cost Management — Trends for the 21st Century," *Cost Engineering,* May 1996, p. 5.

Carr, Lawrence P., "How Xerox Sustains the Cost of Quality," *Management Accounting,* Aug. 1995, pp. 26–33.

Rust, Kathleen G., "Measuring the Costs of Quality," *Management Accounting,* Aug. 1995, pp. 33–39.

II

PROCESSES AND QUALITY TOOLS

In Part II, the tools and techniques needed to conduct analytic studies for the purpose of quality improvement are discussed. When used within the framework of the Deming cycle (Plan-Do-Check-Act), the techniques can serve as a vehicle for the pursuit of quality.

In Chapter 12, the concept of a process is discussed, and a number of examples of a process are offered. Chapter 13 discusses the two types of data and various sampling methodologies. In Chapter 14, the basic quality improvement tools — check sheets, flowcharts, graphs, histograms, Pareto charts, cause-and-effect diagrams, scatter diagrams, and control charts — are presented. Examples are provided to illustrate the use of these basic tools.

Chapters 15, 16, and 17 discuss the use of the various types of control charts and provide examples of each type. Chapter 18 discusses quality improvement stories, and Chapter 19 discusses quality function deployment.

12

THE CONCEPT OF A PROCESS

Everything is a process. Whether it is admitting a patient to a hospital, handling customers at a checkout counter, opening a new account for a bank's customer, or packaging a product for shipment to a customer — all involve a series of activities that are interrelated and must be managed.

WHAT IS A PROCESS?

A *process* is a series of activities or steps used to transform input(s) into output(s). An input or output may exist or occur in the form of data, information, raw material, partially finished units, purchased parts, a product or service, or the environment. It is the steps used by an individual or a group to perform work or complete a task. It is sometimes referred to as a technique, method, or procedure. The absence of a clearly defined process makes any activity subject to an arbitrary mode of execution and its outcome or output subject to unpredictable performance. In order to "do it right the first time" and "do the right things right," processes must be effectively managed. When processes are not adequately managed, quality will regress to mediocrity. An organization is a collection of subprocesses. A customer is affected by one or more processes at any given time. Every process has customers (those who depend on it or are affected by it) and suppliers (those who provide the necessary input for that process). Consequently, everyone in an organization serves a customer or serves someone who is serving a customer.

EXAMPLES OF PROCESSES

The process of providing higher education is complex. For the sake of simplicity, let us assume that it starts with certain inputs such as SAT scores, a completed application for admission, high school grades, letters of reference, or extracurricular activities. Several resources are consumed in the process of transforming input to output. The process of transformation includes the activities of teaching, advising, financial aid processing, residence hall assignment and dwelling, library resources, laboratory experience, class group projects, etc. The output is a student who graduates and is competent, drops out, or graduates but is ill-equipped to perform in a competitive work environment. Important feedback can come from industry employers who may criticize the curriculum or simply refuse to hire the graduate. Within this macro process are many subprocesses. One important subprocess is registration. Key inputs would include a list of previous courses taken by the student, the student's classification (freshman, sophomore, junior, or senior standing), the course offerings for that semester, the course prerequisites, etc. Central to this process are academic advising and payment processing (bursar and financial aid). The output is a registered student.

Another well-known process involves packing and shipping a finished product to a customer. The input to the shipping department includes the finished product, the customer's name and address, invoice information, etc. The transformation process entails inspection, making the box, labeling, packing the product, and arranging for shipment. The output in this case is a successful delivery. The feedback loop in this case is the customer reporting back on the condition of the product when it was delivered and whether it was the right product, model, brand, color, quantity, and performance. Everything is a process — whether opening a new account at a bank, taking a customer's order at the drive-through window of a fast-food restaurant, processing an engineering design change, generating a purchase order, or admitting a patient to a hospital.

One of the primary objectives of total quality management (TQM) is to create processes in which individuals or groups will "do it right the first time" and "do the right things right." As suggested in Figure 12-1, individuals or groups can do the right things right or wrong and the wrong things right or wrong. The manner in which individuals do their work (process) can also be right or wrong. The following examples illustrate each of the four quadrants in Figure 12-1.

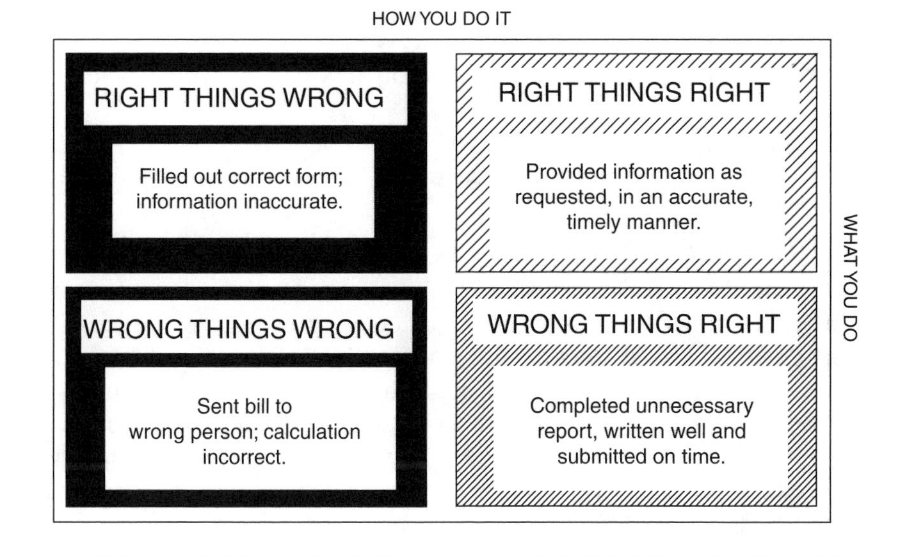

Figure 12-1 The Quality Grid

1. Doing the right things wrong —
   - You have filled out the correct form, but the information is inaccurate.
   - Using the right equipment but not operating it correctly.
   - A nurse provides the necessary explanation to a sick patient, but in an unprofessional manner.
2. Doing the wrong things wrong —
   - The accounts department sends an invoice to the wrong customer, and the calculations are incorrect (two processes are affected here — the billing process and the costing process).
   - Filling out the wrong expense reimbursement form and filling it out incorrectly.
   - Picking up the wrong work order and performing the work incorrectly.
   - Purchasing department orders the wrong parts and orders them several weeks late.
3. Doing the wrong things right —
   Using the examples in #2:
   - The accounts department sends an invoice to the wrong customer, but the calculations are right.
   - Filling out the wrong expense reimbursement form, but filling it out correctly.

- Violating the optimal job sequence by picking up the wrong work order, but performing the work correctly.
- Completing an unnecessary report that is well written and submitted on time.

4. Doing the right things right —
   - Providing the information as requested, in an accurate and timely manner.
   - Ordering the right parts in the right quantity, from the right vendor, within the lead time allowed.

The right things done right means meeting or exceeding the expectations of customers, both internal and external. It also means the elimination of waste, rework, and defects and conformance to valid requirements. If and when it is determined that the customer is incapable of knowing what *right* means, as in the right treatment for a disease, some education and/or explanation would be necessary to bridge the gap between customer expectation and what the service provider delivers.

TYPES OF PROCESSES

There are three types of processes, as follows:

- ***Management process*** — This entails the method(s) used by management in executing its management functions. Three key functional areas used by management are planning, organizing, and controlling.
- ***Functional process*** — A functional process consists of the methods used to achieve functional objectives within a group or by an individual.
- ***Cross-functional process*** — This includes the method(s) used to achieve objectives that require participation or input from more than one group or individual. For example, the problem of minimizing breakage of a fragile product might require input from the shipping, design, marketing, packaging, and manufacturing departments. Similarly, the problem of an adverse drug reaction in a hospital may require the involvement of the pharmacist, the ordering physician, a registered nurse, and a unit secretary. Each group or individual controls one or more of the subprocesses affecting the problem.

THE TOTAL PROCESS

TQM calls for an evaluation of the total system, not just the subsystems. The danger of suboptimization always exists in that subprocesses instead of the total process are optimized. In many instances, the total process is not defined, and therefore, accountability is conspicuously absent. Each internal customer provides intermediate inputs or receives intermediate outputs throughout the process. These intermediate inputs and outputs are used to achieve the final outcome of the organization. The external customers provide an initial input to or receive final output from the process. Because everyone in an organization serves a customer or serves someone who does, everyone is part of a customer–supplier chain. No worker's task is isolated. Consequently, no worker is expected to either accept or pass on defective work or product.

The Feedback Loop

The need to constantly improve processes makes it imperative that a feedback loop be introduced into every process. This feedback loop becomes the link between the output or outcome and the input. It provides the system with an opportunity to evaluate the gap between the expectation of the customer (internal or external) and what is produced or delivered by the supplier. The real value of feedback lies in its usefulness in analyzing the process of transformation.

EXERCISES

12-1 What is a process?
- Give an example of a macro process in either service or manufacturing.
- Give examples of at least two subprocesses within the macro process defined above.

12-2 Define the feedback loop for the macro process and the subprocesses in Exercise 12-1. How can feedback be used to improve the processes you have identified?

12-3 What are the three types of processes? Give examples to illustrate each type.

12-4 Give an example of each of the following in a service and manufacturing sector:
- Doing the right things wrong
- Doing the wrong things wrong
- Doing the wrong things right
- Doing the right things right

REFERENCES

1. Gitlow, H., S. Gitlow, A. Oppenheim, and R. Oppenheim, *Tools and Methods for the Improvement of Quality,* Homewood, Ill.: Irwin Publishers, 1989.
2. Omachonu, V. K., *Total Quality and Productivity Management in Health Care Organizations,* Milwaukee: Quality Press, American Society for Quality Control, and Norcross, Ga.: Industrial Engineering and Management Press, Institute of Industrial Engineers, 1991.

13

UNDERSTANDING DATA

INTRODUCTION

One of the most significant failures in organizations today is the inability to convert raw data into information. Organizations collect hundreds of thousands of data sets on a weekly, monthly, and quarterly basis, with no clear plans as to how to extract information from them. Managers and supervisors are often bombarded by enormous volumes of reports that are aimed at tracking everything from productivity to quality. Often times, the data are presented poorly and the essence of the information is lost. In some cases, the decision to collect the data was made at a time in the history of the organization when it seemed reasonable to track or monitor that activity. However, no one revisits that effort to determine its continued relevance. Successful implementation of the quality management process depends to a large extent on the quality of data and the ability to convert the data into information.

DATA AND INFORMATION

Webster's defines data as "figures from which conclusions can be drawn, a basis for reasoning, discussion or calculation." Data are merely a group of numbers which can represent the measurement of something, such as temperature (degrees), or the count of something, such as the number of rotten apples. Data are transformed into information through analysis. This information influences decision-making and the types of actions that result from these decisions. The following simple rules help us make the distinction between data and information:

Data becomes information if:

- It can be used to make inferences.
- It allows for comparisons.
- It requires little or no intermediate reprocessing or recalculation.
- It is presented or displayed effectively.
- It supports decision-making.
- It was derived from analysis.

Leaders of organizations and their managers and supervisors need information, not data. It is obvious that organizations need data to get to information; however, data should not by itself be the end, but rather a means to an end.

The Concept of a Dashboard

The dashboard of an automobile provides the operator of the vehicle dozens of pieces of information while the vehicle is being operated, throughout the life of the vehicle. The value of the information depends on a number of factors, including timeliness, accuracy, completeness, etc. The displays on a dashboard have been carefully designed to provide the driver with mostly relevant information. If we had access to the back of the dashboard, we might find the data from which the information is derived. For example, there might be a device behind the dashboard that captures the distance traveled and the quantity of gasoline consumed in traveling that distance. These two numbers are used to arrive at the average miles per gallon displayed in some automobiles' dashboard today. Does your organization have a corporate and a departmental dashboard? And if so, what's on the dashboard?

Significance of Data

Organizations collect data for many reasons, good and bad. The following are some of the reasons why organizations should collect data:

- To determine how well it is fulfilling customer requirements
- To determine how close it is to its target
- To track accomplishments
- To recognize when an improvement is made or required
- To track the use of resources and how efficiently they are used
- To provide information that supports efforts to improve

Deciding What to Measure

In general, what to measure depends on whether or not you are referring to a product or a service. In a product environment, for example, we may be interested in measuring degree of conformance to specifications (consistency), whereas in a service environment, we may be interested in attributes such as completeness, accuracy, timeliness, responsiveness, safety, cost effectiveness, and customer satisfaction. Some of these measures are also pertinent for a product environment.

Questions to Ask Prior to Data Collection

- Why are you collecting data? What will it tell you?
- What behavior is the information from this data designed to drive or influence?
- How will the data be collected? How will the information be calculated?
- Who will collect the data and for how long?
- Where will the data be collected?
- What data collection instrument will be used?
- Is the data available? In what format?
- How much data is needed?
- How will the data be analyzed?
- Is the data adequately stratified? If not, how should it be stratified?
- What is the cost of data collection? Is it justified?

Data Collection Methods

There are many approaches for collecting data. The challenge is in knowing which approach is best for your needs, whether you are working in a process improvement team or within the organization. The following data collection methods are widely used in industry:

- Direct observation
- Surveys
- Interviews
- Focus groups
- Experiments
- Manual tracking and reporting
- Computerized tracking and reporting

Data may be collected retrospectively or prospectively. Retrospective data collection refers to the collection of data that is stored somewhere or already gathered, while prospectively collected data refers to the collection of data that does not currently exist and must be collected from this point forward.

Types of Data

Facts concerning a process, service, product, person, or machine are considered data. There are two main types of data: subjective and objective. ***Subjective data*** are based on experiences, opinions, or observations. Subjective data are typically used when making personal decisions like what to order for lunch, what to wear to work, etc. Subjective data can be difficult to quantify and should be used carefully when making decisions that affect many people. In the past, many managerial decisions were based on subjective data. Unfortunately, these decisions were often wrong, which cost the organization time and money. In today's competitive environment and with limited resources, decisions need to be based on facts that come from objective data.

Objective data are typically expressed in numerical form. There are two types of objective data: attribute and variable. An attribute is a quality characteristic of a product, process, etc., that can be counted. Attribute data, therefore, consist of information that can be counted, and so is also referred to as countable or discrete data. It commonly follows yes/no, go/no go criteria. Often, it is not possible to obtain data that can be directly measured; for example, the taste of soup, the feel of carpet, and the ease of driving cannot be measured in numerical terms and would therefore be classified as subjective. However, they can be compared and given a quantitative score, or they can be classified, like good taste vs. bad taste, and the number of occurrences for each classification can be counted. Examples of attribute data include whether or not a pen writes, or whether or not a payment is late.

Variable data involve measurement, reflecting criteria such as quantity, size, or length. The time it takes to assemble a chair, the length of a ruler and the diameter of a shaft are all examples of variable data. This type of data exists on a scale that can be divided into an infinite number of increments, thus it is continuous data. If the data being collected can be measured and expressed as a number on some continuous scale, then the data are considered variable data.

The computation of a value such as miles per gallon is considered variable data since it comes from a measured value, the number of miles driven and the number of gallons of gasoline in the car. In the same way, any computation, such as a proportion or ratio, that comes from attribute data is itself considered attribute data. Therefore, in general, data type can be determined from the original data.

Data Reliability

Even if the correct data are collected, a wrong decision or action can occur if the data are unreliable. There are many causes of unreliable data. One cause is improperly calibrated instruments. Another is improper use of measuring instruments. To minimize potential problems with unreliable data, it is important to check the calibration of measurement equipment before and after the collection of data. Similarly, the people collecting the data should be trained in the use of data collection instruments before and after data collection.

Since variable data require the use of measurement instruments, the data collector needs special skills and training. The skills may be simple, like telling time or using a ruler, or more complex, like determining hardness or measuring viscosity. It is important in all cases that the person collecting the data is trained with regard to data collection techniques and that all measuring devices used are calibrated. Both training and calibration increase the manpower and collection time needed. This also increases the cost.

For attribute data, classification and counting are based on sight, feel, taste, etc. For example, the counting of defects per item is based mainly on sight. The evaluation of wine is based on color, odor, and taste. Since these are individual interpretations, the differences in individual inspectors need to be noted and accounted for when appropriate.

Sometimes data may be reliable but unusable. This usually occurs when the origins of data are not recorded properly. Exactly when, where, who, what, and how the data were collected must be recorded. *When* includes the exact days, the time of day, etc. *Where* includes the plant/line/machine/employee location and where in the plant or on the line the data were collected. *Who* includes who collected the data and who was working on the line/machine when the data were collected. *What* includes the data type and the model number or type of item being produced. *How* includes the instruments (identification number and storage location are helpful if calibration becomes a question) and any other special instructions given for the data collection. All of this information can be easily collected if proper care is taken when designing a check sheet. Without this information, even good data has little meaning.

Stratification

When deciding what data to collect, it is always a good idea to think about the possibility of data stratification. Stratification is merely taking the original set of data and breaking it down into smaller, related subgroups. By allowing the team to determine what effect each subgroup has on the total population, stratification enables a more precise analysis

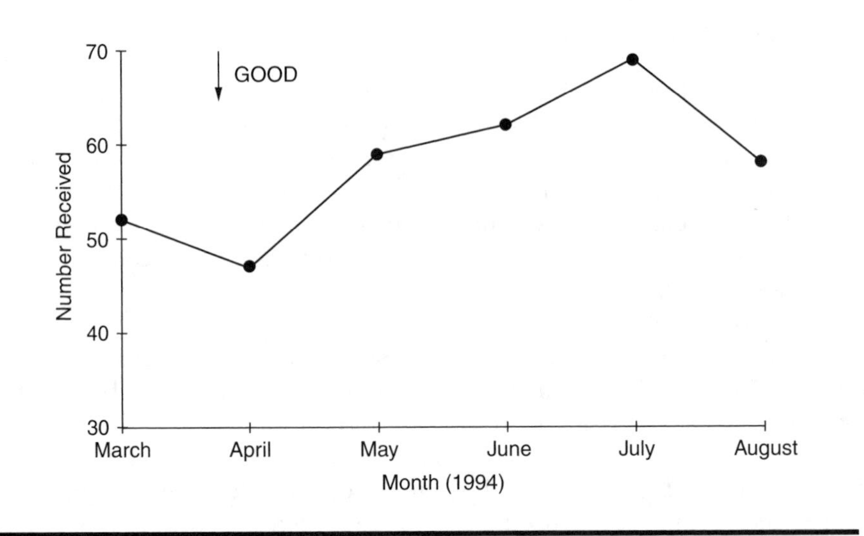

Figure 13-1 Damaged and Spoiled Oranges Received March–August 1994

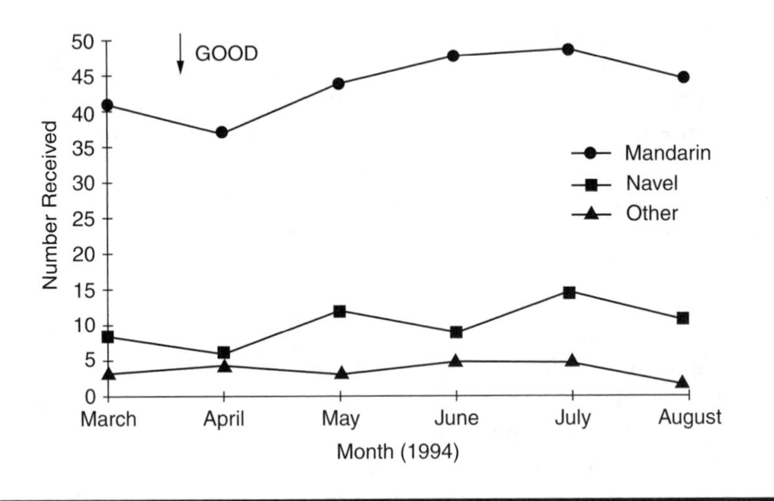

Figure 13-2 Damaged and Spoiled Oranges Received by Type March–August 1994

of the data. It is extremely helpful when doing root cause analysis. How stratification is used when doing root cause analysis with graphs is illustrated in Figure 13-1 and Figure 13-2. The total number of damaged and spoiled oranges received for the last six months is charted in Figure 13-1. The number of damaged and spoiled oranges received for the last six months is graphed by orange type in Figure 13-2. Clearly, there are more damaged and spoiled mandarin oranges per month than any other type. Even further stratification is provided in Figure 13-3 by graphing the

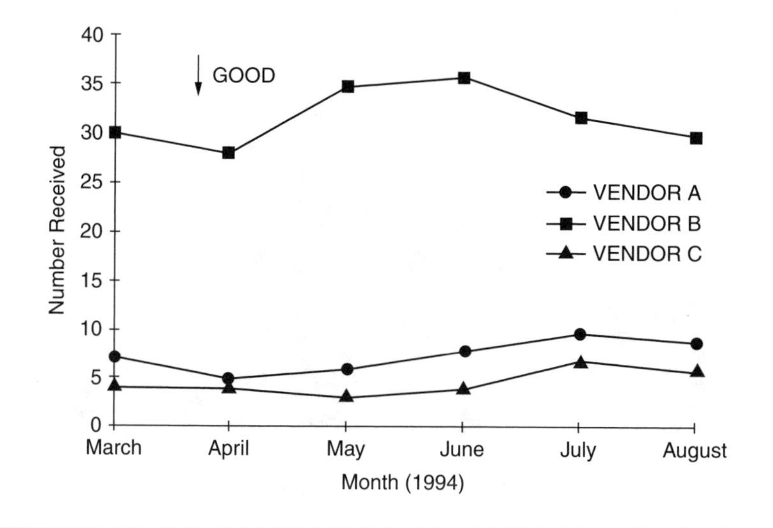

Figure 13-3 Damaged and Spoiled Mandarin Oranges Received by Vendor March–August 1994

number of damaged and spoiled mandarin oranges per month by vendor. This figure indicates that a problem exists with vendor b. These three figures illustrate the power of stratification and the value of collecting all of the data. This can only be accomplished if the data collection is first carefully planned.

HOW TO PRESENT/DESCRIBE DATA

Visual Description: Tabular Displays

Data can be described in tabular form by means of frequency distributions or cumulative frequency distributions. A frequency distribution shows the frequency, or number of times, a given value or values occurs. Figure 13-4 and Table 3-1 illustrate how to present and display data.

Time to complete ER patient registration process for sample 30 patients selected at random.

11.8	3.6	16.6	13.5	4.8	8.3
8.9	9.1	7.7	2.3	12.1	6.1
10.2	8.0	11.4	6.8	9.6	19.5
15.3	12.3	8.5	15.9	18.7	11.7
6.2	11.2	10.4	7.2	5.5	14.5

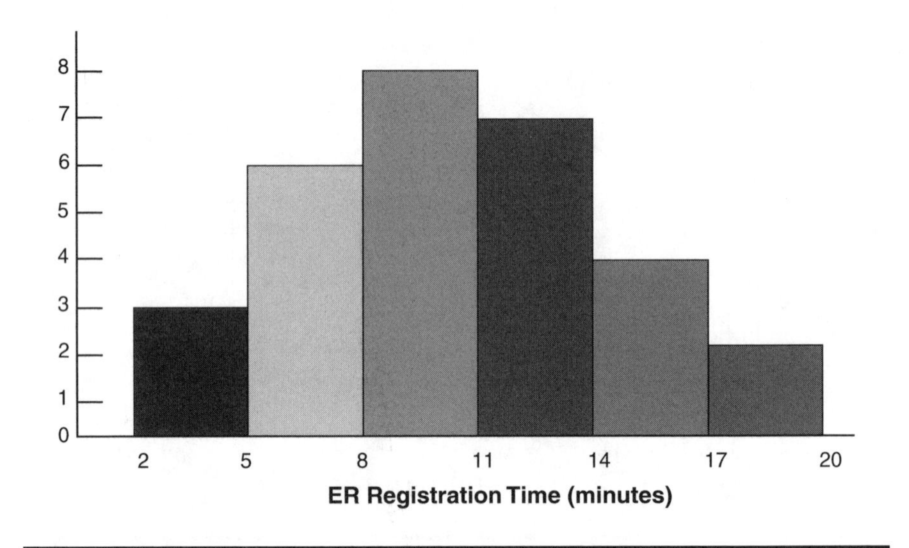

Figure 13-4 Histogram Showing ER Registration Times

Class Limits	*Frequency*
2 up to 5*	3
5 up to 8	6
8 up to 11	8
11 up to 14	7
14 up to 17	4
17 up to 20	2
Total	30

* class contains all measurments from 2 up to but not including 5.

Inferences

10% of ER patients complete their registration in less than 5 minutes; 30% in less than 8 minutes; 57% in less than 11 minutes; 80% in less than 14 minutes; 93% in less than 17 minutes, and 100% in less than 20 minutes.

Tabular displays of data are very useful in providing frequency information. It is important, however, to note that these displays do not provide any insight as to the way the data was collected, or any possible trends that may exist in the data.

Visual Description: Graphical Displays

Data is often presented graphically in order to illustrate relationships and trends that are not visible from tabular displays. Variable data are usually

Table 13-1 Frequency Table Showing ER Registration Time

Class limits	*Relative frequency*	*Culultive relative frequency*
2 up to 5	3/30 = 0.100	3/30 = 0.100
5 up to 8	6/30 = 0.200	9/30 = 0.300
8 up to 11	8/30 = 0.267	17/30 = 0.567
11 up to 14	7/30 = 0.233	24/30 = 0.800
14 up to 17	4/30 = 0.133	28/30 = 0.933
17 up to 20	2/30 = 0.067	30/30 = 1.00

presented in histograms. Attribute data are usually presented in bar charts, although histograms can be used for either type of data. Bar charts are very similar to histograms except that the bars do not enclose an interval; instead, they are centered about each of the attribute categories.

A Run chart is another important tool for displaying data graphically. It is a graph that plots the values of the characteristic being studied versus time, thereby allowing the detection of trends over time.

Numerical Description

To numerically describe data, two aspects or properties of it are usually presented: its central tendency and its dispersion. Central tendency refers to the central portion of the data, while dispersion refers to the spread of the data.

The main measures of central tendency are mean, median, and mode. The mean is simply the arithmetic mean of the values. It is denoted by X and is calculated by ΣX/n, where X is the data value and n is the total number of observations. Consider the following data set, for example:

10 18 12 17 19 16 12 11 14

$$\bar{X} = \frac{10+18+12+17+19+16+12+11+14}{9} = 14.3$$

The mean is the most common of the three measures of central tendency. It is mathematically the strongest, since it takes into account every data point.

The median is the middle value when the data is arranged on ascending order. Its advantage over the mean is that it is less affected by the extreme values. Using the previous data set:

10 11 12 12 **14** 16 17 18 19

median = **14**

In this data set, the number of observations was odd, which made it easier to find the median. Had the number of observations been even, the median would have been calculated by taking the mean of the middle two values. Usually, though, the median is used when there is an odd number of observations.

The third measure of central tendency is the mode. It is the value that occurs most frequently in the data set. Again following the previous example:

10 18 **12** 17 19 16 **12** 11 14

mode = **12**

This example has only one mode. This need not always be the case. A data set with two modes is regarded as bimodal, and one with more than two is known as multimodal. The mode can also be found by creating a frequency table and choosing the value(s) with the highest frequency.

The dispersion or variability of the data are measured by the range and the standard deviation. The range, R, measures the spread of the data by subtracting the lowest value from the highest value. In this data set, the range is:

$$R = 19 - 10 = 9$$

Therefore, this data has a spread of 9 units.

The standard deviation is another measure of spread, but it takes into account all of the data. To calculate the sample standard deviation S, use:

$$S = [\ (\Sigma(X_i - \overline{X})^2)/(n - 1)]^{1/2}$$

1. The Mean, $\overline{X}$, must be calculated. From the previous calculation, we know the mean, $\overline{X} = 14.3$.

2. $(X_i - \overline{X})$ is calculated for each reading X_i, and is then squared:

$$\begin{aligned}
10 - 14.3 &= -4.3 & -4.3^2 &= 18.49 \\
18 - 14.3 &= 3.7 & 3.7^2 &= 13.69 \\
12 - 14.3 &= -2.3 & -2.3^2 &= 5.29 \\
17 - 14.3 &= 2.7 & 2.7^2 &= 7.29 \\
19 - 14.3 &= 4.7 & 4.7^2 &= 22.09 \\
16 - 14.3 &= 1.7 & 1.7^2 &= 2.89 \\
12 - 14.3 &= -2.3 & -2.3^2 &= 5.29 \\
11 - 14.3 &= -3.3 & -3.3^2 &= 10.89 \\
14 - 14.3 &= -0.3 & -0.3^2 &= 0.09
\end{aligned}$$

3. These numbers are then added:

$$18.49 + 3.69 + 5.29 + 7.29 + 22.09 + 2.89 + 5.29 + 10.89 + .09 = 86.01$$

4. This value is then divided by the number of observations minus one, $(n - 1)$:

$$\frac{86.01}{9-1} = 10.75$$

This value is called the variance.

5. Finally, taking the square root of the variance gives the standard deviation of the sample:

$$(10.75)^{1/2} = 3.28$$

NOTE: Data is obtained from a sample. A distinction must be made between a sample and a population. For example, if a few different nurses take the vital signs of a patient, the time values would constitute a sample and yield a certain mean and standard deviation. These values differ from the true mean and standard deviation, which would have to be calculated from the population. In this case, the population would be all the nurses who take patients' vital signs, not just a group of them.

This difference between sample and population is reflected in the notation and calculation of the mean and standard deviation. The true (population) parameters are denoted by Greek letters: μ (mu) the mean and σ (sigma) the standard deviation. For both samples and the population, the mean is still simply an arithmetic average; only the notation differs (μ and X). The standard deviation of the population is calculated as follows:

$$= [\ (\Sigma(X_i - \overline{X})^2)/n]^{1/2}$$

When the true parameters μ and σ are not known, $\overline{X}$ and S are used as estimators. This is denoted by a circumflex (^) over the parameter, so that $\hat{\mu} = \overline{X}$ and $\hat{\sigma} = S$.

SAMPLING

Because data provide the basis for all decisions and actions, it is important that the data collected accurately represent the situation, process, lot, etc.

being studied. It would be ideal if every item in the population of items could be measured or tested. However, because most lots, processes, customer interactions, etc. that are typically studied contain large numbers of items, sampling is required.

Basic Definitions

Some basic terms need to be defined before sampling can be discussed further:

Population — A population is the totality of items being studied. This could be a single lot, output from one machine or person, or even a process line. Due to time, personnel, and money constraints, collecting population data is often impossible or impractical.

Sample — A sample is one or more items taken from a population and used to reflect the distribution of the population and to estimate the parameters of the distribution.

Population distribution — Because every item cannot be produced exactly the same each time, variability about the expected mean occurs. This variability has a pattern associated with it, which is referred to as the population distribution. In natural events, such as weights, heights, etc., the population distribution is usually the normal distribution. However, in the workplace, the distribution could be normal, exponential, Poisson, etc. By knowing the population distribution of the process being studied, proper corrections or improvements to the process can be made.

Parameters of a distribution — Distribution parameters are the characteristics that describe the position and the variability of the population distribution. For a normal distribution, the parameters of interest are the mean and the variance.

Symbols — The distribution parameters are represented by symbols. The symbols used to represent the normal distribution parameters are shown in Table 13-2.

Table 13-2 Parameter Symbols Representing the Normal Distribution

	Population		
Parameter	*Known*	*Estimated*	*Sample*
Mean	μ	$\hat{\mu}$	$\overline{X}$
Variance	σ^2	$\hat{\sigma}^2$	S^2

Table 13-2 shows that two types of parameters for the population exist, known or estimated (unknown). If perfect knowledge exists about a population or if you can sample every item in the population, the parameters of the population distribution are said to be known. If the parameters are known, then Greek letters are used to represent this fact. However, perfect knowledge seldom exists, and it is usually impossible to sample every item in the population. Therefore, the parameters must be estimated. Estimated parameters are represented by Greek letters with a circumflex (^); alphabetic characters are used to represent the sample parameters. Whenever the sample is used to estimate population parameters, the estimated population parameters will be the same as the sample parameters.

Types of Sampling

The purpose of sampling is to acquire accurate knowledge about a given population and to take appropriate action for improvement. Therefore, it is very important that the sample be truly representative of the population. In order to be truly representative of the population, the sample must be random.

Random sampling is taking a sample in such a way that every item in the population has an equal and independent chance of being included in the sample. ***Simple random sampling*** is random sampling without replacement. It is primarily used when the population being studied is thought to be homogeneous (the variance is due to chance causes instead of assignable causes). The steps for collecting a simple random sample are as follows:

Step 1 Determine the size of the population being studied and assign sequential numbers to the items. For example, if the population is 75 items, the items would be numbered from 1 to 75.

Step 2 Determine what the sample size will be. For this example, the sample size will be 10.

Step 3 Use a random numbers table. This is typically done by closing your eyes and putting your finger on the table. (Many books provide a single random numbers table. If there is more than one page to the table being used, then a die can be rolled to determine which page of the table to start on.) The number your finger points to is where you start.*

---

*There are a multitude of ways to use a random numbers table. This is just one of them.

Step 4 Determine the number of each item that will be in the sample. In this case, 10 numbers with values less than 75 have to be obtained. If the table being used provides numbers in groups of 4, take either the first 2, the last 2, or the middle 2. It doesn't matter. The table being used for this example provides numbers in groups of 5. The first 15 numbers in the table (starting with the number my finger landed on and moving across the row**) are given here, in order to ensure that there will be 10 usable numbers.

14346	09172	30168	90229	04734
59193	54164	58492	22421	74103
47072	25306	76468	26384	58151

The last two digits are used. The numbers then become:

46 72 68 29 34 93 64 92 21 03
72 06 68 84 51

Because there only 75 items are being studied, any number greater than 75 is eliminated. Similarly, any number that appears twice is eliminated because the sampling is being done without replacement. The numbers then left are:

46 72 68 29 34 64 21 03 06 51

These 10 numbers represent the number of the items within the population that will be included in the sample. If there are fewer than the required 10 numbers left, then the procedure is to return to the random numbers table at the point where the last number was obtained and collect additional random numbers. This is done until the number required are obtained.

Step 5 Collect the sample. In this example, a random numbers table is used. This is the most convenient means of generating a set of random numbers, since these tables are easily available. However, there are other ways of obtaining random numbers, such as rolling dice, tossing coins, and spinning a roulette-type wheel. These methods obviously have limitations and are not really practical on the job. One method that can be used on the job is to have a computer generate a set of random numbers.

**Moving down the column is also acceptable.

Stratified random sampling is done by dividing the population into several mutually exclusive and exhaustive strata or regions. The division of the population into the different strata is done in such a way that the units are as similar as possible within each stratum. For example, the strata could be the top, middle, and bottom portions of a lot or they could be groupings by like items (see Figure 13-5). Once the population is stratified, simple random samples are then taken from each strata (use Steps 1 to 5 to take each sample). This sampling is done independently within each stratum.

Clustered sampling is done by subdividing the population into groups or clusters and taking a sample of these clusters. Clustered sampling is performed when it is not feasible (and perhaps not possible) to exhaustively define a sampling region. For example, a survey of individual expenditures is to be done in Dade County, Florida. It is not possible to exhaustively define a sampling region due to daily births, deaths, etc. Therefore, the population is divided into several groups from which samples can be defined. In this example, Dade County has 27 incorporated cites within it. Therefore, the clusters would be the 27 incorporated cities and the remaining unincorporated portion of the county. Individuals within each cluster are chosen randomly and studied.

Selected sampling is done by taking the sample from only one special part of the population. For example, one form of selected sampling is to take the samples from the end of the roll or edge of the plate. Other types of selected sampling include sampling at specific times or sampling only one of the ingredients in a mixture (e.g., sampling only the milk in cake batter).

Systematic sampling is done by sampling at fixed intervals (e.g., every 25th transaction or every 10th item produced). The specific interval is determined by dividing the population by the sample size. Items are then selected at this interval throughout the population until the specified

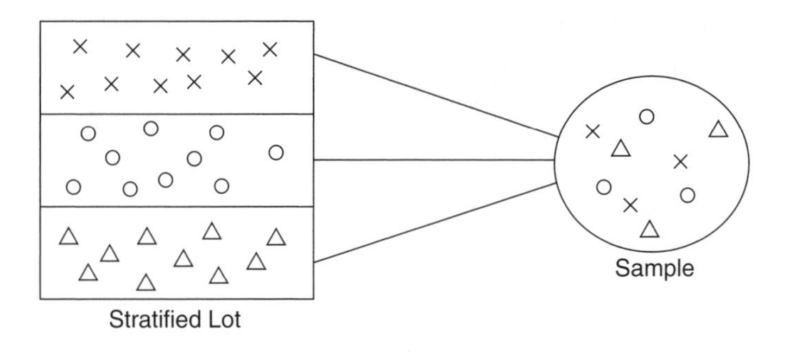

Figure 13-5 Stratified Sampling

sample size has been reached. For example, let the population be 150 items and the sample size be 10:

$$\text{sample interval} = 150/10 = 15$$

Therefore, the fixed interval is every 15th item or items 15, 30, 45, 60, 75, 90, 105, 120, 135, and 150. A degree of randomness is added to this method by selecting the starting point at random. This is done by randomly selecting a number from the interval range. In this example, the interval range is 15. Therefore, a number from 1 to 15 is chosen at random. If 5 is the number chosen, the items included in the sample become 5, 20, 35, 50, 65, 80, 95, 110, 125, and 140.

The collection of data using selected sampling and systematic sampling is more precise than simple random samples. It is also easier and more economical. However, there is always some bias. Also, because the sample is not truly random, it may not accurately represent the population.

Acceptance sampling is primarily done during incoming inspection for the purpose of accepting or rejecting a lot. Acceptance sampling plans consist of tables that are indexed according to different criteria. There are many standard acceptance sampling plans in use today. The most common is the MIL-STD 105D1 (or ANSI/ASQC Z1.4) acceptance sampling plan for attributes and the MIL-STD 4142 (ANSI/ASQC Z1.9) acceptance sampling plan for variables. Both of these plans are designed to ensure that the producer makes lots with a quality level as good or better than the specified acceptable quality level (AQL).

Sampling Error

Sampling error has occured if the sample statistics differ from the population statistics after the entire population has been examined. There are two types of sampling error: bias and dispersion. ***Bias*** or "lack of accuracy" occurs when the sample mean is different than the population mean. Bias can result from factors such as:

- Sampling only from the surface of a liquid at rest
- Sampling only from one edge of rolls or sheets
- Sampling from only one segment of the lot
- Instruments out of calibration

The result of a bias error is illustrated in Figure 13-6. Dispersion or "lack of precision" occurs when the measurements taken and recorded vary around the true measurement. ***Dispersion error*** is the result of variability in the sample standard deviation (see Figure 13-7). This type of error is typically due to improper reading or use of an instrument or an instrument that cannot read to the specified precision. The key to eliminating dispersion error is to choose the proper measuring instruments and make sure that the people using them are adequately trained in their use. The key to eliminating bias error is to make sure the instruments are calibrated and that the sample chosen accurately represents the population.

Summary

Remember that data provide the basis for all decisions and actions. It is important to know why data are needed and what they will be used for. Therefore, careful planning should precede every data collection effort. Be sure to collect only the data needed, and be sure that all information associated with the data and the data collection process is accurately recorded. When collecting the data, measure as accurately as possible within the given time and cost constraints. Also, all data collection should be done in such a way that the data can be easily used and understood.

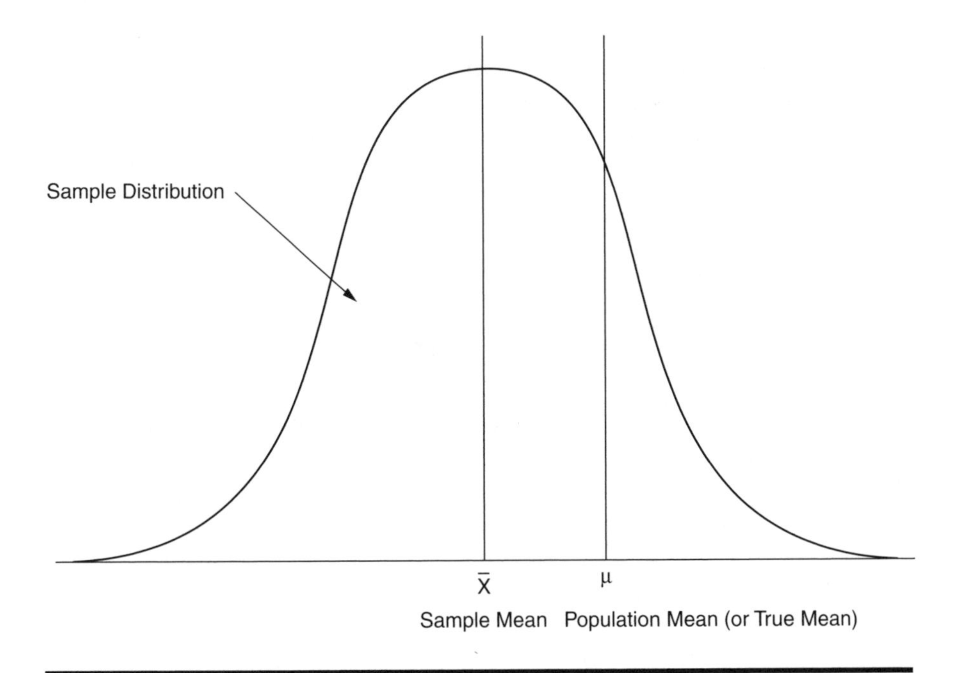

Figure 13-6 Bias Error

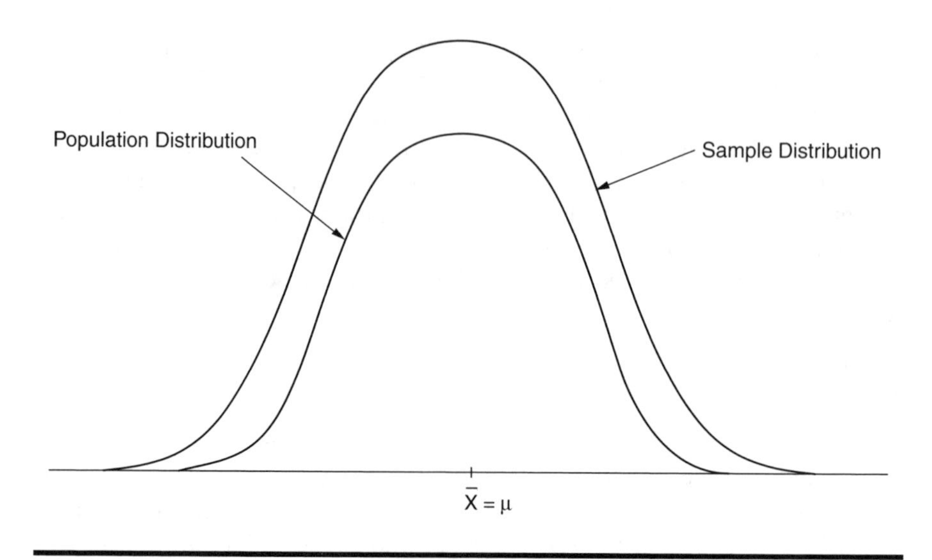

Figure 13-7 Dispersion Error

Data collected from samples are used to make decisions. Therefore, it is critical that the sample be carefully chosen. When choosing a sampling scheme, several aspects should be considered: (1) the accuracy and reliability the scheme provides, (2) the additional cost in time and personnel that will be incurred, and (3) the timeliness with which the sample can be taken. If sampling is done properly, the data will accurately represent the population, and correct decisions can be made and appropriate actions taken. If sampling is not done properly, the data will not truly represent the population and the decisions made and actions taken may be the wrong ones. Therefore, whatever sampling method is used, it should be carefully designed.

EXERCISES

13-1 List several possible reasons why you might need to collect data in your job.

13-2 What are the two main types of data? Give an example of each.

13-3 What is the difference between attribute data and variable data? Give several examples of each.

13-4 What is stratification?

13-5 Suppose an airline had over 6000 missing pieces of luggage last year, and this number represents 0.2% of the total pieces of luggage handled. What types of data would you request from the airline to help you understand the nature of the problem and how to solve it?

13-6 Suppose a bicycle manufacturer is experiencing a high rate of returned merchandise from its customers. The most frequently cited reason for the return (95%) is "difficulties with assembly." What types of data would you request from the bicycle manufacturing company to help you understand the nature of the problem and how to solve it? Be sure to state why you believe the data you request would help you, and state how it would help you.

13-7 What is the purpose of sampling?

13-8 What is sampling error?

13-9 What is dispersion error and bias error?

13-10 What factors should be considered when choosing a sampling scheme?

ENDNOTES

1. U.S. Department of Defense, "Sampling Procedures and Tables for Inspection by Attributes, MIL-STD 105D," Washington, D.C.: U.S. Government Printing Office, 1963.
2. U.S. Department of Defense, "Sampling Procedures for Inspection by Variables for Percent Defective, MIL-STD 414," Washington, D.C.: U.S. Government Printing Office, 1957.

REFERENCES

Cohen, Ruben D., "Why Do Random Samples Represent Populations So Accurately?" *Journal of Chemical Education,* Nov. 1991, pp. 902–903.

Dallal, Gerard E., "The 17/10 Rule for Sample-Size Determinations," *The American Statistician,* Feb. 1992, p. 70.

Dietz, E. Jacquelin, "A Cooperative Learning Activity on Methods of Selecting a Sample," *The American Statistician,* May 1993, pp. 104–109.

Eckblad, James W., "How Many Samples Should Be Taken?" *BioScience,* May 1991, pp. 346–347.

Ishikawa, Kaoru, *Guide to Quality Control,* Hong Kong: Asian Productivity Organization, 1982 (available in North America, the U.K., and Western Europe from Unipub, a division of Quality Resources).

Levy, Paul S. and Stanley Lemeshow, *Sampling of Populations — Methods and Applications,* New York: John Wiley & Sons, 1991.

Taylor, John Keenan, *Statistical Techniques for Data Analysis,* Chelsea, Mich.: Lewis Publishers, 1990.

Thompson, Steven K., *Sampling,* New York: John Wiley & Sons, 1992.

14

THE SEVEN BASIC QUALITY CONTROL TOOLS

BACKGROUND

In 1968, Dr. Kaoru Ishikawa wrote a book entitled *Gemba no QC Shuho* to introduce quality control techniques and practices to the workers of Japan. It was designed to be "used for self-study; training of employees by foremen; or in QC reading groups"[1] in the Japanese workplace. It is in this book that the seven basic quality control tools were first presented. (Dr. Ishikawa did not call them the seven basic quality control tools. This descriptor came later.) In 1971, an English translation of Dr. Ishikawa's book, entitled *Guide to Quality Control,* was published by the Asian Productivity Organization.[2] This book has been widely used and is still a valuable resource when using the seven basic tools.

The seven basic quality control tools, as originally identified by Dr. Ishikawa, are:

- Check sheets
- Graphs
- Histograms
- Pareto charts
- Cause-and-effect diagrams
- Scatter diagrams
- Control charts

These seven are considered the traditional tools because they are the ones presented in Dr. Ishikawa's book. However, another basic tool, the flowchart, is considered to be just as valuable. Because the flowchart is such a valuable tool, it sometimes replaces a lesser used tool (like scatter diagrams) in the list of seven. Depending on what book or article you read, a listing of the seven basic tools may exclude one or more of those listed above and include a personal favorite of the particular author. Regardless of which tools are listed, the fundamental criterion is that the tool be a structured technique for collecting and analyzing data.

The remainder of this chapter provides an introduction to and the basics of how to use the traditional seven tools. A section on flowcharts is also included because they are so popular. For better understanding, these tools can be divided into three distinct categories: tools for identifying, tools for prioritizing and communicating, and tools for analyzing. The identifying tools are the check sheet and the flowchart. Both are used to help identify and quantify where and what problems exist. Once a problem area has been identified, the prioritizing tools can be used. The prioritizing tools consist of histograms, Pareto charts, and graphs. These tools help the user organize, understand, interpret, and present the data gathered. With this information, the user can now prioritize which problems to work on and in what order they should be addressed. Because these tools provide charts and graphs that are very easy to understand, they can also be considered the major communication tools of the group. With a specific problem identified, the analyzing tools can be used. The analyzing tools are the cause-and-effect diagram, the scatter diagram, and control charts (histograms can also be considered an analyzing tool). These tools are used to examine and investigate the causes of the problem. They can also suggest possible corrective actions. It should be noted that 70 to 80% of all problems can be solved by using check sheets, Pareto diagrams, and cause-and-effect diagrams.

CHECK SHEETS

Check sheets are forms that are used to systematically collect data. They give the user a "place to start" (a major stumbling block for some) and provide a structure for collecting the data. They also aid the user in organizing the data for use later. (The data gathered in a check sheet can be used in building histograms, Pareto charts, control charts, etc.) The primary benefits of check sheets are that they are very easy to use and understand and can provide a clear picture of the situation. Check sheets essentially allow the user to speak with facts (a fundamental tenet of total quality management).

There are many types of check sheets that can be and are being used. Three major types are presented here: defect-location check sheets, tally check sheets, and defect-cause check sheets.

Defect-Location Check Sheets

The defect-location check sheet is usually a sketch, drawing, or picture of the product being made. The location and nature of problems or defects are marked on the diagram. An example is provided in Figure 14-1, which is a sketch of an automobile door. It should be noted that the sketch is not to scale. The important thing is that it represents the part being studied and the defects can be easily stratified. This check sheet was used to examine paint blemishes on a car door. From this check sheet, it was found that the majority of paint blemishes occurred on the lower right corner. Upon investigation, it was discovered that the shape of this door differed from the shape of the door on the previous model and the programming for the spray guns had not been properly changed. This type of check sheet typically leads to fast corrective action.

Tally Check Sheet

The tally check sheet is used to count the number of occurrences of different types of defects. By knowing which type of defect occurs most frequently, appropriate action can be taken to reduce the total number of defects. Figure 14-2 is an example of a tally check sheet to collect data

ABC MOTOR COMPANY
SOUTH PLANT

Defect location check sheet for examining paint blemishes occurring on passenger and driver front door on 1991 Model 480si

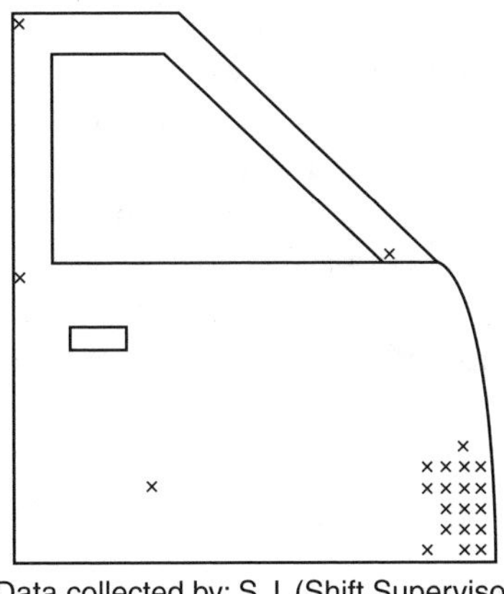

Data collected by: S.J. (Shift Supervisor)
Period of study: Shift 1, February 11, 1991

Figure 14-1 Defect-Location Check Sheet

COMPLAINT ANALYSIS
XYZ BREAD
DECEMBER 1992

COMPLAINT TYPE	TALLY	TOTAL
QUALITY	卌 卌 卌 卌 卌 卌 卌 IIII	39
PACKAGING	卌 卌 卌 卌 卌 卌 卌 卌 卌 II	47
INFESTATION	卌 卌 I 卌 卌	21
FOREIGN MATERIAL	卌 卌 卌 IIII	19
OTHER	卌 I 卌	11

137 Total Complaints

Data tabulated by Andrew Thomas January 27, 1993

Data tabulated from records collected by customer service department for the month of December 1992

Figure 14-2 Tally Check Sheet

on the reasons for customer complaints about a particular brand of bread. The check sheet suggests that the main cause of complaints was due to packaging. The team felt that this was not an accurate representation of the problem, because the data were collected over a limited time span. To verify the main cause, the team decided to collect data for the previous six months. (This was relatively easy since the customer service department had been keeping very accurate records on all complaints received during the past year.) Figure 14-3 is a summary tally check sheet for this six-month time frame and indicates that the main problem was really the quality of the bread.

Defect-Cause Check Sheet

The previous check sheets are used to determine certain aspects of defects, such as location or general cause. However, when more information about the cause of a defect is required, a defect-cause check sheet is used. A defect-cause check sheet that was used to determine the reasons for poor

COMPLAINT ANALYSIS
XYZ BREAD
JULY 1992-DECEMBER 1992

COMPLAINT TYPE	TOTAL
Quality	1309
Packaging	498
Infestation	261
Foreign Material	192
Other	112
	2372 Total Complaints

Data tabulated by Andrew Thomas   Februrary 18, 1993

Data tabulated from records collected by customer service department for the months July-December 1992

Figure 14-3 Summary Tally Check Sheet

DEFECT CAUSE ANALYSIS
XYZ BREAD

Purpose: Investigate occurrances of poor quality bread
Location: Oven center
Study period: March 15–19, 1993
Data collected by: F.W. (Supervisor)

Operator	Oven	Monday 3/15/93	Tuesday 3/16/93	Wednesday 3/17/93	Thursday 3/18/93	Friday 3/19/93
1	A	xxxx	xx o	oo	x	
	B	xxx oo	x	xxx o	xx o	xxx oo
2	C	xxxxxx xxx ooooo o	xxxxxxx xxx ooooo	xxxxxxxx xxx oooooo oo	xxxxxxx xx ooooo	xxxxxxx ooo
	D	xxx o	x	xx	x	x

x Burned   o Undercooked

Figure 14-4 Defect-Cause Check Sheet

quality bread is illustrated in Figure 14-4. The check sheet indicates that the highest percentage of poor quality bread was being produced by operator 2 in oven C. Because operator 2 also used oven D, which had the lowest percentage of poor quality bread, operator error was ruled out as the main cause. Further study showed that proper maintenance had not been done on oven C because this oven was older than the other three ovens and replacement parts were difficult to get.

FLOWCHARTS

Flowcharts are graphical representations of a process which detail the sequencing of the materials, machinery, and operations that make up that process. They are an excellent means of documenting what is going on in a process and communicating that information to everyone.

There are many benefits to using a flowchart. First, it clearly identifies the components of a process. This helps the people who work in the process understand where they fit in and what the overall objective is. Second, it also can be used as a training tool for new workers who are brought into the process or for existing workers who change locations within the process. Third, it can serve as a guide for identifying problems or areas of improvement within the process. It also helps identify where and when in the process measurements can be made. Fourth, it can be used to document a simple operation such as a cash sales transaction, as shown in Figure 14-5. A flowchart can also be used to document a complex concept such as the running of a corporation. Finally, and most importantly, if used consistently, everyone will understand the process in the same terms. In other words, everyone will speak the same language.

GRAPHS

Graphs are visual displays of data that are used to organize and summarize data. They are typically the simplest and best way of analyzing, understanding, and communicating data. Therefore, they can easily be used for illustrating the current situation, identifying a problem area, or for illustrating the new, improved situation.

There are many different types of graphs, ranging from simple to complex. The three major types of graphs most commonly used are the line graph, bar graph, and circle graph.

Line Graph

A line graph is a visual display of the pattern of data. It is primarily used for comparing data, identifying problem areas, and outlining the pattern of data. A typical line graph is shown in Figure 14-6, which outlines the allowed employee exposure time for a wide range of noise levels.

A special type of line graph is the run chart. A run chart plots a given variable as a function of time. This type of graph is very useful in that it shows the variability of a variable over time. With enough data, patterns such as trends and cycles can be identified. (Note that the run chart merely displays the nature of the data. No statistical conclusions can be drawn from this type of chart.) An example of a run chart is provided in Figure 14-7, which shows the number of airline deaths per year for 1982 to 1994.

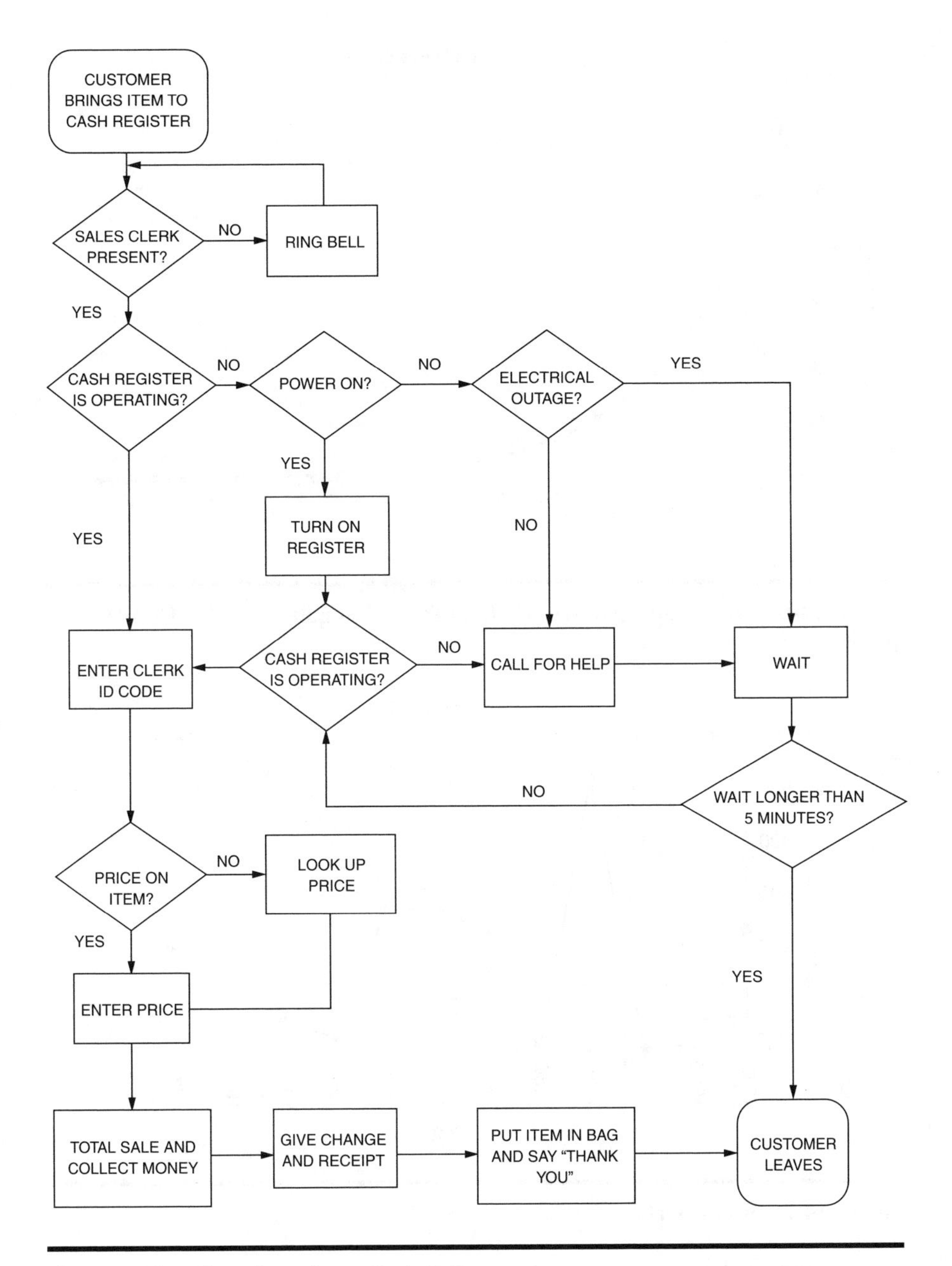

Figure 14-5 Flowchart for a Cash Sale

Bar Graphs

A bar graph, better known as a bar chart, is a visual illustration of data in which rectangular bars are used to represent the quantity of the variable

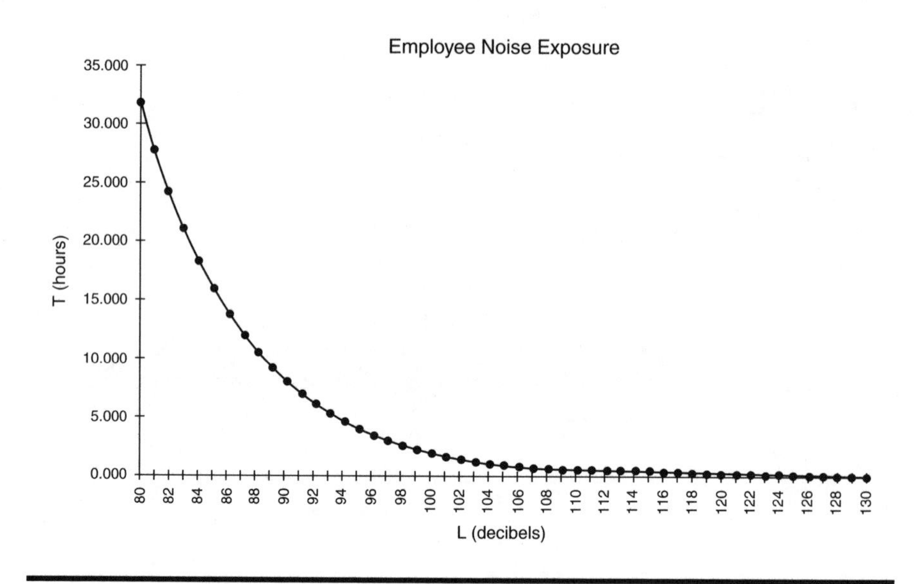

Figure 14-6 Line Graph (Source: Code of Federal Regulations 29 CFR 1910.95, p. 184)

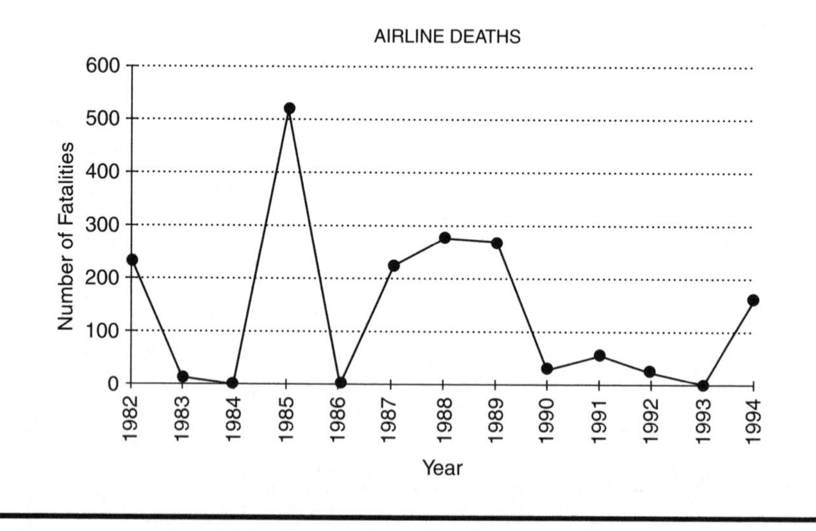

Figure 14-7 Run Chart

being studied. This chart is used primarily for comparison purposes. There are special types of bar charts available for use. The two most common are the histogram and the Pareto chart. Each has a specific purpose and will be discussed later.

A bar chart is displayed in Figure 14-8, which illustrates the serving speed of the top ten servers in the IBM/ATP tennis tour.

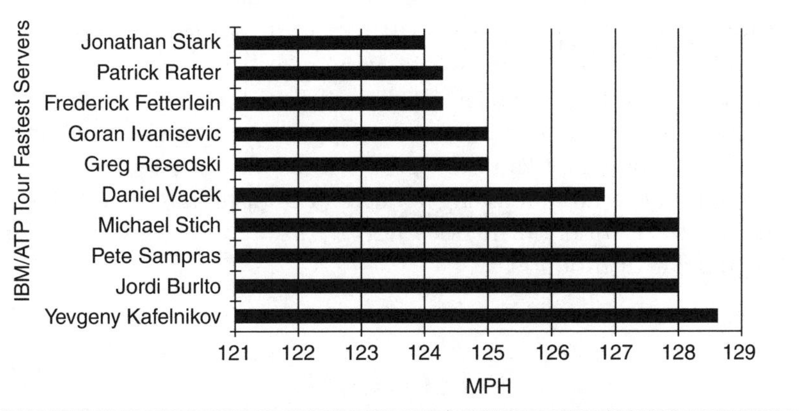

Figure 14-8 Bar Graph

Circle Graph

Circle graphs are more commonly known as pie charts. Pie charts represent data as slices of a pie. The larger the slice, the larger the percentage that item is of the whole. Figure 14-9 illustrates how a particular four-year-old spends his day. The whole pie represents 24 hours. Other than sleeping, this four-year-old spends most of his time playing, which is to be expected.

The pie chart is a very effective tool for comparing relative magnitude or frequency and how it contributes to the whole. This is true only if the number of categories being compared is kept low. If there are too many categories, the user spends most of his or her time trying to determine what the categories are and misses the whole point of the chart.

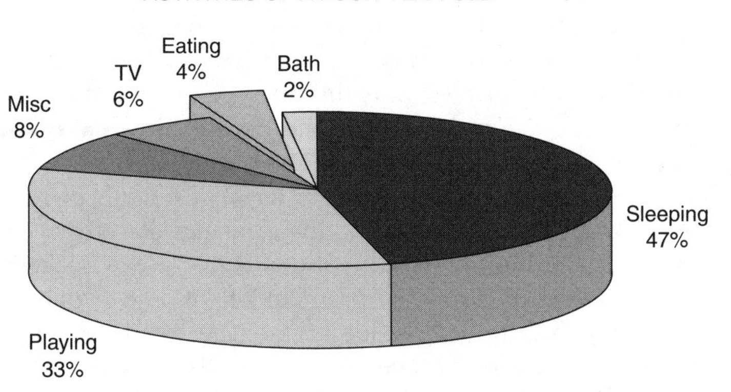

Figure 14-9 Pie Chart

HISTOGRAMS

The histogram is a type of bar chart that visually displays the variability of a product or process. It shows the various measures of central tendency (mean, mode, and average). It can be used to illustrate whether product specifications are being met by drawing the specifications on the histogram. A histogram can also be used to study and identify the underlying distribution of the variable being studied. (The histogram merely illustrates the nature of distribution. It does not, by itself, provide statistical proof of a particular distribution.)

PARETO CHARTS

The Pareto principle was first defined in an article written by Dr. Joseph M. Juran in 1950.[3] While studying quality defects as a young engineer in the 1920s, Dr. Juran noted a phenomenon that he called "the vital few and the trivial many." He discovered that if quality defects were arranged in order of frequency of occurrence, relatively few of these defects accounted for the bulk of the defectiveness. Later in his works, he again noted that a similar phenomenon existed in employee absenteeism, causes of accidents, and other managerial areas. In the late 1930s, while on a temporary assignment at General Motors, one of the executives revealed to Dr. Juran that this phenomenon existed in other fields as well. During this time, Dr. Juran discovered the work of Vilfredo Pareto, a 19th-century economist who had made extensive studies on the unequal distribution of wealth. Pareto observed that 80% of the wealth was owned by only 20% of the population. (Pareto developed several mathematical models to quantify this unequal distribution.) Pareto's observation with respect to economics was similar to Juran's observation.

By the late 1940s, Dr. Juran had recognized that the concept of "the vital few and the trivial many" was truly universal in management and in fact was universal in nature. He was the first person to reduce to writing this universal phenomenon. He coined the phrase "the vital few and the trivial many" and called it the Pareto principle as a shorthand notation to convey the concept of maldistribution. In the first edition of his *Quality Control Handbook,* he used the shorthand name Pareto principle to identify this idea, and the universality of this concept was picked up and used by writers who enthusiastically promoted it with the erroneous name.

Later, Dr. Juran was forced to admit that he had made a mistake in attributing so much to the 19th-century economist Pareto and admitted that Pareto's work dealt only with the unequal distribution of wealth. Dr. Juran also acknowledged that the cumulative frequency distribution curves used in the first edition of the *Quality Control Handbook* should have been attributed to Lorenz instead of Pareto.[4] Dr. Juran also modified his phrase to "the vital few and the *useful* many" when he later learned that all problems found deserve attention and therefore are not *trivial*.[5]

The power of the Pareto principle comes from how it is illustrated via the Pareto chart and the ease with which this chart can be understood. A Pareto chart is basically a bar graph in which the bars are arranged in descending order of height, starting at the left. This "picture" quickly highlights the "vital few" problems that should be worked on first. Thus, it aids in identifying and prioritizing what needs to be done. It also provides a common knowledge base founded on facts, instead of hunches, which results in gaining the cooperation of all involved.

Pareto charts have a variety of applications. In addition to providing a means for studying and improving quality, they also provide a means for studying and improving efficiency, material waste, energy conservation, safety issues, cost reductions, etc. Virtually any area a team wants to study can benefit from the use of a Pareto chart.

A good example of the use of the Pareto chart can be shown by analyzing the data provided in Figure 14-3. The Pareto chart developed from these data is shown in Figure 14-10. As the chart illustrates, the biggest problem is complaints related to the category "quality." At this point, more information is needed before improvements can be made, because quality encompasses such a broad area. Stratification of the quality category is illustrated in Figure 14-11. From this chart, the "vital few" areas are "stale" and "burned." Because "burned" was an in-house problem, the team decided to work on this problem first. (Note that this was not the major quality problem, but it was one that the team could directly address and would have a major impact overall.) The results are shown in Figure 14-12, which illustrates the overall improvement by placing the before-improvement and after-improvement Pareto diagrams side by side. (This can be done because both the before and after diagrams had the same relative time frame, six months.) Also note that by solving the "burned" problem, the "undercooked" problem was also corrected. The main quality problem then appeared to be stale bread. Because this problem involved vendors and other departments not represented on the team, the current team was disbanded and a new team was formed to address the new quality problem. In other words, the improvement cycle began all over again (hence, continuous improvement, another tenet of total quality management).

As can be seen from Figures 14-10 to 14-12, Pareto charts can be used to study an overall problem area (Figure 14-10), study one specific cause of the overall problem area (Figure 14-11), and to provide a means of measuring the impact of any changes made (Figure 14-12).

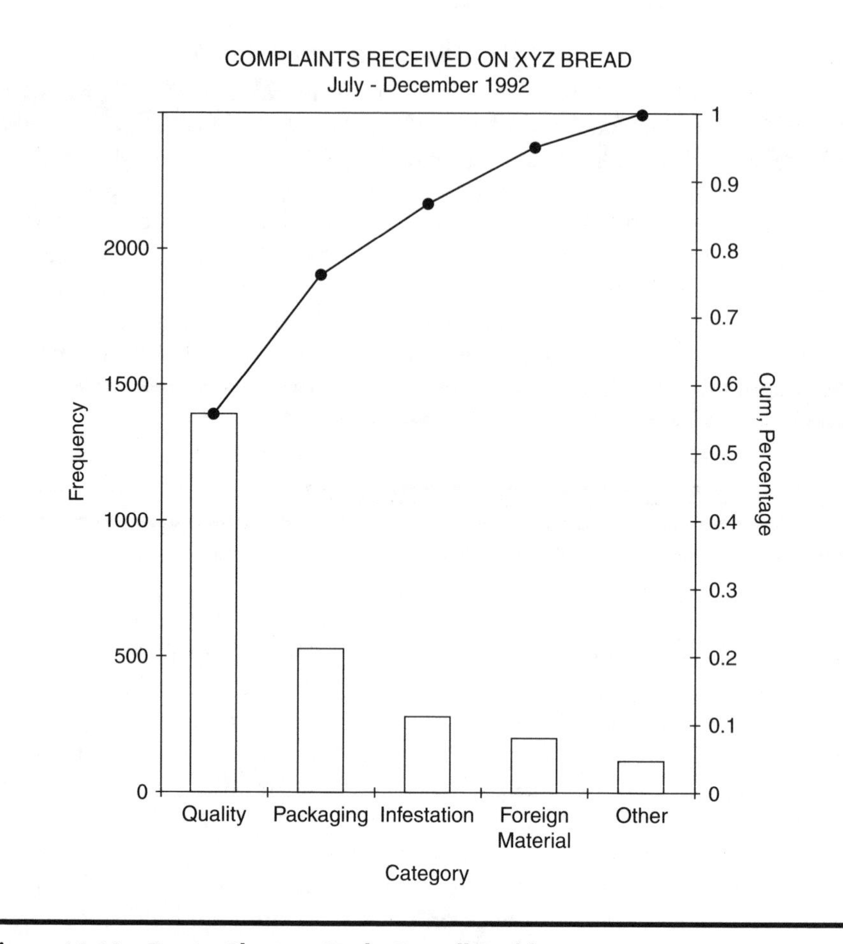

Figure 14-10 Pareto Chart to Study Overall Problem Area

CAUSE-AND-EFFECT DIAGRAMS

The cause-and-effect (CE) diagram was "developed by Dr. Kaoru Ishikawa of the University of Tokyo in the summer of 1943, while he was explaining to some engineers at the Kawasaki Steel Works how various factors can be sorted out and related."[6] For this reason, this diagram is also known as the Ishikawa diagram. Its third name, the fishbone diagram, stems from the fact that a completed diagram resembles the skeleton of a fish.

The primary purpose of the CE diagram is to show the relationship between a given effect and all identified causes of that effect. There are typically several major causes for any given effect. Therefore, a CE diagram assists the team in (1) gathering and organizing the possible causes, (2) reaching a common understanding of the problem, (3) exposing gaps in

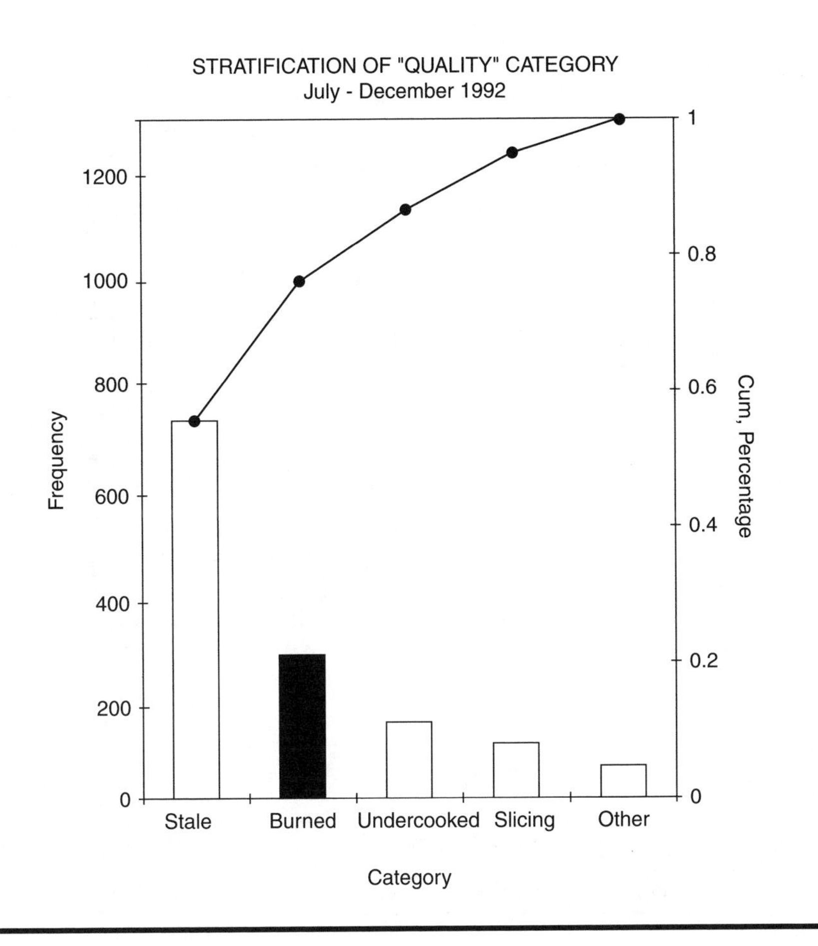

Figure 14-11 Stratified Pareto Chart

existing knowledge, (4) ranking the most probable causes, and (5) studying each cause.

In his book *Guide to Quality Control,* Dr. Ishikawa describes three types of CE diagrams:

1. Dispersion analysis
2. Production process classification
3. Cause enumeration

Even though dispersion analysis is by far the most popular, all three will be discussed here.

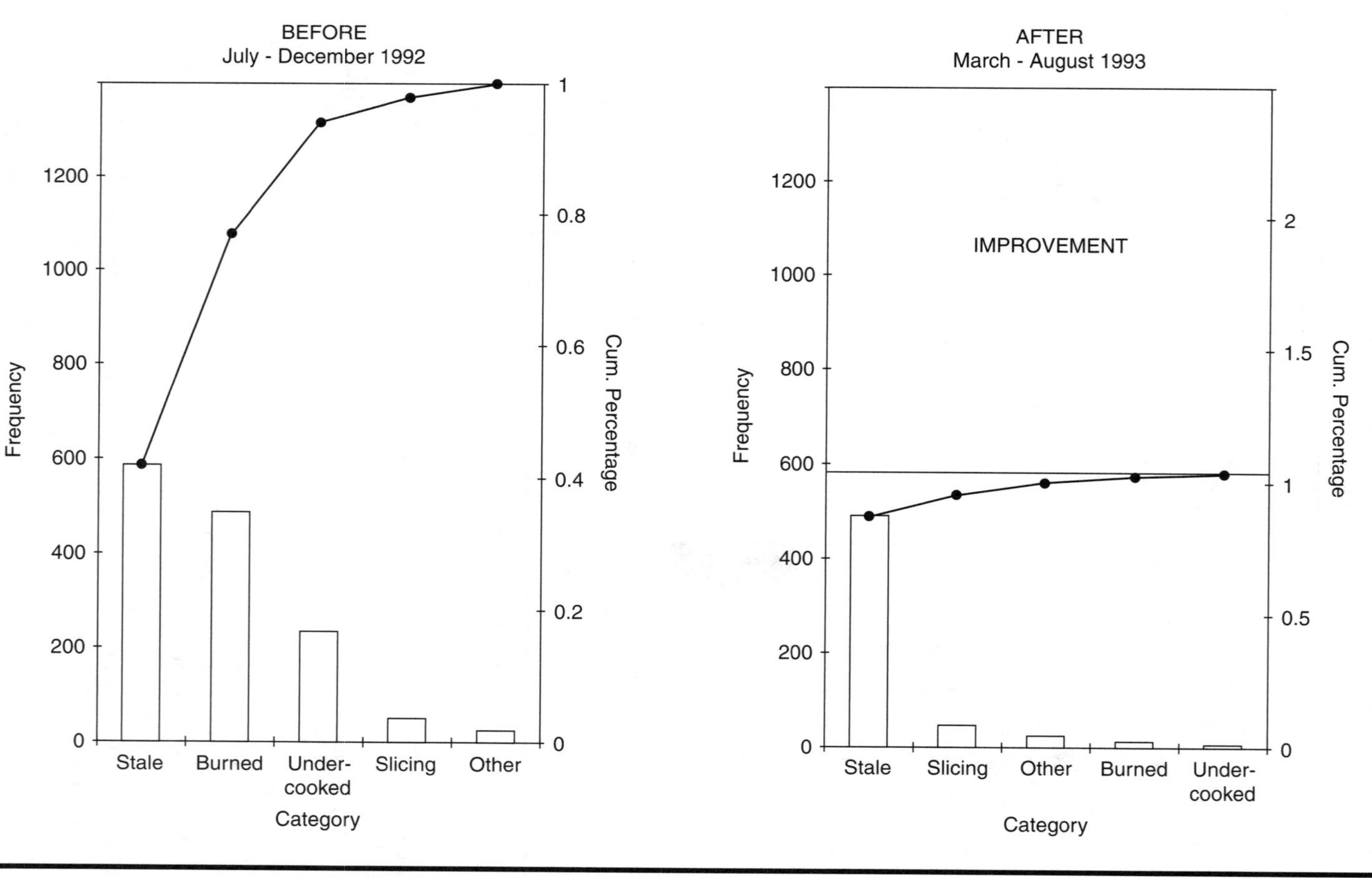

Figure 14-12 Before and After Pareto Charts

Dispersion Analysis Cause-and-Effect Diagram

The dispersion is the quality characteristic or effect being studied. In Figure 14-13, the dispersion, or effect, is damaged gowns. It should be noted that the six major "bones" or causes correspond to what are called the "standard six": manpower (or employees), machines, methods, materials, measurement, and environment. These six are obviously geared for the manufacturing environment. However, if an analysis is being performed in a service or office environment, employees, equipment, procedures, policies, and workplace can be used as generic causes. (Any set of causes can be used; these two sets are just suggestions to get the team started.)

The key to a dispersion analysis CE diagram is to continuously ask "Why does this cause produce this dispersion?" This question is reiterated for each major cause. For example, when developing the "employee" cause in Figure 14-13, the team asked, "Why do employees produce damaged gowns?" Inadequate training and no experience became the "bones" or subcauses of the major cause category. This is done until all major causes have been addressed.

The major benefit of this type of CE diagram is that it helps organize and relate the factors of the dispersion (or effect). It also gives structure to team discussions or brainstorming sessions. A major drawback is that it might fail to identify minor causes (this is typically due to the type of people on the team).

Production Process Classification Cause-and-Effect Diagram

In this type of diagram, the main line, or "backbone," sequentially follows the process flow. The major bones represent the different stages of the process. Anything that has an influence on the effect during the different process stages is represented as a "bone" on that respective stage. The process classification diagram for the damaged gown problem is illustrated in Figure 14-14. The same type of question is used to develop the "bones" (for example, "Why does the cutting stage produce damaged gowns?").

This type of CE diagram can also be made to resemble an assembly line. Figure 14-15 shows how Figure 14-14 would have looked if the assembly line approach had been used.

The major benefit of this type of diagram is that it is easy to assemble and understand because it follows the process sequence. The major drawback is that similar causes (such as employees) appear over and over again.

Note that because this type of diagram can also be used outside of the manufacturing environment, the term "production" is usually dropped from the type name.

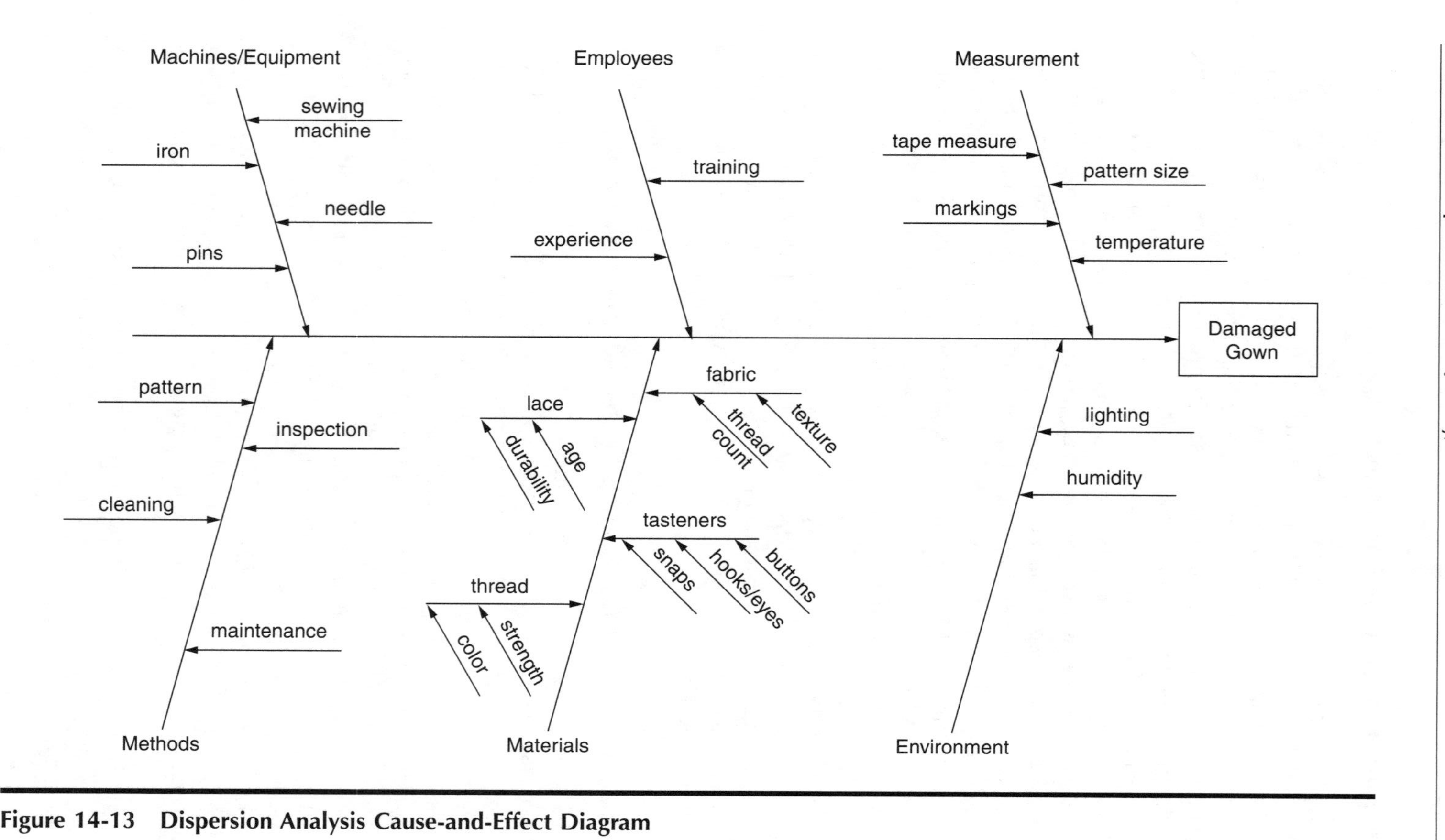

Figure 14-13 Dispersion Analysis Cause-and-Effect Diagram

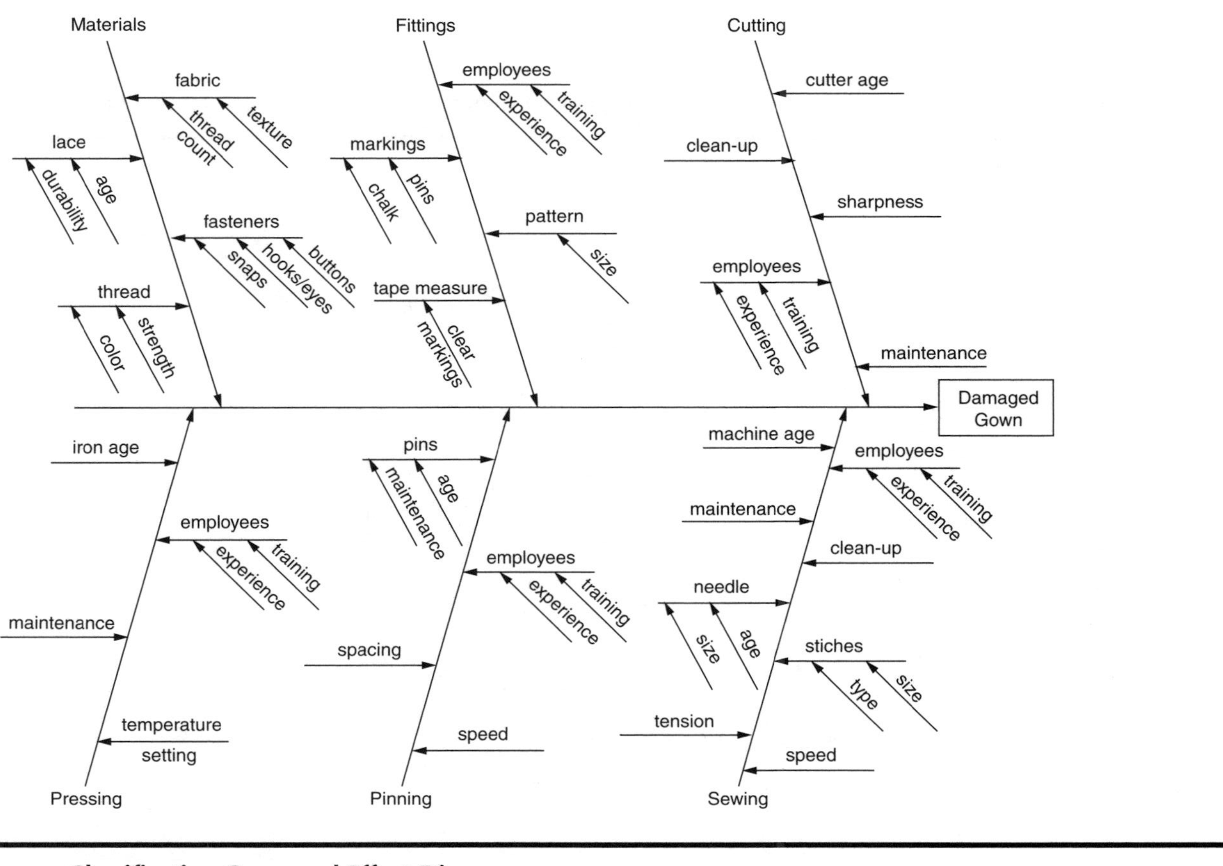

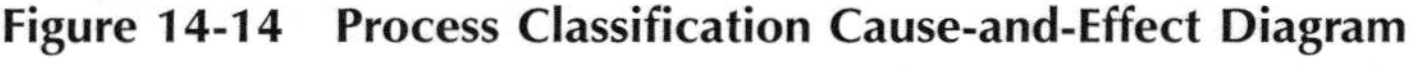
Figure 14-14 Process Classification Cause-and-Effect Diagram

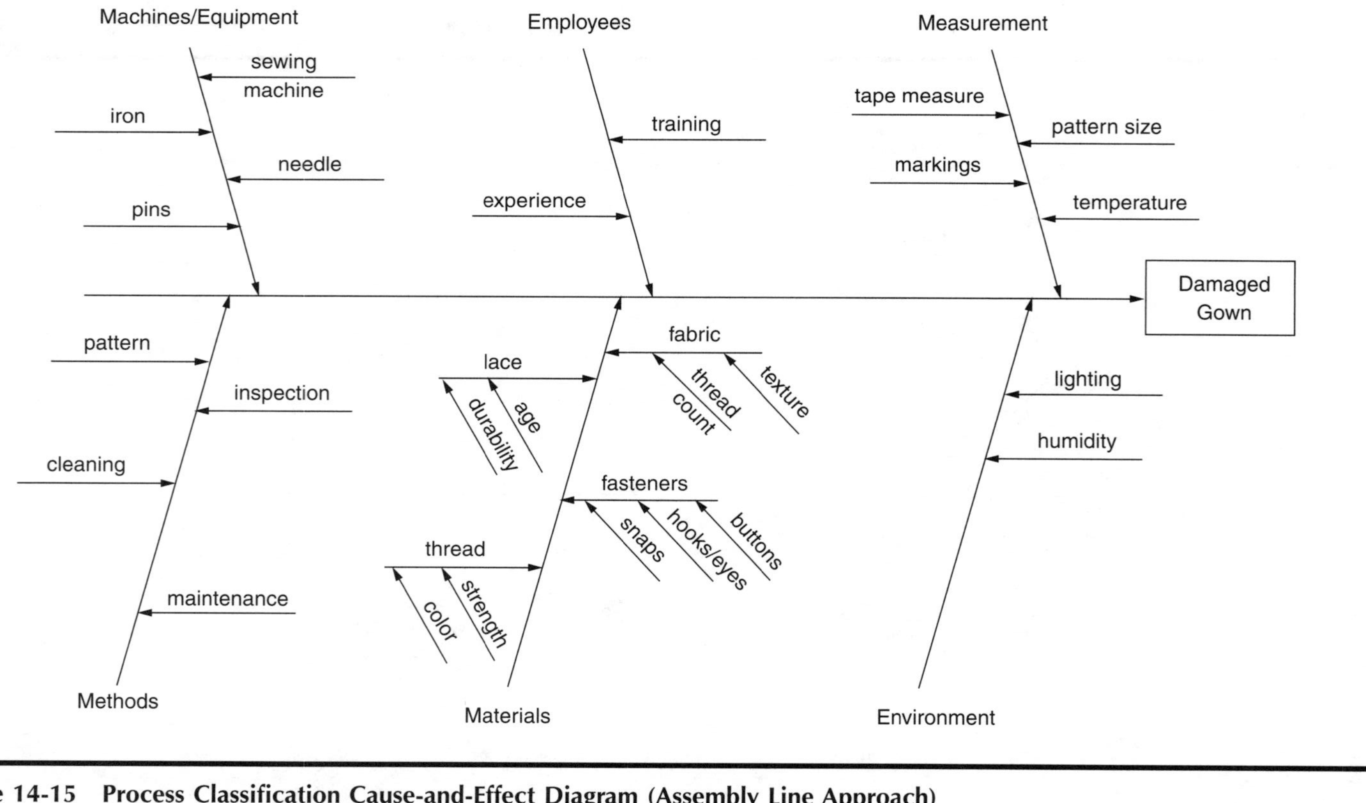

Figure 14-15 Process Classification Cause-and-Effect Diagram (Assembly Line Approach)

Cause Enumeration Cause-and-Effect Diagram

In this type of diagram, all identified causes are listed (as in a brainstorming session). Once all identified causes have been listed, they are placed in major categories. The resulting diagram would resemble Figure 14-13.

The major benefit of this type of diagram is that all identified causes are listed. The major drawback is that it is sometimes difficult to relate all the causes listed, which makes the diagram difficult to draw and can easily frustrate team members.

SCATTER DIAGRAMS

A scatter diagram is a graph of point plots that is used to compare two variables. The distribution of the points indicates the cause-and-effect relationship (or lack thereof) between two variables. In order to use a scatter diagram, paired data must be available for the two variables being studied.

Scatter diagrams are very useful in that they (1) can clearly indicate whether or not a cause-and-effect relationship exists and (2) give an idea of the strength of that relationship. Five different scatter diagrams are displayed in Figure 14-16. Figure 14-16a shows that there is a strong positive relationship between $x1$ and $y1$ and indicates that an increase in $y1$ depends on increases in $x1$. Figure 14-16b shows that there is a positive relationship between $x2$ and $y2$. However, other factors seem to be influencing $y2$. Figure 14-16c shows that there is no relationship between $x3$ and $y3$. Figure 14-16d shows a negative relationship between $x4$ and $y4$, but that other factors are affecting $y4$. Figure 14-16e shows a strong negative relationship between $x5$ and $y5$.

A scatter diagram by itself does not imply statistical significance of the observed relationship. Additional analysis, in the form of probability plotting or calculation of the correlation coefficient, is required for statistical correlation. It should also be noted that the conclusion drawn from a given scatter diagram is only valid over the range of values that were actually observed.

CONTROL CHARTS

A control chart is a special type of run chart with limits. It shows the amount and nature of variation in the process over time. It also enables pattern interpretation and detection of changes in the process.

There are three main reasons for using a control chart. First, it is used to monitor a process in order to determine if the process is operating with only chance causes of variation. If it is, then the process is said to be in statistical control. If it is not, then the process is said to be out-of-

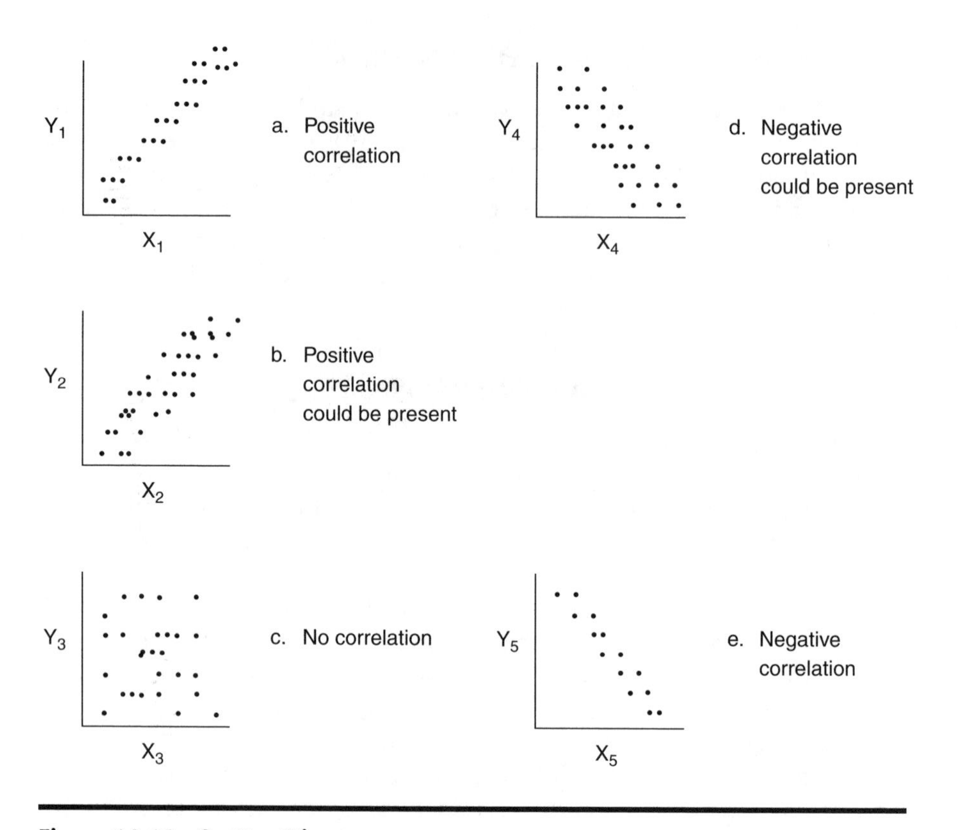

Figure 14-16 Scatter Diagrams

control. If the process is out-of-control, then the control chart can be used to help identify the assignable causes of variation and correct the process. Second, control charts are used to estimate the parameters of a process. Third, control charts are used in reducing the variability of a process.

The type of control chart depends on the type of data used in its construction. If the data are based on measurements (such as pounds, inches, etc.), then the data are said to be ***continuous*** and a variables control chart is used. If the data are based on counting (such as the number of defects in a product), then the data are said to be ***discrete*** and an attributes control chart is used. There are two types of variables control charts. One, based on averages, is called the X bar, ($\overline{X}$), chart. An $\overline{X}$ chart is accompanied by either a range (R) chart or a standard deviation (S) chart. The second type of variables control chart is based on the individual measurements and is called an X chart. It is accompanied by a moving range (MR) chart. There are four types of control charts for attributes: the p chart, np chart, c chart, and u chart. The p chart plots the fraction non-conforming; the np chart plots the number non-conforming; the c chart plots the number of non-conformities the u chart plots the number of non-conformitites per unit.

Because there are so many different types of control charts, separate chapters are devoted to control charts for variables and control charts for attributes.

EXERCISES

14-1 a. Design a location check sheet that can be used by the local volleyball team to help identify in what areas of the court most of their mistakes are occurring. Mistakes include out-of-bounds spikes, spikes not returned, serves not returned, blocks out of bounds, etc.

b. Design a tally check sheet that summarizes the errors per game that the volleyball team makes during a specific match. Errors include out-of-bounds serves, spikes, blocks, double hits, foot faults, etc.

c. Design a defect-cause check sheet that summarizes the errors made by each player. This check sheet will help the coach identify specific areas that each player needs to work on during practice. Errors include out-of-bounds serves, spikes, blocks, foot faults, out of position, etc.

Note: (A) You may need to attend a volleyball match in order to develop a complete list of possible errors for parts a, b, and c. (B) Parts a, b, and c can be used for any team sport.

14-2 Given the following data, create a bar chart that shows the estimated revenues from sales, property, and income taxes paid by undocumented aliens for the states given:

State	*Estimated revenue*
California	$732 million
Florida	$277 million
Nevada	$175 million
New York	$422 million
Texas	$202 million

14-3 Given the following data, create a pie chart indicating the percentage distribution of substance abuse:

Distribution of substance abuse among regular users	
Beer	40
Wine	7
Liquor	17
Marijuana	49
Hashish	9
Hard drugs	22
Total	144

Note: Hard drugs include speed, acid, cocaine, valium, Quaaludes, inhalants, mescaline, opium, codeine, heroin, Darvon, and peyote.

14-4 Tony has been on a special diet for the past 15 weeks. His goal is to lose 40 pounds. At the end of each week, Tony weighed himself. The following table gives his weight for each week. Draw a line graph that shows Tony's progress.

Week	*Weight (lbs.)*	*Week*	*Weight (lbs.)*
0	245 (starting weight)	8	217
1	241	9	219
2	236	10	216
3	233	11	214
4	231	12	215
5	227	13	212
6	220	14	210
7	218	15	207

14-5 In 1992, over 20 million cars traveled Interstate 10 between Pensacola and Tallahassee. That year there were 112 accidents that resulted in death. The police classified each accident into one of the five categories listed below. Draw a Pareto chart using the five categories. Further stratify the data using the additional information provided.

Cause of accident	*Accidents*
Excessive speed	24
Improper lane change	6
Mechanical failure	47
Incapable driver	22
Weather conditions	13
Total	112

Type of mechanical failure	*Accidents*
Blown tire	32
Lost brakes	9
Lost steering control	5
Other	1
Total	47

14-6 Durgest Woods manufactures a low production volume of chairs. The wood it receives is cut, sanded, painted, and then assembled. At the end of the line, the chairs are inspected. Lately, the number of rejected chairs has increased and the company would like to know why. Analyze the situation using the three types of cause-and-effect diagrams.

14-7 Use a flowchart to describe the operation of pumping gas into your car at a self-service pump. To simplify the process, make the following assumptions:

- There are no power outages
- No waiting in line for a vacant pump
- No waiting in line at the cashier
- Credit cards and ATM cards are accepted
- There is no shortage of gas

ENDNOTES

1. Kaoru Ishikawa, *Guide to Quality Control,* Hong Kong: Asian Productivity Organization, 1982. p. iii.
2. Kaoru Ishikawa, *Guide to Quality Control,* Hong Kong: Asian Productivity Organization, 1982, p. i.
3. Joseph M. Juran, "Pareto, Lorenz, Cournot, Bernoulli, Juran and Others," *Industrial Quality Control,* Oct. 1950, p. 25.
4. Joseph M. Juran, "The Non-Pareto Principle: Mea Culpa," *Quality Progress,* May 1975, p. 8.
5. Joseph M. Juran, *Juran on Leadership for Quality,* New York: McGraw-Hill, 1989.
6. Kaoru Ishikawa, *Guide to Quality Control,* Hong Kong: Asian Productivity Organization, 1982, p. 29.

REFERENCES

Benjamin, Marti and James G. Shaw, "Harnessing the Power of the Pareto Principle," *Quality Progress,* Sep. 1993, pp. 103–107.

FPL Quality Improvement Program. QI Story and Techniques, Miami: Florida Power and Light Company, 1987.

Ishikawa, Kaoru, *Guide to Quality Control,* Hong Kong: Asian Productivity Organization, 1982 (available in North America, the U.K., and Western Europe from Unipub, a division of Quality Resources).

Juran, Joseph M., "Pareto, Lorenz, Cournot, Bernoulli, Juran and Others," *Industrial Quality Control,* Oct. 1950, pp. 25–30.

Juran, Joseph M., "The Non-Pareto Principle; Mea Culpa," *Quality Progress,* May 1975, pp. 8–9.

Juran, Joseph M., *Juran on Leadership for Quality,* 4th ed., New York: McGraw-Hill, 1989.

Kume, Hitoshi, *Statistical Methods for Quality Improvement,* Japan: The Association for Overseas Technical Scholarship, 1985.

Quality Improvement Tools Workbooks, Wilton, Conn.: Juran Institute, 1989.

Seven-Part Series on "The Tools of Quality," *Quality Progress,* June–Dec. 1990:

Burr, John T., "Part I: Going with the Flow(chart)," *Quality Progress,* June 1990, pp. 64–67.

Sarazen, J. Stephen, "Part II: Cause-and-Effect Diagrams," *Quality Progress,* July 1990, pp. 59–62.

Shainin, Peter D., "Part III: Control Chart," *Quality Progress,* Aug. 1990, pp. 79–82.

"Part IV: Histograms," *Quality Progress,* Sep. 1990, pp. 75–78.

"Part V: Check Sheets," *Quality Progress,* Oct. 1990, pp. 51–56.

Burr, John T., "Part VI: Pareto Charts," *Quality Progress,* Nov. 1990, pp. 59–61.

Burr, John T., "Part VII: Scatter Diagrams," *Quality Progress,* Dec. 1990, pp. 87–89.

Swift, J. A., *Introduction to Statistical Quality Control and Management,* Delray Beach, Fla.: St. Lucie Press, 1995.

The Memory Jogger: A Pocket Guide of Tools for Continuous Improvement, Methuen, Mass.: GOAL/QPC, 1988.

15

CONTROL CHARTS FOR VARIABLES

BACKGROUND

It was Dr. Shewhart who first suggested the use of control charts. In the late 1920s, Dr. Shewhart suggested that every process exhibits some degree of variation. Since no two things can be produced exactly alike, variation is natural and should be expected. However, Dr. Shewhart discovered that there were two types of variation, chance cause variation and assignable cause variation. Chance cause variation is variation that is inherent in the process. It is random in nature and cannot be controlled. Any process that operates with only chance cause variation is said to be in a state of statistical control. Once a process is in statistical control, adjustments can be made to minimize the random variation, which will improve the process. Assignable cause variation is variation that is controlled by some outside influence or special cause, such as change in material, change in operator, change in tool setting, tool wear, or other phenomena. Any process that operates with assignable cause variation is said to be out-of-control. By using Dr. Shewhart's control charts, outside influence can usually be identified and controlled.

USES OF CONTROL CHARTS

There are three basic uses of control charts. First, they are used to monitor a given process. Because a control chart shows the degree and nature of variation over time, it can be used to determine whether a process is in a state of statistical control or is out-of-control. If it is out-of-control, the chart aids in quickly finding the assignable causes of the out-of-control condition, which enables taking corrective action before too many bad products can be produced. If a process is in-control, continued monitoring allows for quicker detection of process changes. It also allows for process improvement.

Second, control charts are used to estimate the parameters (mean, variation) of a process. By knowing the parameters of a process, the output and the variability of the output can be predicted.

Third, control charts are used to improve a process. Once a process is in a state of statistical control, efforts to reduce process variability can begin. By reducing the variability of the process, the overall quality of the final product increases, which reduces scrap and rework and increases profits.

In short, the emphasis in using control charts is on the early detection and prevention of problems. By preventing problems from occurring, productivity and profits increase.

VARIABLES CONTROL CHARTS

Variables control charts are the more classical type of control chart. They are used to monitor measurable quality characteristics of a process. Measurable quality characteristics include weight, temperature, viscosity, etc. Anything that can be measured can be monitored using a variables control chart. The main restriction is that a variables control chart can monitor only one quality characteristic at a time. If more than one quality characteristic needs to be monitored, then a chart for each characteristic must be created.

A variables control chart monitors the mean value and the variability of the quality characteristic being studied. The mean value is monitored via an X bar ($\overline{X}$) chart or an individuals (X) chart. Variability is measured via a range (R) or moving range (MR) chart or a standard deviation (S) chart.

An $\overline{X}$ chart monitors between-sample variability and is the most common type of control chart. It is used with either an R chart or an S chart. Both the R chart and the S chart measure within-sample variability, and the decision on which to use is based on the size of the sample taken. The R chart is used when the sample size is less than or equal to ten, and the S chart is used when the sample size is greater than ten.

An individuals chart is used to monitor the mean when the sample size is one. A sample size of one typically occurs when one of the following is true:

- Automated inspection is being used and every unit produced is inspected.
- It is uneconomical to take multiple measurements.
- Destructive testing is being used.

Since the sample size is one, a range does not exist. Therefore, a moving range (MR) must be established. The moving range is found by determining the range within a set of successive numbers. If the moving range size is two, then the moving ranges would be found by taking the difference between the first and second numbers, then between the second and third numbers, then the third and fourth numbers, etc. The American Society for Testing Materials (ASTM) recommends that a subgroup size of two be used for determining the moving ranges.

A summary of the control chart formulas used when sampling a process is provided in Table 15-1.

APPLICATIONS OF VARIABLES CONTROL CHARTS

Preparing to Use Variables Control Charts

When preparing to use variables control charts, several things need to be decided:

1. Problem definition
2. Choice of quality characteristic
3. Size and number of samples
4. Sampling frequency
5. Rational subgroups
6. Choice of control limits

Table 15-1 Control Chart Formula Summary[a]

Chart	*3σ control limits*	*Center line*
X	$\bar{X} \pm \frac{3\overline{MR}}{d_2}$	$\bar{X}$
$\bar{X}$ (using R)	$\bar{\bar{X}} \pm A_2 \bar{R}$	$\bar{\bar{X}}$
$\bar{X}$ (using S)	$\bar{\bar{X}} \pm A_3 \bar{S}$	$\bar{\bar{X}}$
R	UCL = $D_4 \bar{R}$ LCL = $D_3 \bar{R}$	$\bar{R}$
MR	UCL = $D_4 \overline{MR}$ LCL = $D_3 \overline{MR}$	$\overline{MR}$
S	UCL = $B_4 \bar{S}$ LCL = $B_3 \bar{S}$	$\bar{S}$

[a] For statistical development of these formulas, refer to J. A. Swift, *Introduction to Modern Statistical Quality Control and Management*, Delray Beach, Fla.: St. Lucie Press, 1995.

Problem Definition

It is very important to determine the goal of monitoring a particular quality characteristic or a group of characteristics. It is not sufficient to say that we want to improve quality. A good problem definition would be as follows:

> The difficulties associated with fitting parts 00146A with part 00146B is due to inconsistencies in the diameter of one or both of the parts. We wish to monitor the dimensions of both parts for uniformity.

Choice of Quality Characteristic

The choice of quality characteristic is based mainly on two factors. First, the quality characteristic to be examined must be measurable (for example, weight, temperature, viscosity, tensile strength, etc.). The second consideration is whether studying a particular quality characteristic will lead to reduced costs. Typically, quality characteristics that are currently exhibiting high scrap or rework rates are ideal candidates for study.

Size and Number of Samples

Several factors affect the determination of the sample size. Because a variables chart is being used, the units within each sample must be measured. The measurement taken can be as simple as weighing an item or as complex as reading a vernier caliper. The time to perform each measurement is different. Therefore, when deciding on sample size, the amount of time needed to take each sample must be considered. The larger the sample, the longer it takes to gather and measure it, and the longer it takes, the higher the costs. This would seem to imply that a small sample size should always be chosen. However, the trade-off that must be considered is that the ability of variables control charts to accurately monitor a process is decreased by collecting smaller sample sizes. This implies that larger sample sizes are preferable. The decision must be made after considering both aspects and making trade-offs. In industry, the typical sample size is four or five. Also, the sample size must remain constant.

When first setting up a control chart, enough samples need to be collected to accurately estimate the process mean and standard deviation. Also, enough samples should be collected so that any unusual source of variation has an opportunity to appear. A good rule of thumb is to collect 20 to 25 samples.

Sampling Frequency

The main purpose of using a control chart is to detect changes in a process over time. Therefore, how often to take a sample is a real concern. Taking small samples at short intervals provides quicker feedback. However, it costs more, and the process may not produce enough in a short interval to gather a random sample. Taking large samples at longer intervals provides better feedback. However, if the interval is too long, problems can occur, causing unnecessary losses. Because every process is different, there are no guidelines; there are only trade-offs to be made.

Rational Subgroups

Data should be collected in rational subgroups. This means that the subgroups should be selected so that each subgroup is as homogeneous as possible. It also means that samples are selected so that if a problem does exist, the chances for differences between subgroups are maximized and the chances for differences within subgroups are minimized.

Choice of Control Limits

The standard practice for choosing the width of the control limits is to use a multiple of the standard deviation, typically $\pm 3\sigma$.

Collecting the Samples

Once all the preparations have been made, collection of the samples begins. Collection of the samples is simplified if a standardized form is used. The form can have any appearance. The key point is that all pertinent information is recorded on the form. Pertinent information includes:

- Housekeeping items such as department, operation, specs, part number, machine number, etc.
- A place for recording the data
- A place for graphing the charts
- Action instructions
- A place for process information

The most important information that must be recorded is the process information. This information is recorded on the back of most forms, because there is more room. The process information is a record of any changes that occur to the process while it is being monitored. This includes changes in people, material, environment, methods, or machines. These changes and the exact times they occur are recorded. If the control charts indicate that a problem exists with the process, it is the process information that will aid in identifying and correcting the problem. If the process information is not taken properly, then the control charts cannot be evaluated properly. In other words, the control charts are only usable if the process information is recorded. This underscores the need for and the value of collecting the process information.

EXAMPLES OF VARIABLES CONTROL CHARTS

The general terminology used in control charting is as follows:

m = number of samples or subgroups

n = number of observations in each sample

x_i = value (measurement) of an individual item i

$\bar{x}$ = sample mean = $\dfrac{x_1 + x_2 + x_3 + \ldots +}{n}$

$\bar{\bar{X}}$ = average of the sample average = $\dfrac{\sum_{i=1}^{m} \bar{x}_i}{m}$

R = sample range = $x_{max} - x_{min}$

$\bar{R}$ = average range = $\dfrac{(R_1 + R_2 + R_3 + \ldots + R_m)}{m} = \dfrac{\sum_{t=1}^{m}}{r}$

S = sample standard deviation = $\sqrt{\dfrac{\sum_{i=1}^{n} (x_i - \bar{x})^2}{n-1}}$

$\bar{S}$ = average of the sample standard deviations

$$= \frac{(S_1 + S_2 + S_3 + \ldots + S_m)}{m} = \frac{\sqrt{\sum_{i=1}^{m} S_i}}{m}$$

UCL = upper control limit

LCL = lower control limit

Example 1: $\bar{X}$ and R Charts

A line foreman wants to establish statistical control on shaft lengths being cut. He decides to use $\bar{X}$ and R charts. The foreman collects 25 samples, each of size 5. The data collected are shown in Table 15-2.

Table 15-2 Shaft Length Measurements (Inches)

Sample #	*Sample observations*					$\overline{X}$	*R*
1	9.26	10.44	10.39	9.87	10.26	10.04	1.18
2	10.92	10.08	9.97	10.16	9.30	10.09	1.62
3	10.10	10.61	8.62	10.24	10.17	9.95	1.99
4	10.17	9.24	10.60	10.08	10.51	10.12	1.35
5	10.29	10.36	10.39	10.58	9.96	10.32	0.63
6	9.84	9.52	10.11	9.65	10.18	9.86	0.66
7	9.84	9.77	10.47	10.25	10.28	10.12	0.70
8	9.94	9.35	9.61	9.09	10.09	9.62	0.99
9	10.18	11.18	10.16	10.68	10.80	10.60	1.02
10	9.46	10.15	10.80	9.57	9.20	9.84	1.61
11	9.64	9.71	10.38	10.23	9.92	9.98	0.74
12	10.54	10.76	10.83	9.97	9.91	10.40	0.92
13	10.41	9.67	9.88	10.28	9.77	10.00	0.75
14	9.72	8.70	9.81	9.39	9.68	9.46	1.12
15	9.35	10.28	10.86	11.11	10.05	10.33	1.76
16	9.11	10.22	9.50	9.82	9.65	9.66	1.11
17	9.53	10.12	10.03	9.71	9.72	9.82	0.60
18	10.14	8.98	9.84	9.74	9.68	9.68	1.16
19	9.81	10.18	9.95	10.40	9.71	10.01	0.69
20	10.12	10.01	10.01	9.75	9.25	9.83	0.87
21	10.30	9.95	9.55	9.84	10.46	10.02	0.91
22	10.73	9.78	9.27	11.03	9.99	10.16	1.76
23	9.89	9.82	9.40	10.67	9.43	9.84	1.27
24	9.99	10.20	9.37	9.05	10.27	9.78	1.21
25	10.10	10.25	9.51	9.16	10.17	9.84	1.09
					Total	249.35	27.70
						= 9.97	= 1.11

When setting up the $\overline{X}$ and R charts, begin with the R chart to ensure that the variability within samples is in-control. Using the data in Table 15-2, we find that the center line for the R chart is

$$\overline{R} = \frac{27.70}{25} = 1.1$$

For samples of size 5, the values for D_3 and D_4 are found from Appendix A to be 0.00 and 2.114, respectively. Therefore, the control limits for the R chart can be determined from the equations in Table 15-1:

$$\text{UCL} = \overline{R}D_4 = (1.11)\ (2.114) = 2.35$$

$$\text{LCL} = \overline{R}D_3 = (1.11)\ (0.00) = 0.00$$

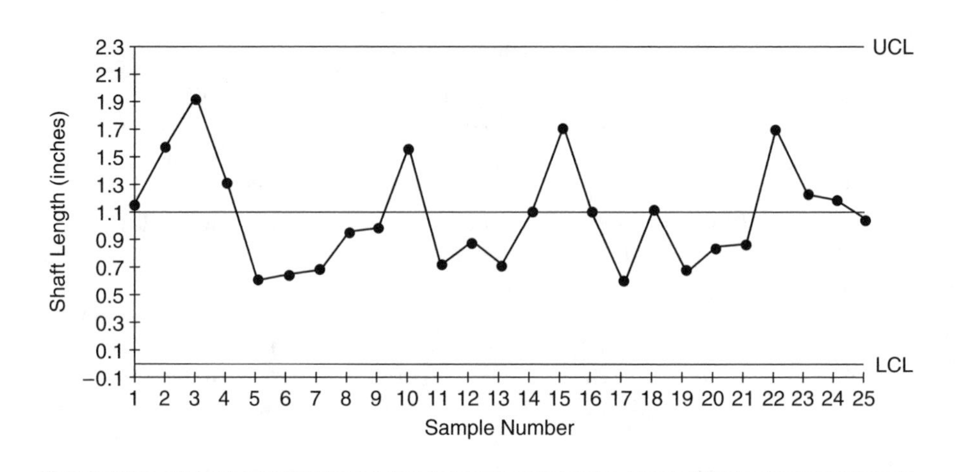

Figure 15-1 R Chart for Example 1

The R chart is shown in Figure 15-1. As can be seen, there is no indication of any out-of-control conditions. Since the within-sample variability is in control, the $\overline{X}$ chart can now be constructed. From Table 15-2, the center line for the $\overline{X}$ chart is

$$\overline{\overline{X}} = \frac{249.35}{25} = 9.97$$

From Appendix A, the value of A_2 is found to be 0.577 for a sample size of 5. Using the equations from Table 15-1, the control limits can be determined:

$$\text{UCL} = \overline{\overline{X}} + A_2 \; \overline{R} = 9.97 + (0.577)\,(1.11) = 10.61$$

$$\text{LCL} = \overline{\overline{X}} - A_2 \; \overline{R} = 9.97 - (0.577)\,(1.11) = 9.33$$

The $\overline{X}$ chart is shown in Figure 15-2. The chart shows no indication of an out-of-control condition. Therefore, both the R chart and the $\overline{X}$ chart indicate that the process is in-control.

Example 2: $\overline{X}$ and S Charts

The construction of the $\overline{X}$ and S charts will be illustrated using the shaft length measurements from the previous example. The data for the $\overline{X}$ and S charts are summarized in Table 15-3.

The parameters for the S chart are determined as follows:

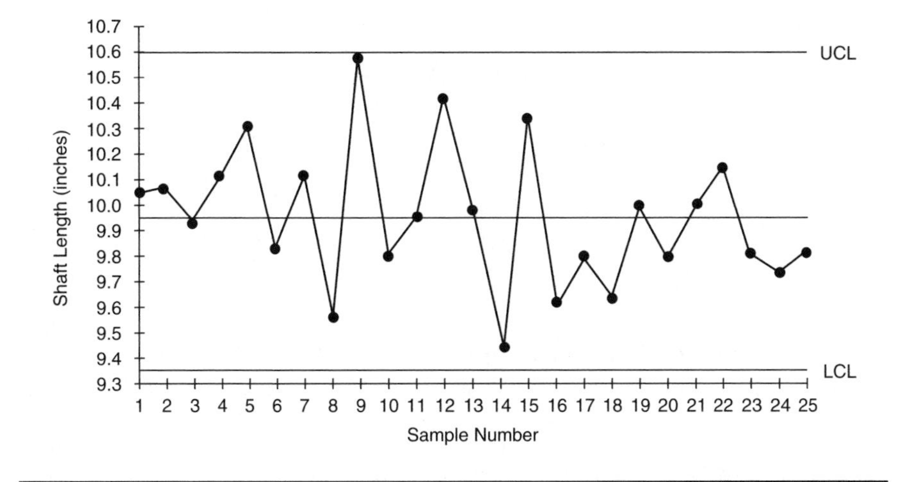

Figure 15-2 $\overline{X}$ Chart for Example 1

$$\text{center line } = \overline{S} = \frac{11.23}{25} = 0.45$$

$$\text{UCL} = B_4 = (0.45)\ (2.089) = 0.94$$

$$\text{LCL} = B_3 = (0.45)\ (0.00) = 0.00$$

The S chart is shown in Figure 15-3. As can be seen, the S chart is in-control. Therefore, the $\overline{X}$ chart can be developed using the equations from Table 15-1:

$$\text{center line} = \overline{\overline{X}} = \frac{249.35}{25} = 9.97$$

$$\text{UCL} = + A_3 = 0.97 + (1.427)\ (0.45) = 10.61$$

$$\text{LCL} = - A_3 = 0.97 - (1.427)\ (0.45) = 9.33$$

The $\overline{X}$ chart is shown in Figure 15-4. Both the $\overline{X}$ chart and the S chart indicate that the process is operating in-control.

Notice that the control limits for the $\overline{X}$ chart in Example 1 are identical to the $\overline{X}$ chart control limits for this example. This will not always be the case. Many times, the $\overline{X}$ chart control limits based on $\overline{R}$ will differ slightly from those based on $\overline{S}$.

Table 15-3 Shaft Length Measurements (Inches)

Sample #	*Sample observations*						*s*
1	9.26	10.44	10.39	9.87	10.26	10.04	0.49
2	10.92	10.08	9.97	10.16	9.30	10.09	0.58
3	10.10	10.61	8.62	10.24	10.17	9.95	0.77
4	10.17	9.24	10.60	10.08	10.51	10.12	0.54
5	10.29	10.36	10.39	10.58	9.96	10.32	0.23
6	9.84	9.52	10.11	9.65	10.18	9.86	0.29
7	9.84	9.77	10.47	10.25	10.28	10.12	0.30
8	9.94	9.35	9.61	9.09	10.09	9.62	0.41
9	10.18	11.18	10.16	10.68	10.80	10.60	0.43
10	9.46	10.15	10.80	9.57	9.20	9.84	0.64
11	9.64	9.71	10.38	10.23	9.92	9.98	0.32
12	10.54	10.76	10.83	9.97	9.91	10.40	0.44
13	10.41	9.67	9.88	10.28	9.77	10.00	0.33
14	9.72	8.70	9.81	9.39	9.68	9.46	0.45
15	9.35	10.28	10.86	11.11	10.05	10.33	0.69
16	9.11	10.22	9.50	9.82	9.65	9.66	0.41
17	9.53	10.12	10.03	9.71	9.72	9.82	0.25
18	10.14	8.98	9.84	9.74	9.68	9.68	0.43
19	9.81	10.18	9.95	10.40	9.71	10.01	0.28
20	10.12	10.01	10.01	9.75	9.25	9.83	0.35
21	10.30	9.95	9.55	9.84	10.46	10.02	0.36
22	10.73	9.78	9.27	11.03	9.99	10.16	0.72
23	9.89	9.82	9.40	10.67	9.43	9.84	0.51
24	9.99	10.20	9.37	9.05	10.27	9.78	0.54
25	10.10	10.25	9.51	9.16	10.17	9.84	0.48
					Total	249.35	11.23
						= 9.97	= 0.45

Example 3: X and MR Charts

Tensile strength is an important quality characteristic for bridge bolts. The bolts are made of a mild steel (ASTM A36). Because a tensile-strength test destroys the bolt and each bolt is expensive, a sample of size 1 was decided upon. Twenty-five bolts were tested. The results are given in Table 15-4.

The moving range chart can be constructed using the equations from Table 15-1:

$$\text{center line} = \overline{MR} = \frac{275.23}{24} = 11.47$$

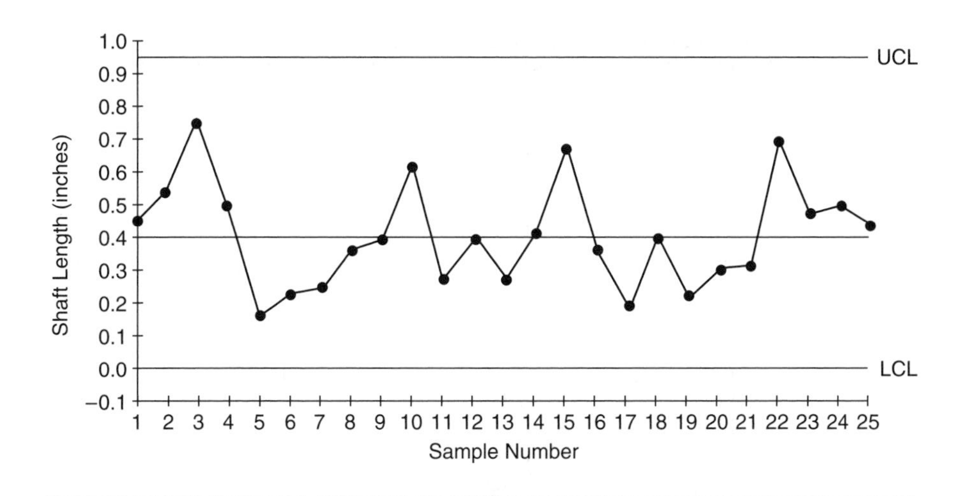

Figure 15-3 S Chart for Example 2

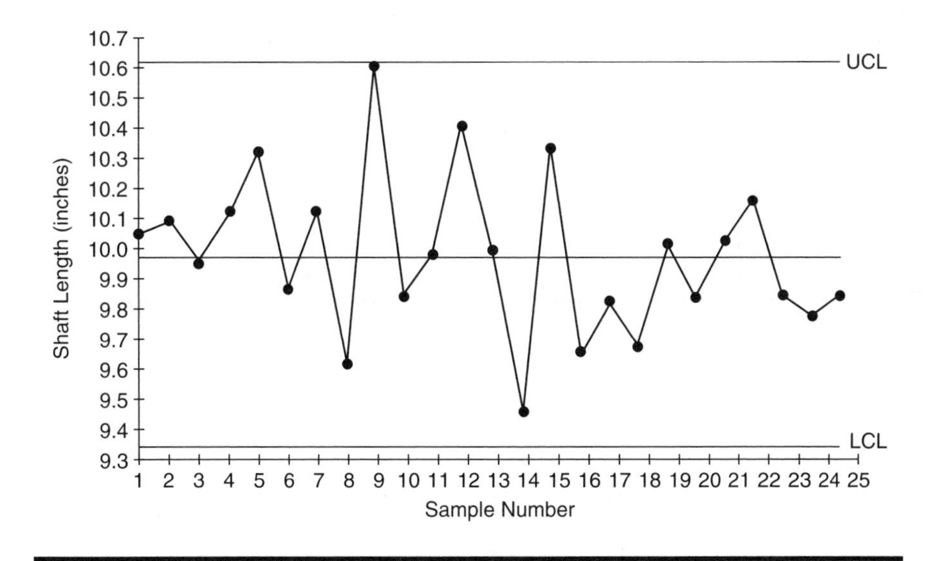

Figure 15-4 $\overline{X}$ Chart for Example 2

$$\text{UCL} = D_4\,\overline{MR} = (3.267)\ (11.47) = 37.47$$

$$\text{LCL} = D_3\,\overline{MR} = (0.00)\ (11.47) = 0.00$$

The moving range chart is shown in Figure 15-5. The moving range chart appears to be in-control. Therefore, the X chart can be constructed.

Table 15-4 Tensile Strength of Bridge Bolts (ksi)

Bolt	*Tensile strength*	*Moving range*
1	70.87	—
2	60.27	10.60
3	69.51	9.24
4	66.84	2.67
5	63.94	2.90
6	70.01	6.07
7	58.73	11.28
8	80.23	21.50
9	63.26	16.97
10	77.29	14.03
11	77.01	0.28
12	62.32	14.69
13	69.54	7.23
14	63.62	5.92
15	74.95	11.33
16	73.37	1.58
17	74.13	0.76
18	76.66	2.52
19	52.95	23.71
20	64.95	12.00
21	73.93	8.99
22	53.26	20.67
23	74.19	20.93
24	93.69	19.50
25	63.82	29.87
TOTAL	1729.34	275.23
	$\overline{X} = 69.17$	$\overline{MR} = 11.47$

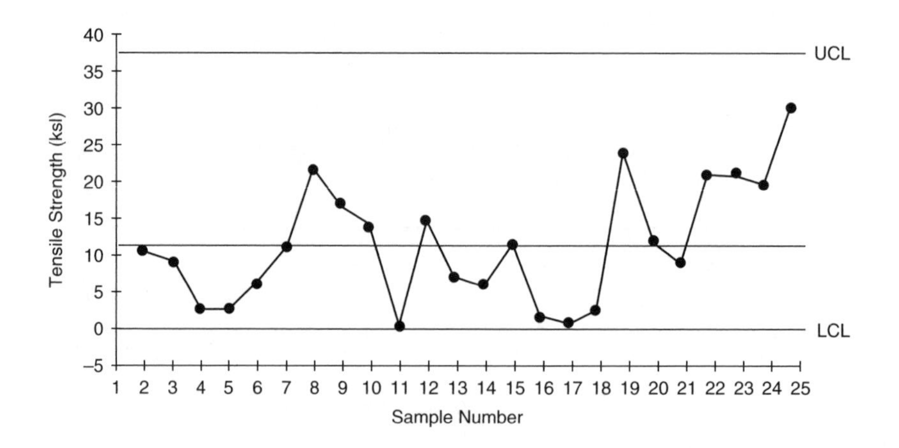

Figure 15-5 MR Chart for Example 3

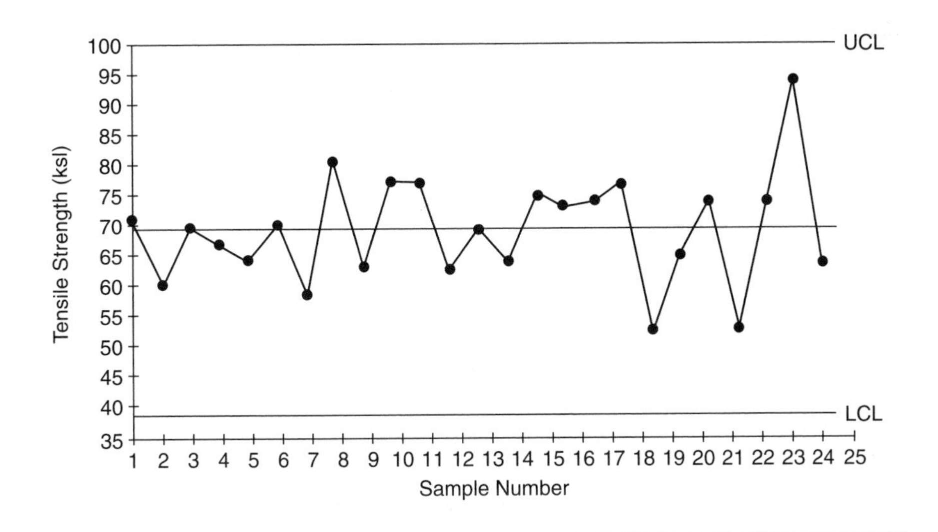

Figure 15-6 X Chart for Example 3

$$\text{center line} = \overline{X} = \frac{1729.34}{25} = 69.17$$

$$\text{UCL} = \overline{X} + 3\frac{\overline{MR}}{d_2} = 69.17 + \frac{3(11.47)}{1.128} = 99.68$$

$$\text{LCL} = \overline{X} + 3\frac{\overline{MR}}{d_2} = 69.17 + \frac{3(11.47)}{1.128} = 38.66$$

Figure 15-6 gives the X chart for the bridge bolts. Since there are no apparent out-of-control conditions present in either chart, the process producing the bridge bolts is considered to be operating in-control.

Interpreting Control Charts

Thus far, we have discussed the fact that a process exhibits a lack of statistical control if a subgroup statistic falls outside of either of the control limits. It is also possible for a process to exhibit a lack of statistical control even when all the subgroup statistics are within the control limits. There are other factors that indicate a lack of control and should be investigated. A process that is in a state of statistical control exhibits random patterns of variation that obey the laws of chance. A stable process should exhibit the following characteristics:

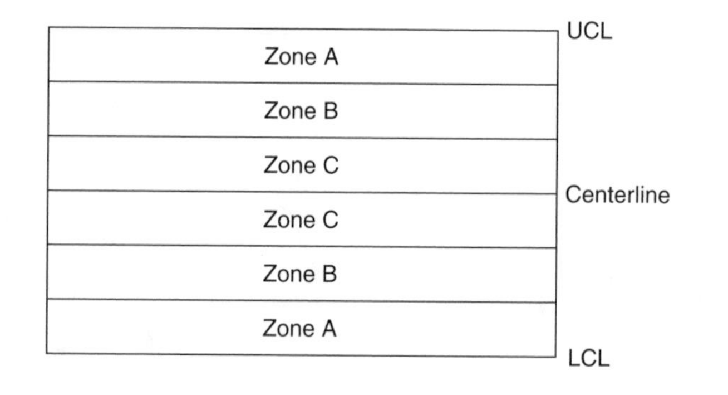

Figure 15-7 AT&T Run Rule Test Zones

1. Approximately 68% of the points should be within $\pm 1\sigma$ of the centerline, hence, most of the points are close to the centerline.
2. A few of the values will lie close to the control limits. In fact, approximately 5% of the points will lie between $\pm 2\sigma$ and $\pm 3\sigma$.
3. Rarely would a point fall outside of the control limits. Approximately 0.3% of the points would fall beyond the $\pm 3\sigma$ limits.
4. There will seldom be prolonged runs upward or downward for a large number of subgroup statistics.

The most widely used method for determining if a process is not in statistical control is to test for instability. One of the best know methods for checking for instability is the AT&T run rules, which were developed based on the preceding characteristics. The AT&T run rules require that the area between the control limits be divided into six bands. Each band represents one standard error. As Figure 15-7 shows, bands within one standard error of the centerline are called the *C zones*; bands between one and two standard errors from the centerline are called *B zones*; and the outermost bands (between two and three standard errors from the centerline) are called the *A zones*.

The following four rules, based on the six bands, are commonly used to determine if a process is exhibiting a lack of statistical control.

Rule 1: A process is not in statistical control if any subgroup statistic falls outside of the control limits. This point is marked with an "X" directly on the control chart.

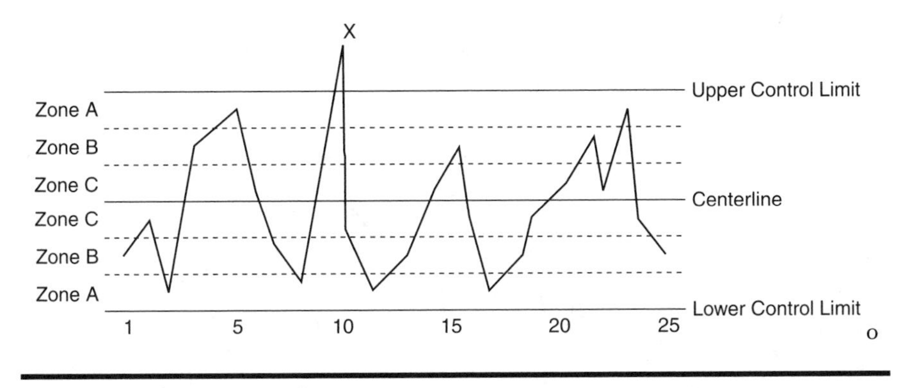

Figure 15-8 Rule 1 — Lack of Statistical Control

Rule 2: A process is not in statistical control if any two out of three successive subgroup statistics fall in one of the A zones or beyond on the same side of the centerline. The second of the two points in or beyond zone A is marked with an "X."

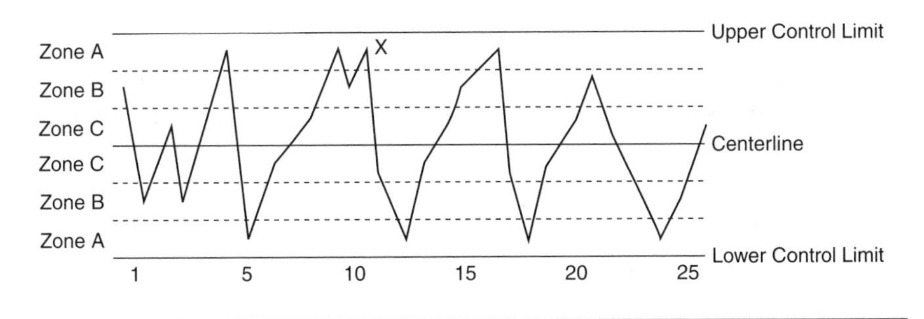

Figure 15-9 Rule 2 — Lack of Statistical Control

Rule 3: A process is not in statistical control if four out of five successive subgroup statistics fall in one of the B zones or beyond on the same side of the centerline. Only the fourth point is marked with an "X."

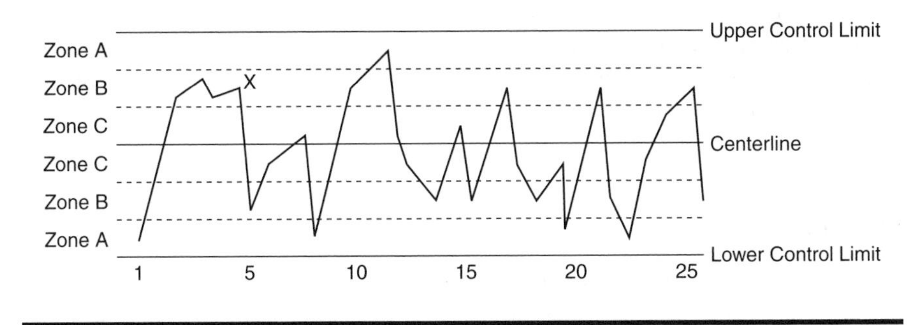

Figure 15-10 Rule 3 — Lack of Statistical Control

Rule 4: A process is not in statistical control if eight successive points fall in zone C on either side of the centerline. Only the eight point is marked with an "X."

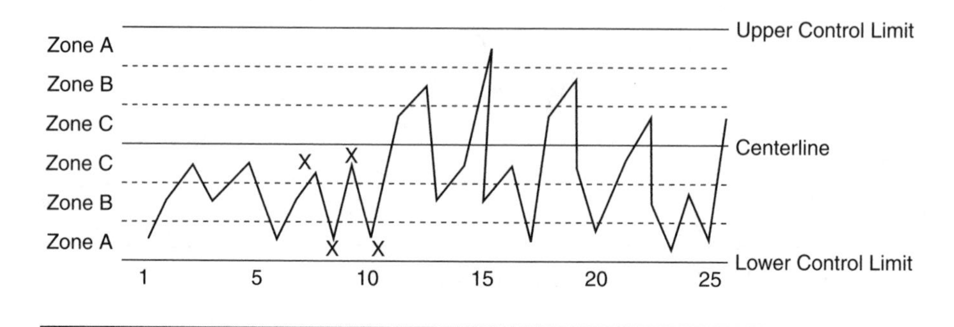

Figure 15-11 Rule 4 — Lack of Statistical Control

SUMMARY

Remember that variables control charts are used to monitor both the process mean and process variability. Therefore, for every quality characteristic being studied, two control charts must be done. Thus, an $\overline{X}$ and R chart, an $\overline{X}$ and S chart, or an X chart and MR chart should be used.

EXERCISES

Questions

15-1 What does the $\overline{X}$ chart measure? Why is this important?
15-2 What does the R chart measure? Why is this important?
15-3 What is the underlying distribution of the $\overline{X}$ chart?
15-4 Who first suggested the use of control charts?
15-5 What are the three basic uses of control charts?
15-6 What is the special notation used to denote the normal distribution?
15-7 What factors need to be decided when preparing to use variables control charts?
15-8 What is a rational subgroup?

Problems

15-1 Consider the following 20 samples:

Sample	*Observations*				
1	42	59	38	28	24
2	32	22	22	25	41
3	38	40	31	52	40
4	22	52	33	27	37
5	46	32	20	50	43
6	27	29	24	15	24
7	31	4	34	60	37
8	32	46	30	32	40
9	35	20	34	46	39
10	55	25	33	54	41
11	22	44	51	42	36
12	14	24	12	33	22
13	36	52	19	47	50
14	29	21	17	9	21
15	33	31	26	18	7
16	40	34	17	27	23
17	23	41	21	29	20
18	28	22	35	21	45
19	32	27	16	30	16
20	23	29	31	42	13

a. Determine the control limits for the $\overline{X}$ and R charts.
b. Plot the $\overline{X}$ and R charts. What can you conclude about the process?

15-2 A process is currently operating in-control. The control limits for the $\overline{X}$ bar and R charts are:

$\overline{X}$ *chart*	*R chart*
UCL = 47.69	UCL = 54.99
LCL = 15.23	LCL = 0
= 31.46	= 22.26

The sample size is four. What would the appropriate parameters of the S chart be?

15-3 A process has a mean of 100.0 and a standard deviation of 2.5. The process is in-control. An $\overline{X}$ and an R chart with sample sizes of six are being used to monitor the process. What are the values of the $\overline{X}$ and R chart control limits?

15-4 A particular part has specifications of 110.50 $\pm$ 0.25. Parts produced outside of specifications are scrapped. Two different machines produce this particular part at a rate of 150 parts per hour each. Items from both machines are discharged into the same collection box. Every half hour, the inspector selects a sample of six parts from the collection box. A single $\overline{X}$ and R control chart is kept using the inspector's samples.

a. After 40 samples have been taken, $\Sigma \overline{X} = 4{,}162.33$ and $\Sigma R = 8.66$. Determine the control limits for the $\overline{X}$ and R charts.
b. Calculate the natural tolerance limits of the process.
c. What serious mistake is being made in the situation described in the statement of this problem?

15-5 $\overline{X}$ and S control charts are maintained for a certain quality characteristic. The sample size is nine. After 50 samples, we have $\Sigma \overline{X} = 9{,}810$ and $\Sigma S = 1{,}350$.
a. Calculate the control limits for the two charts.
b. Assuming that both charts exhibit control, what are the natural tolerance limits?
c. Suppose an R chart were to be substituted for the S chart. What would the control limits for the R chart be?

15-6 Tensile strength is an important quality characteristic for bridge bolts. The bolts are made of a mild steel (ASTM A36). Because a tensile strength test destroys the bolt and each bolt is expensive, a sample of size 1 was selected, and 20 bolts were tested. The results are as follows:

Bolt	*Tensile strength (psi)*	*Bolt*	*Tensile strength (psi)*
1	57,500	11	74,000
2	77,750	12	60,000
3	85,000	13	77,000
4	57,560	14	59,900
5	62,000	15	64,500
6	69,900	16	71,000
7	79,800	17	80,500
8	75,300	18	69,700
9	64,500	19	78,000
10	70,500	20	68,900

a. Set up the X chart and the MR chart. Is the process in-control?
b. If the specification limits are 58,000 and 80,000 psi, what can you say about the process?

REFERENCES

AT&T Statistical Quality Control Handbook, Charlotte, N.C.: Delmar, 1985.
Case, Kenneth E., David H. Brooks, and James S. Bigelow, "Proper Use of Process Capability Indices in SPC," in *1987 IIE Integrated Systems Conference Proceedings,* Norcross, Ga.: Institute of Industrial Engineers, 1987, pp. 107–111.
Grant, Eugene L. and Richard S. Leavenworth, *Statistical Quality Control,* 7th ed., New York: McGraw-Hill, 1996.

McMillen, Nevin, *Statistical Process Control and Company-Wide Quality Improvement,* Bedford, England: IFS Publications, 1991.
Montgomery, Douglas C., *Introduction to Statistical Quality Control,* 3rd ed., New York: John Wiley & Sons, 1996.
Omachonu, V. K. *Healthcare Performance Improvement,* Norcross, GA. Engineering and Management Press, 1999.
Swift, J. A., *Introduction to Statistical Quality Control and Management,* Delray Beach, Fla.: St. Lucie Press, 1995.

Appendix A

	Factors for $\overline{X}$ charts			Factors for R Charts					Factors for S charts				Factors for center line		
n	A	A_2	A_3	D_1	D_2	D_3	D_4	d_3	B_3	B_4	B_5	B_6	c_4	d_2	n
2	2.121	1.880	2.659	0.000	3.686	0.000	3.267	0.853	0.000	3.267	0.000	2.606	0.7979	1.128	2
3	1.732	1.023	1.954	0.000	4.358	0.000	2.574	0.888	0.000	2.568	0.000	2.276	0.8862	1.693	3
4	1.500	0.729	1.628	0.000	4.698	0.000	2.282	0.880	0.000	2.266	0.000	2.088	0.9213	2.059	4
5	1.342	0.577	1.427	0.000	4.918	0.000	2.114	0.864	0.000	2.089	0.000	1.964	0.9400	2.326	5
6	1.225	0.483	1.287	0.000	5.078	0.000	2.004	0.848	0.030	1.970	0.029	1.874	0.9515	2.534	6
7	1.134	0.419	1.182	0.204	5.204	0.076	1.924	0.833	0.118	1.882	0.113	1.806	0.9594	2.704	7
8	1.061	0.373	1.099	0.388	5.306	0.136	1.864	0.820	0.185	1.815	0.179	1.751	0.9650	2.847	8
9	1.000	0.337	1.032	0.547	5.393	0.184	1.816	0.808	0.239	1.761	0.232	1.707	0.9693	2.970	9
10	0.949	0.308	0.975	0.687	5.469	0.223	1.777	0.797	0.284	1.716	0.276	1.669	0.9727	3.078	10
11	0.905	0.285	0.927	0.811	5.535	0.256	1.744	0.787	0.321	1.679	0.313	1.637	0.9754	3.173	11
12	0.866	0.266	0.886	0.922	5.594	0.283	1.717	0.778	0.354	1.646	0.346	1.610	0.9776	3.258	12

16

CONTROL CHARTS FOR ATTRIBUTES

It is often inconvenient, impractical, or impossible to take numerical measurements of the type necessary to set up variables control charts. In these cases, the quality characteristic of a unit is judged, or classified, as either conforming or non-conforming based on whether or not it has certain attributes (leaks/does not leak, works/does not work, etc.). Another means of judging the item is to count the number of non-conformities that appear on the unit (number of scratches, dents, holes, etc.). These types of quality characteristics are referred to as ***attributes***. Attribute data can have only two values, such as pass/fail, conforming/non-conforming, present/absent, etc. Even though attribute data cannot be measured, they can be counted. There are four special control charts for analyzing attribute data:

- ***p chart*** — Plots the fraction non-conforming per sample
- ***np chart*** — Plots the number non-conforming per sample
- ***c chart*** — Plots the number of non-conformities per inspection unit
- ***u chart*** — Plots the average number of non-conformities per inspection unit

The following are several basic terms that are used in connection with attributes charts. It is important to understand them.

- ***Non-conformity*** — A characteristic that does not meet requirements or specification. Non-conformities provide reasons for not accepting (rejecting) the unit. A non-conformity is also called a ***defect***.

- ***Non-conforming*** — A unit that is rejected. A non-conforming unit contains more than the allowable number of non-conformities. However, in many cases, the allowable number of non-conformities is one. A non-conforming unit is also called a ***defective*** unit.

One advantage of using attributes charts is that they can handle multiple characteristics. Therefore, attributes charts typically require more inspection. The inspection is less precise (no measurements to be taken and recorded) and is usually cheaper (no special training needed).

CONTROL CHART FOR FRACTION NON-CONFORMING (p CHART)

The control chart for fraction non-conforming is also known as the p chart. The "p" stands for proportion because it measures the proportion of non-conforming units in a group of units being inspected. The p chart monitors the fraction non-conforming of a process by plotting sample fraction non-conforming over time. The p chart is the most versatile and widely used attribute chart. It can be used for the following purposes:

1. To determine the average proportion (or fraction) of non-conforming units over a given time span
2. To signal a change in the average fraction non-conforming
3. To identify out-of-control points that call for immediate action
4. To suggest places to implement $\overline{X}$ and R charts

The p chart is based on the ***binomial distribution***. The binomial distribution assumes that:

1. For every trial there are only two possible outcomes (e.g., pass/fail, conforming/non-conforming).
2. The same trial is repeated any number of times.
3. The repeated trials are independent of one another. For example, the outcome of the second trial is not affected by (or dependent on) the first trial, and the outcome of trial n is not affected by the outcomes of trial 1 through trial $n - 1$.
4. The probability of a specific outcome remains constant from trial to trial.

The p chart plots the fraction non-conforming per sample. This means that the unit either conforms or does not conform. There are only two possible outcomes. The probability of pass/fail of each item remains constant from unit to unit. Also, the probability of a unit passing/failing is independent of how previous units tested. Therefore, the binomial distribution describes the p chart.

Typically, p is not known. It is estimated from:

$$\bar{p} = \frac{\sum_{i=1}^{m} D_i}{\sum_{i=1}^{m} n_i} \tag{16-1}$$

where D_i = number of non-conforming units in sample i
n_i = number of units in sample i
m = number of samples

This allows the control limits for the p chart to be determined from:

$$\text{center line} = \bar{p} \tag{16-2a}$$

$$\text{UCL} = \bar{p} + 3\sqrt{\frac{\bar{p}(1-\bar{p})}{n}} \tag{16-2b}$$

$$\text{LCL} = \bar{p} - 3\sqrt{\frac{\bar{p}(1-\bar{p})}{n}} \tag{16-2c}$$

Equation 16-2c may give a value less than zero for the lower control limit. Whenever this occurs, a lower control limit of zero is used.

Example 1

A manager wants to keep track of the number of non-conforming circuit testers being produced. There are six types of defects that can cause a circuit tester to be considered defective or non-conforming:

1. Mechanical defect
2. Short
3. Open
4. Peak inverse voltage (PIV)
5. Voltage forward (VF)
6. Reverse polarity (RP)

Table 16-1 Data Collected on Non-Conforming Circuit Testers

Day	*Mechanical defect*	*Short*	*Open*	*PIV*	*VF*	*Total RP*	*Total defective*	*inspected*	*Fraction non-conform-ing*
1	38	50	67	78	7	1	241	2000	12.05
2	47	61	78	90	3	2	281	2000	14.05
3	42	51	89	99	5	0	286	2000	14.30
4	50	50	76	103	4	0	283	2000	14.15
5	12	47	72	98	2	1	232	2000	11.60
6	11	64	88	87	8	0	258	2000	12.90
7	5	49	71	93	12	1	231	2000	11.55
8	15	52	69	92	14	2	244	2000	12.20
9	13	63	70	98	7	3	254	2000	12.70
10	9	72	76	87	9	5	258	2000	12.90
11	8	67	77	86	13	0	251	2000	12.55
12	12	59	71	80	11	0	233	2000	11.65
13	22	62	74	82	9	1	250	2000	12.50
14	9	61	63	90	6	2	231	2000	11.55
15	17	83	64	97	8	1	270	2000	13.50
16	19	65	68	98	7	1	258	2000	12.90
17	18	50	79	104	4	0	255	2000	12.75
18	14	51	64	118	3	1	251	2000	12.55
19	20	57	77	96	7	0	257	2000	12.85
20	17	60	63	98	5	0	243	2000	12.15
21	16	68	79	114	3	1	281	2000	14.05
22	14	61	71	92	11	0	249	2000	12.45
23	10	68	78	89	8	1	254	2000	12.70

The manager sets up a data collection sheet and begins to collect the information on the defective circuit testers being produced. The manager sets the sample size at 2000. The data collected for a period of 23 days are given in Table 16-1. The average fraction non-conforming can be determined using Equation 16-1:

$$\bar{p} = \frac{\sum_{i=1}^{m} D_i}{\sum_{i=1}^{m} n_i} = \frac{5{,}851}{46{,}000} = 0.1272$$

The control limits for the p chart can be found from Equations 16-2b and 16-2c:

$$\text{UCL} = \bar{p} + 3\sqrt{\frac{\bar{p}(1-\bar{p})}{n}} = 0.1272 + 3\sqrt{\frac{0.1272(1-0.1272)}{2{,}000}} = 0.1496$$

$$\text{LCL} = \bar{p} - 3\sqrt{\frac{\bar{p}(1-\bar{p})}{n}} = 0.1272 + 3\sqrt{\frac{0.1272(1-0.1272)}{2{,}000}} = 0.1496$$

The p chart for the fraction non-conforming is displayed in Figure 16-1. The process appears to be in-control, and the average fraction non-conforming is 12.72%.

In Example 1, the sample size was constant. Many times, however, the sample size is not constant, especially if 100% inspection is being performed. With 100% inspection, inspection is typically performed on units produced during some defined sampling period, such as a shift, a day, or a production run. Many factors influence the number of units produced during a sampling period. Therefore, it is very difficult to produce exactly the same number of items every time.

Since the control limits for the p chart are a function of the sample size, some modifications need to be made to ensure that the chart is properly interpreted. The three most common modifications are:

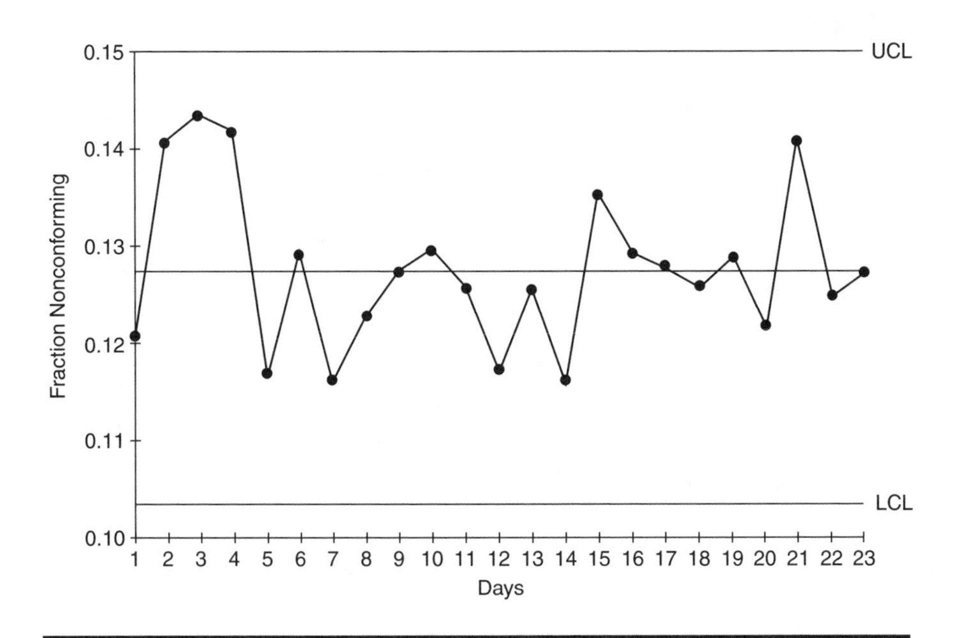

Figure 16-1 Chart for Circuit Testers

1. Compute separate control limits for each sample based on its specific sample size.
2. Compute the control limits based on an average sample size.
3. Compute the control limits based on a standardized fraction nonconforming.

Because computers are so readily available these days and because it is the easiest to interpret, the first modification is recommended.

Example 2

A novelty shop owner feels that a particular supplier is delivering a high percentage of unusable product. The owner gets a shipment of novelties once a week. The size of the shipment varies from week to week. The shop owner decides to inspect 100% of each shipment he gets from this particular supplier. The data the shop owner collected over a five-month period are displayed in Table 16-2, which also provides the control chart

Table 16-2 Development of p Chart with Variable Sample Size

Week	*Number nonconforming*	*Number inspected*	*Fraction nonconforming*	*LCL*	*UCL*
1	44	250	0.176	0.121	0.271
2	40	200	0.200	0.112	0.280
3	52	275	0.189	0.124	0.268
4	37	225	0.164	0.177	0.275
5	41	250	0.164	0.121	0.271
6	49	250	0.196	0.121	0.271
7	40	225	0.178	0.117	0.275
8	46	200	0.230	0.112	0.280
9	55	250	0.220	0.121	0.271
10	61	275	0.222	0.124	0.268
11	58	275	0.211	0.124	0.268
12	39	250	0.156	0.121	0.271
13	46	225	0.204	0.117	0.275
14	57	250	0.228	0.121	0.271
15	53	250	0.212	0.121	0.271
16	41	225	0.182	0.117	0.275
17	58	275	0.211	0.124	0.268
18	49	250	0.196	0.121	0.271
19	42	225	0.187	0.117	0.275
20	54	275	0.196	0.124	0.268
21	47	250	0.188	0.121	0.271

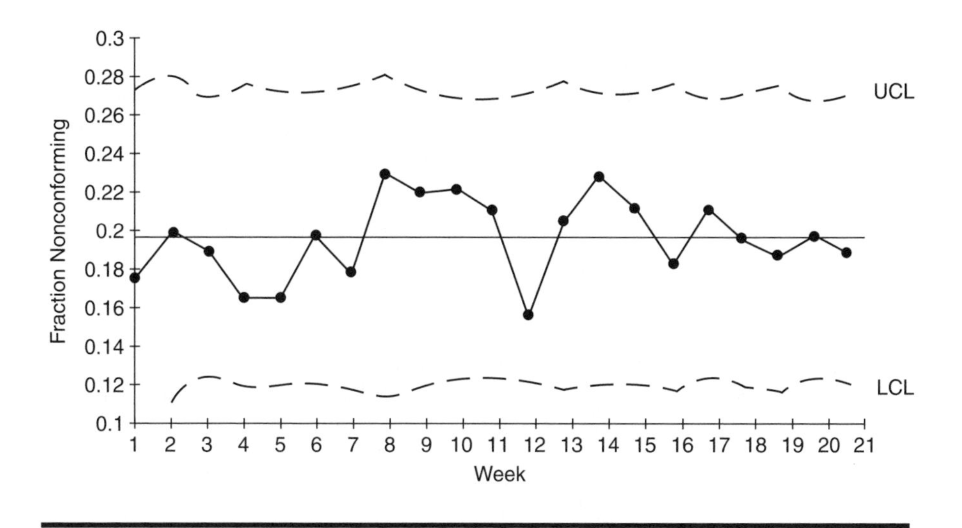

Figure 16-2 Chart with Variable Sample Size

limits for the variable sample sizes. The control chart limits were calculated using Equations 16-2b and 16-2c. The plotted p chart is shown in Figure 16-2, which indicates that the process appears to be in-control. However, the average fraction non-conforming is 19.59%, which is very high. The shop owner has decided to give the supplier three months to significantly reduce the fraction non-conforming.

CONTROL CHART FOR NUMBER NON-CONFORMING (np CHART)

If the sample size can be kept constant, then the p chart can be simplified. With a constant sample size, there is really no need to convert the number non-conforming to fraction non-conforming. Simply plot the number non-conforming. In actual practice, plotting the number non-conforming is more easily understood than the plotting of fractions. It is also a good chart to use when first introducing the work force to control chart techniques.

The np chart is also based on the binomial distribution. The control limits for the np chart can be found using the following equations:

$$\text{center line} = \bar{p} \qquad (16\text{-}3a)$$

$$\text{UCL} = n\bar{p} + 3\sqrt{n\bar{p}(1 -} \qquad (16\text{-}3b)$$

$$\text{LCL} = n\bar{p} - 3\sqrt{n\bar{p}(1 - \bar{p})} \qquad (16\text{-}3c)$$

As with the p chart, if the LCL equation gives a value less than zero for the lower control limit, then the lower control limit is set to zero. Equations 16-3a to 16-3c are used if p (the process fraction non-conforming) is known. If p is unknown, then Equation 16-1 is used to estimate p. Once an estimate of p has been obtained, Equations 16-3a to 16-3c can then be used to set up the control limits. The key to being able to use the np chart is that the sample size must remain constant.

Example 3

A small sheet-metal part is shaped through a series of processes. When the shaping is completed, the part is inspected. If any defect is found, the part is considered defective and is scrapped. The shop foreman has decided to track the number of defective sheet-metal parts produced. The parts are produced and shaped in batches of size 100. Since the sample size is constant, the foreman decides to use an np chart. The inspection results of the last 26 batches are shown in Table 16-3.

According to Table 16-3, there were a total of 417 non-conforming pieces out of 2600 sheet-metal pieces inspected. The estimated fraction non-conforming is determined from Equation 16-1:

$$\bar{p} = \frac{\sum_{i=1}^{m} D_i}{\sum_{i=1}^{m} n_i} = \frac{417}{2,600} = 0.1604$$

The parameters for the np chart can be determined from Equations 16-3a to 16-3c:

$$\text{center line} = n\bar{p} = 100(0.1604) = 16.04$$

$$\text{UCL} = n\bar{p} + 3\sqrt{n\bar{p}(1-\bar{p})} = 16.04 + 3\sqrt{16.04(1-0.1604)} = 27.05$$

$$\text{LCL} = n\bar{p} - 3\sqrt{n\bar{p}(1-\bar{p})} = 16.04 - 3\sqrt{16.04(1-0.1604)} = 5.03$$

The np chart is given in Figure 16-3. The chart appears to be in-control as no points fall outside the control limits. However, an average of 16 non-conforming pieces per batch is too high. The foreman has decided further investigation is needed.

Table 16-3 Number of Non-Conforming Sheet-Metal Parts (Batch Size = 100)

Batch number	*Number non-conforming*	*Batch number*	*Number non-conforming*
1	14	14	20
2	18	15	16
3	21	16	11
4	32	17	10
5	30	18	25
6	18	19	23
7	21	20	12
8	11	21	14
9	18	22	12
10	12	23	8
11	17	24	13
12	15	25	15
13	17	26	13
Total			417

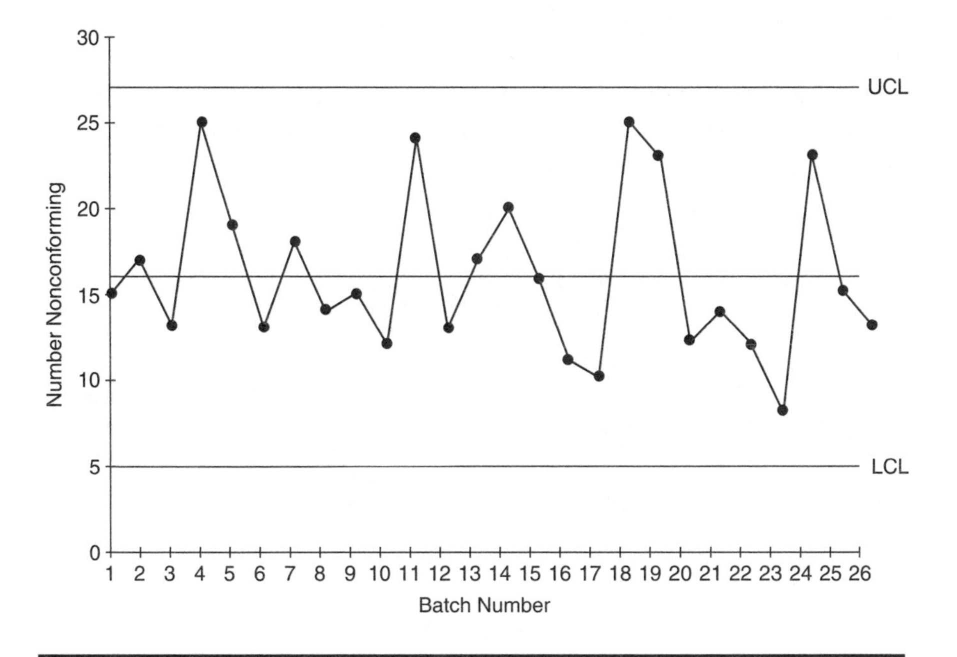

Figure 16-3 np Chart for Sheet-Metal Parts

CONTROL CHART FOR NON-CONFORMITIES

While the p and the np charts monitor the non-conformance of a process, the c and u charts monitor the non-conformities of a process. The c and u charts monitor the number of times a particular characteristic (or non-conformity) appears in a given area of opportunity (or inspection unit). More specifically, the c chart monitors the number of non-conformities per inspection unit and the u chart monitors the average number of non-conformities per inspection unit. With respect to c and u charts, a sample is an area of opportunity. Within an area of opportunity, there can be one or many inspection units. With a c chart, the area of opportunity and the inspection unit are usually one and the same. With the u chart, the area of opportunity varies, thus making it difficult for the area of opportunity and the inspection unit to be the same.

c Chart

The c chart plots the number of non-conformities that occur in a ***constant*** area of opportunity or inspection unit. The constant means that the area of opportunity and the inspection unit are equal. Examples of a constant inspection unit are fixed length, area, and quantity. Fixed length could be a section of road, a roll of paper, etc. Fixed area could be the hood of a car, a defined portion of a circuit board, etc. Fixed quantity could be a work week, three computers, etc.

In most applications, the center line, or mean, of the c chart is based on the average number of non-conformities per inspection unit. This can be determined by:

$$\mu = \bar{c} = \frac{\sum_{i=1}^{k} c_i}{k} = \frac{\text{total number of non-conformities found}}{\text{number of inspection units}} \quad (16\text{-}4)$$

where ci is the observed number of non-conformities found in inspection unit i. The control limits are determined by

$$\text{control limits} = \mu \pm 3\sigma$$

Therefore, the control limits for the c chart are determined from

$$\text{center line} = \bar{c} \quad (16\text{-}5a)$$

$$\text{UCL} = \bar{c} + 3\sqrt{\bar{c}} \quad (16\text{-}5b)$$

$$LCL = \bar{c} - 3\sqrt{\bar{c}} \tag{16-5c}$$

As with the p and np charts, if the calculation of the lower control limit is less than zero, then the lower control limit is set to zero.

Example 4

A production line supervisor decides to count the number of non-conformities that exist on the 3.5-inch floppy disks being produced by process line 2. The production rate is approximately 500 per hour. For reasons of convenience, the supervisor takes a sample of 75 per hour. In other words, the inspection unit is 75 floppy disks. The supervisor collects data for three days. The number of non-conformities observed over the three days is displayed in Table 16-4.

Table 16-4 shows that there were a total of 472 non-conformities in the 24 samples. From Equation 16-4, the mean can be determined by:

$$\bar{c} = \frac{\sum_{i=1}^{k} c_i}{k} = \frac{472}{24} = 19.67$$

The control limits can be determined using Equations 16-5b and 16-5c:

$$UCL = \bar{c} + 3\sqrt{\bar{c}} = 19.67 + 3\sqrt{19.67} = 32.97$$

$$LCL = \bar{c} - 3\sqrt{\bar{c}} = 19.67 - 3\sqrt{19.67} = 6.36$$

Table 16-4 Number of Non-Conformities for 3.5-Inch Floppy Disks

Sample number	*Number non-conformities*	*Sample number*	*Number non-conformities*
1	14	13	24
2	18	14	22
3	21	15	23
4	32	16	32
5	30	17	20
6	18	18	28
7	21	19	8
8	11	20	15
9	18	21	12
10	12	22	16
11	17	23	24
12	15	24	21
Total			472

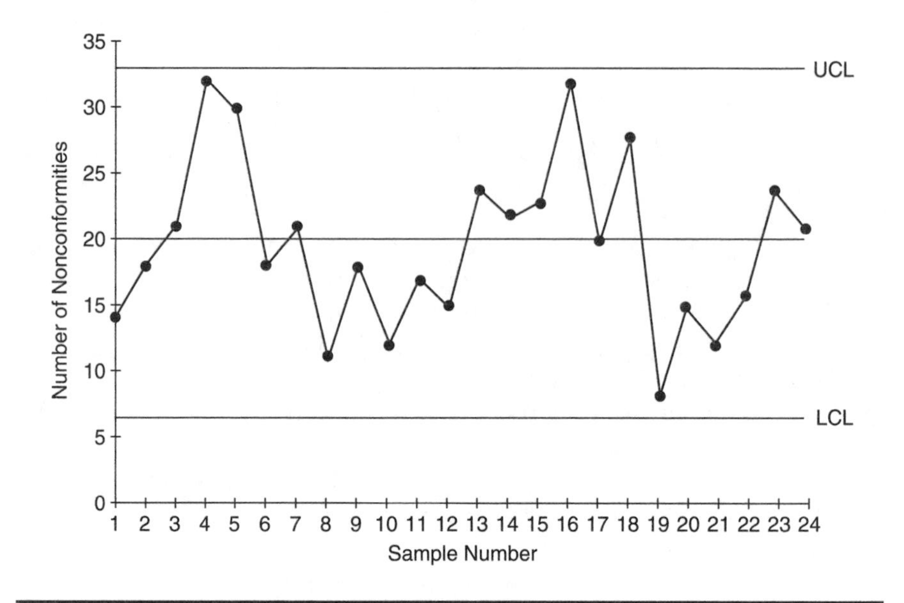

Figure 16-4 c Chart for 3.5-Inch Floppy Disks

The c chart is shown in Figure 16-4. As can be seen, all points fall within the control limits and the pattern appears to be random. Therefore, no lack of control is indicated.

u Chart

In many applications, the area of opportunity varies. This violates the constant opportunity space assumption upon which the c chart is based. Therefore, it becomes necessary to create a standardized statistic. This standardized statistic is defined as u and is the average number of non-conformities per inspection unit. It is determined by:

$$u = \frac{c}{n} \qquad (16\text{-}6)$$

where c = number of non-conformities per area of opportunity
n = number of inspection units in the area of opportunity inspected

In this case, the inspection unit is not equal to the area of opportunity. This means that there will typically be multiple inspection units for a given area of opportunity. It also means that n will not always be an integer. For example, let's say that the inspection unit is ten feet of paper. The process that makes the paper produces the paper in rolls that vary in size from 100 to 125 feet. Because the paper is produced in rolls, it is inspected for non-conformities by rolls. Thus, the area of opportunity is one roll. The number of inspection units in a given area of opportunity (a roll of paper) would vary from 10 to 12.5 units.

The characteristic that is plotted for this type of inspection is u determined from Equation 16-6. This gives the u chart. As with all control charts, a center line and control limits must be established. The mean or center line is found by:

$$\bar{u} = \text{mean} = \text{center line} = \frac{\sum_{i=1}^{k} c_i}{\sum_{i=1}^{k} n_i} \tag{16-7}$$

where c_i = observed number of non-conformities in opportunity area i
n_i = number of inspection units in opportunity area i
k = number of opportunity areas inspected

The control limits are obtained from $\bar{u}$. Therefore, the control limits for the u chart are

$$\text{UCL} = \bar{u} + 3\sqrt{\frac{\bar{u}}{n_i}} \tag{16-8a}$$

$$\text{LCL}^* = \bar{u} - 3\sqrt{\frac{\bar{u}}{n_i}} \tag{16-8b}$$

where n = number of inspection units in the given area of opportunity

If the number of inspection units varies from opportunity area to opportunity area, the control limits will also vary, because the control limits are a function of the number of inspection units.

Example 5

A copy shop manager notices that the number of complaints on completed jobs has increased. The manager realizes that the total number of completed jobs per day has also increased. The manager wants to determine if the increase in complaints should be expected due to the increase in total number of jobs or if the quality of work has really decreased. The manager decides to inspect all completed jobs for defects on a daily basis and does so for three weeks. The results of the manager's inspections are shown in Table 16-5.

* If the lower control limit is less than zero, then it is set equal to zero.

Table 16-5 Development of u Chart for Complaints Received

Date	Number of completed jobs, n_i	Number of complaints, c_i	Non-conformities per unit, $u_i = c_i/n_i$	LCL	UCL
1	21	375	17.86	19.57	14.19
2	17	244	14.35	19.87	13.89
3	28	421	15.04	19.21	14.55
4	25	478	19.12	19.34	14.41
5	20	388	19.40	19.63	14.12
6	29	511	17.62	19.17	14.59
7	22	345	15.68	19.50	14.25
8	25	477	19.08	19.34	14.41
9	31	566	18.26	19.09	14.66
10	28	414	14.79	19.21	14.55
11	23	364	15.83	19.45	14.31
12	27	444	16.44	19.25	14.50
13	26	435	16.73	19.29	14.46
14	29	533	18.38	19.17	14.59
15	23	433	18.83	19.45	14.31
16	30	497	16.57	19.13	14.63
17	23	356	15.48	19.45	14.31
18	26	388	14.92	19.29	14.46
19	22	403	18.32	19.50	14.25
20	28	467	16.68	19.21	14.55
21	24	355	14.79	19.39	14.36

Since the size of each completed job varies only slightly, the manager lets one completed job be the inspection unit. Since the area of opportunity (or completed jobs) varies, the manager develops a u chart. This means that the control limits vary from day to day. Table 16-5 also shows the development of the u chart. Equation 16-7 was used to calculate average number of defects per unit. Equations 16-8a and 16-8b were used to calculate the individual control limits. The plotted u chart is illustrated in Figure 16-5.

The average number of defects per completed job is 16.88. All points fall within control limits and the pattern appears random. Therefore, the manager concludes that the process is operating in-control. However, the manager feels that an average of 16.88 defects per completed job is too high. Therefore, the manager decides to form a task team to lower the average number of defects per completed job.

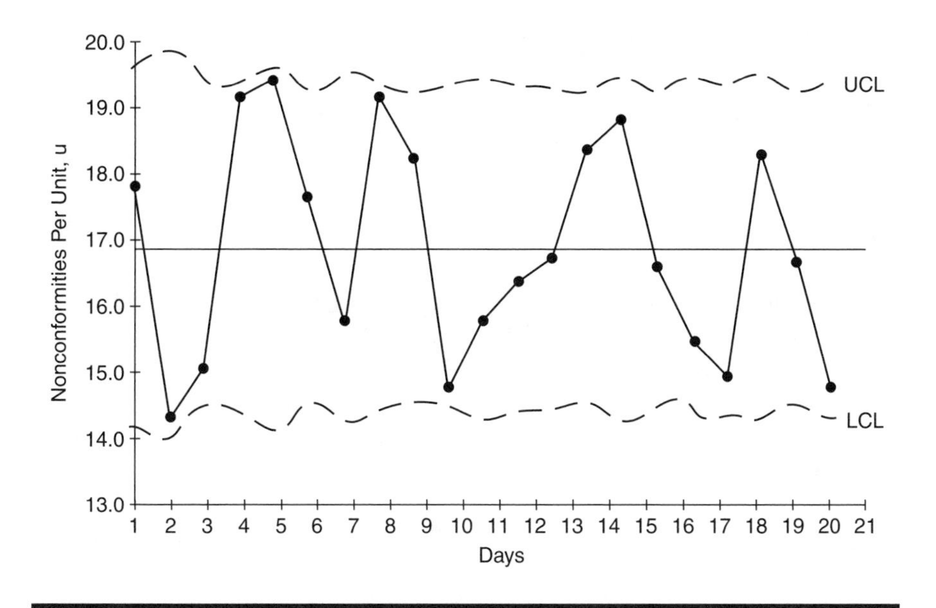

Figure 16-5 u Chart for Complaints Received

SUMMARY

Control charts for attributes are widely used in non-manufacturing applications. They are very easy to use. Also, the concept of number nonconforming or number of defects is readily understood by most people, thus making these charts easy to understand.

EXERCISES

Questions

16-1 List the four types of attributes control charts and what each chart plots.

16-2 What is attribute data?

16-3 What is the difference between non-conforming and non-conformity?

16-4 What is/are the advantage(s) of using attributes charts?

16-5 Which control chart(s) is/are used when the sample size is not constant?

16-6 What happens to the control limits when the sample size is not constant?

16-7 When the sample size varies, the np chart and the c chart are not appropriate for use. Explain why.

Problems

16-1 A process produces silicon wafers. Each wafer is 0.01 inch thick and 10 cm in diameter. There are 250 integrated circuits or chips on each wafer. At the end of the process, the wafer is inspected. The following data give the number of non-conforming chips per wafer. Construct a fraction non-conforming control chart for these data. What can you conclude about the process?

Wafer number	*# non-conforming*	*Wafer number*	*# non-conforming*
1	17	19	18
2	14	20	13
3	11	21	15
4	13	22	19
5	16	23	11
6	18	24	17
7	20	25	21
8	15	26	13
9	12	27	16
10	17	28	12
11	16	29	15
12	22	30	19
13	11	31	17
14	19	32	11
15	15	33	13
16	11	34	20
17	14	35	14
18	15	36	16

16-2 Construct an np chart for the data in Problem 16-1. Which type of chart do you prefer? Why?

16-3 The following data represent the number of dishes that are broken at a restaurant while loading and unloading the dishwasher. The total number of dishes washed varies daily. Construct a p chart for these data. What can you conclude about the situation? Do you have any recommendations?

Day	Total number of dishes washed	Number of broken dishes	Day	Total number of dishes washed	Number of broken dishes
1	465	51	13	465	39
2	425	42	14	425	41
3	500	55	15	500	51
4	500	53	16	500	47
5	425	44	17	425	38
6	425	41	18	425	44
7	425	39	19	425	41
8	465	47	20	465	48
9	500	56	21	465	42
10	500	43	22	500	47
11	425	46	23	500	51
12	465	49	24	465	36

16-4 A process produces overhead projectors. Each projector is inspected for defects or non-conformities. Typical defects found are scratches, dents, paint blemishes, improper labeling, etc. The supervisor wants to keep track of the defects being produced and decides to set up a u chart. The inspection unit the supervisor uses is 10% of all projectors produced per day. Since the number of projectors produced per day varies, so does the inspection unit (or sample size). The following data were collected over the first four weeks. Develop a table similar to Table 16-5 given in the chapter. Plot the u chart. Does this process appear to be in-control? What are your recommendations?

Day	Sample size	Number of defects	Day	Sample size	Number of defects
1	100	654	11	110	667
2	120	688	12	125	777
3	95	544	13	120	644
4	115	592	14	95	466
5	120	721	15	110	632
6	110	652	16	100	584
7	90	502	17	120	669
8	125	756	18	115	623
9	115	570	19	100	567
10	100	633	20	100	678

16-5 For the situation in Problem 16-4, assume the supervisor used a constant sample size of 110 projectors. Use the following data to set up and plot a c chart. Does the process appear to be in-control?

Day	*Sample size*	*Number of defects*	*Day*	*Sample size*	*Number of defects*
1	110	632	11	110	627
2	110	644	12	110	717
3	110	548	13	110	648
4	110	579	14	110	586
5	110	701	15	110	632
6	110	652	16	110	554
7	110	582	17	110	659
8	110	655	18	110	613
9	110	578	19	110	667
10	110	639	20	110	618

16-6 Set up and plot a u chart for the data in Problem 16-5.

REFERENCES

Farnum, Nicholas R., *Modern Statistical Quality Control and Improvement,* Belmont, Calif.: Duxbury Press, 1994.

Gitlow, Howard, Shelly Gitlow, Alan Oppenheim, and Rosa Oppenheim, *Tools and Methods for the Improvement of Quality,* Boston: Richard D. Irwin, 1989.

Grant, Eugene L. and Richard S. Leavenworth, *Statistical Quality Control,* 7th ed., New York: McGraw-Hill, 1996.

Hines, William W. and Douglas C. Montgomery, *Probability and Statistics in Engineering and Management Science,* 2d ed., New York: John Wiley & Sons, 1980.

Montgomery, Douglas C., *Introduction to Statistical Quality Control,* 3rd ed., New York: John Wiley & Sons, 1996.

Omachonu, V. K. *Healthcare Performance Improvement*, Norcross, GA. Engineering and Management Press, 1999.

Swift, J. A., *Introduction to Statistical Quality Control and Management,* Delray Beach, Fla.: St. Lucie Press, 1995.

17

WHEN TO USE THE DIFFERENT CONTROL CHARTS

INTRODUCTION

How to develop and interpret the different control charts was discussed in detail in the previous two chapters. In this chapter, a simplified aid is provided for deciding which control chart is appropriate for a given circumstance. In the past, this decision had to be made by someone experienced in the use of control charts. However, the increased awareness of quality in the workplace has meant that more and more people are being introduced to the various tools of quality, including control charts. Thus, the need exists for a basic aid in choosing the correct control chart.

The flowchart in Figure 17-1 is a very simple and easy-to-use aid for determining which control chart is needed for a given situation. There are several flowcharts of this nature. However, the distinguishing features of this flowchart are:

- It is simple and easy to use.
- Only a basic understanding of statistics is required.
- It can be used (and, more importantly, understood) by personnel at all levels.

EXAMPLE 1

A ceramic tile manufacturing company has just secured a contract with NASA to supply the tiles for the new space shuttle. The manufacturing process is long and detailed, and only 20 to 25 tiles can be manufactured per day. NASA requires that the tiles be subjected to specific measured tests to prove that they are capable of withstanding repeated exposure to

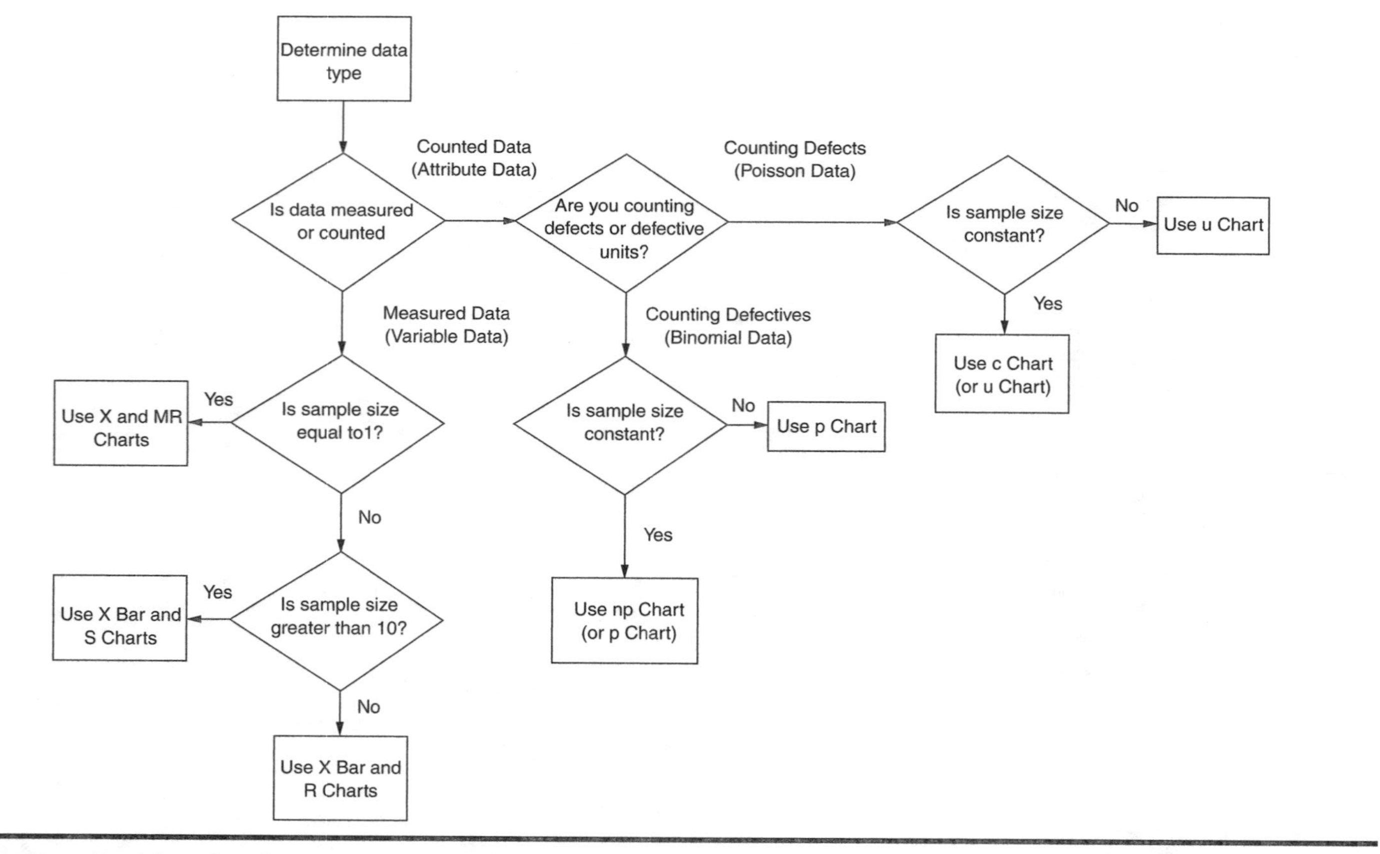

Figure 17-1 Decision Flowchart (©1994 J. A. Swift)

extreme high temperatures. The tests required are destructive tests. Because the tests are destructive tests and the production output per day is low, the manufacturing company has decided to use a sample size of one. Which chart should be used?

Solution (refer to Figure 17-1)

data are measured → sample size = 1 → use X and MR charts

EXAMPLE 2

Mr. Fence runs a small alterations shop. Recently, there has been an increase in the number of complaints about the work done in his shop. He has decided that at the end of each day, he will inspect all the work completed that day for defects. Which chart should Mr. Fence use?

Solution

data are counted → counting defects →
sample size varies (a different number of alterations
are completed each day) → use a u chart

EXAMPLE 3

A boot manufacturer wants to check a certain style of boot for possible defects in the sole stitching. The defects include missed stitches, loose threads, and any other observed defects. This particular style of boot is produced at a rate of 100 pairs per hour. The manager suggests checking 10 pairs per hour. Which chart should be used?

Solution

data are counted → counting defects →
sample size is constant (10 parts/hour) → use a c chart

SUMMARY

As seen from the previous examples, using Figure 17-1 makes the decision as to which chart to use in a given situation relatively easy. Once an appropriate chart has been identified for use, the previous two chapters can be used to aid in setting up the chart and interpreting it.

EXERCISES

For the following situations, use Figure 17-1 and the solution formats given in the examples in the chapter to determine the appropriate control chart.

17-1 A process that packages a ready-to-make cake mix automatically weighs each bag of mix before placing it into its respective box. The specification for each bag of mix is 8 ± 0.01 ounces. If a bag weighs outside of specs, it is automatically separated from the rest. Which chart should be used?

17-2 A hardware store has recently changed its supplier of wood products. In order for the store to sign a permanent contract with this new supplier, the products must pass some quality standards. At the present time, the products are ordered weekly and order quantity varies from week to week. All items ordered are received at the beginning of the week following placement of the order. The company tracks quality by inspecting 20% of all items ordered and tracking the number of defects (scratches, dents, blemishes, etc.). Which chart should be used?

17-3 A T-shirt outlet in Miami Beach wants to keep track of the number of unsellable T-shirts it receives from its supplier. A T-shirt is considered unsellable if the design is faded, peeling, crooked, etc. The number of T-shirts delivered per week varies. The T-shirt outlet does a 100% inspection of every delivery. Which chart should be used?

17-4 Mrs. Green is a gymnastics coach. She wants to keep track of the number of technical errors made by one of her students during a certain floor routine. Due to time constraints, the student can only perform the floor routine twice a day. Which chart should Mrs. Green use?

17-5 FWS Water Company produces bottled water. Because the company is located in the Middle East, it must use salt water as its source. The company has recently developed a new and significantly cheaper process to remove the salt. The new process produces 200 1-liter bottles per hour. The process manager inspects 15 bottles every hour. For each bottle inspected, he determines the percentage of salt remaining in the water. Which chart should the process manager use?

17-6 An airline manager in Memphis wants to track the number of late plane arrivals from Tampa. The manager wants to track the late arrivals on a weekly basis. Due to the air route licensing agreement this airline has, the number of planes leaving Tampa for Memphis per week is constant. Which chart should be used?

17-7 A furniture company produces kitchen tables. The length of the table legs has very tight tolerances in order to keep the finished table from wobbling. The process produces 80 table legs per hour and 8 per hour are inspected. Which chart should be used?

17-8 A local delivery company is interested in determining how many deliveries are not made on time. The dispatcher randomly checks on 30 deliveries a day. Which chart should the dispatcher use?

18

QUALITY IMPROVEMENT STORIES

WHAT IS A QUALITY IMPROVEMENT STORY?

A quality improvement (QI) story is a step-by-step guide for problem solving (or process improvement). It is called a story because it organizes the work a team does in such a way that its story is told. It tells who the team is, when and why it got together, where and what it worked on, and how it solved its problem. By having every team use the same storytelling technique, communication is standardized. This standardization allows for easier transfer of ideas between teams, departments, and even companies. It also provides a framework for training all employees in the application of the basic quality control tools.

A QI storyboard is a visual display of the QI story. The storyboard is typically mounted (or hung) in the area where the problem is occurring. This allows everyone to see what problem is being worked on and how far along in the problem-solving process the team has advanced.

There are four main reasons for using a QI story. First, it helps the team organize, gather, and analyze the data in a logical fashion. In this sense, all team members are speaking the same language and working on the same problem. Second, it monitors the team's progress. When a QI storyboard is visible, team members (as well as non-team members) can see exactly where the team is in the problem-solving process. Third, it facilitates understanding by non-team members. This is helpful in that non-team members may be able to provide useful feedback that the team may have missed or not thought of. Because everyone in the company is interested in problem/process improvement, feedback is desirable. Fourth, it standardizes presentations to management. Because management is (or should be) already familiar with the use of QI stories, management can concentrate on the problem being presented, which saves valuable time. It also lets the team speak with facts (something that strongly influences management).

A QI story follows several basic steps for solving a problem:

1. Identify the problem area
2. Observe and identify causes of the problem
3. Analyze, identify, and verify root cause(s) of the problem
4. Plan and implement preventive action
5. Check effectiveness of action taken
6. Standardize process improvement
7. Determine future actions

The relationship between the QI process and the PDCA cycle is depicted in Figure 18-1.

Within each step, certain tasks are performed to ensure that the team does not miss an important point. In the remainder of this chapter, each step is discussed.

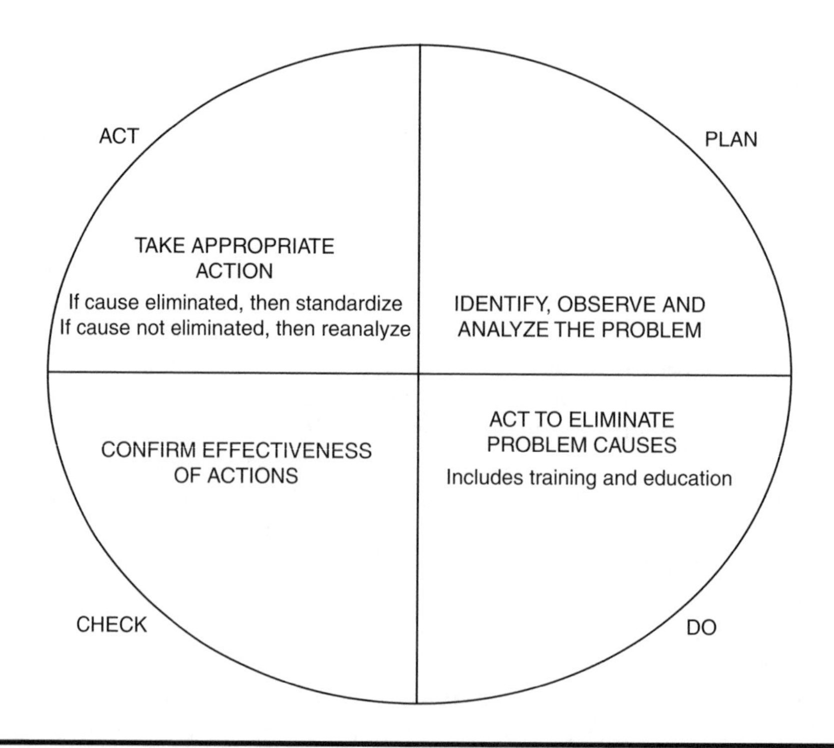

Figure 18-1 Relationship between the Quality Improvement Process and the PDCA Cycle

STEP 1: IDENTIFY THE PROBLEM AREA

The purpose of this step is to identify the general problem to be worked on and to recognize the importance of the problem. There are many problems, both large and small, that could be worked on. However, because resources and time are limited, only the most important problems can be addressed. The best way to identify the most important problems is through the use of "good data." With good data, it is easy to quantify the significance of the problem and consequently solve it.

The reasons for working on a problem can stem from many sources. The problem could have arisen from a process that was being tracked. The problem could come from customer complaints. It could also come from upper management. Whatever the source or reason, once a problem area has been defined, data must be gathered to substantiate the importance of the problem. One of the best ways to demonstrate the importance of a problem is to show the loss in performance (time, money, etc.) that is occurring at the present time and how much it should improve by correcting the problem. Identifying who the customers of the process are and how the problem affects them also demonstrates the significance of the problem. With the problem identified and its importance substantiated, the team can proceed to the next step.

STEP 2: OBSERVE AND IDENTIFY CAUSES OF THE PROBLEM

The purpose of the observation step is to evaluate the present status of the problem and identify the factors that could be causes of the problem. This is done by investigating the specific features of the problem. This step takes the initial tracking indicator and stratifies it to the point where the team can take action. This stratification is done by investigating the four points that define the features of the problem: time, place, type, and symptom. (Additional data typically must be collected at this point.) Once the initial tracking indicator has been broken down and causes of the problem have been identified, a specific problem statement can be formulated and the team can proceed to the next step.

Step 2 initially appears very similar to Step 1. However, the purpose of Step 1 was to identify the general problem area and recognize its significance. The purpose of Step 2 is to identify the factors that are causes of the problem.

STEP 3: ANALYZE, IDENTIFY, AND VERIFY ROOT CAUSE(S) OF THE PROBLEM

The purpose of the analysis step is to determine the main or root cause(s) of the specific problem identified in the previous step. It is very important that causes, and not symptoms, of the problem are identified because treating or fixing a symptom does not eliminate the problem. The problem can only be eliminated if the cause is eliminated. Much like taking aspirin for a toothache, it may relieve the ache temporarily, but if the cause of the ache is not treated, the ache usually returns. The best way to identify the root cause(s) is to construct a cause-and-effect diagram. The cause-and-effect diagram takes the problem identified in the previous step and makes it the "effect" portion of the diagram. Now the root cause analysis begins. All elements that seem to relate to the effect are identified and put on the cause-and-effect diagram. The potential root cause(s) are compared with the actual causes using the information obtained in the observation step. (The cause-and-effect diagram will typically undergo several iterations at this point.) The root cause(s) that appear to have the greatest impact are highlighted on the cause-and-effect diagram. At this point, the analysis step is only half done. Verification of the main root cause(s) must be done. The biggest mistake teams make is to proceed to the next step without verifying the root cause(s). In order to verify that the correct root cause(s) have been identified, additional new data must be collected. Once the root cause(s) have been verified, the team then proceeds to the next step.

STEP 4: PLAN AND IMPLEMENT PREVENTIVE ACTION

Purpose

The purpose of this step is to plan and implement actions that will eliminate the root cause(s). There are two types of action. The first type is corrective action. ***Corrective action*** is action taken to correct or to temporarily fix the problem, like taking aspirin to relieve a toothache (treating the symptom instead of the cause). This temporarily fixes the problem, but does nothing to prevent it from recurring. The second type of action is preventive action. ***Preventive action*** is action taken to prevent the problem from happening again, in this case treating the cause of the toothache.

Preventive action eliminates the cause of the problem and prevents it from recurring. Therefore, preventive action is the best way to solve a problem. It is also the reason the team came together in the first place. Determining an appropriate preventive action takes time. Therefore, it may be necessary to implement temporary corrective actions in order to minimize the impact of the problem while an appropriate preventive action is being determined.

Care must be taken when developing actions to eliminate the root cause. This is because many actions may cause other problems not foreseen, like side effects from drugs. To prevent the unwanted side effect, the proposed preventive action must be thoroughly evaluated before implementing it. Because people are always part of the process and typically resistant to change, unwanted people side effects need to be anticipated and avoided. Active cooperation of all involved is essential for success.

Because action means change and change means additional money or resources, a cost/benefit analysis needs to be performed. The cost/benefit analysis should take into account both tangible and intangible benefits. A cost/benefit analysis ensures that the preventive action to be implemented does not cost more than just leaving the problem alone.

Once all factors concerning the proposed preventive actions have been investigated, an action plan is developed and the preventive action is implemented. Tracking of the previously selected key indicators is also initiated.

STEP 5: CHECK EFFECTIVENESS OF ACTION TAKEN

The purpose of the check step is to ensure that the main cause(s) of the problem have been eliminated. This is done by comparing the data obtained after the preventive action has been implemented with the data obtained before implementation. The same format (tables, graphs, charts), the same time frame (weeks, months), and the same indicator that was used to show the status prior to implementing the preventive action are used in the comparison.

If the preventive action did not eliminate the problem, then something is happening in the process that was not properly identified, and the problem needs to be reanalyzed. If the results of the preventive action are not as good as expected, then the reasons why need to be investigated and fully documented. In short, the problem-solving process has failed and the team must return to the observation step.

If the preventive action appears to have eliminated the problem, then the team can proceed to the standardization step.

STEP 6: STANDARDIZE PROCESS IMPROVEMENT

The purpose of this step is to eliminate the cause of the problem permanently. This is done by replicating and documenting the preventive action taken in Step 4. Within the new standard, who, what, when, where, why, and how must be clearly identified. This is essential in communicating the reasons for the new standard and ensuring active cooperation from all workers involved. In addition, education and training are needed for all the workers involved so that the new standard becomes part of their thoughts and habits. If proper education and training are not provided, the new standard will not be carried out properly and problems will begin to recur.

STEP 7: DETERMINE FUTURE ACTION

The purpose of this step is to provide a summary of the story. This is done by reviewing the problem-solving process just completed and determining any future action that needs to be taken. By reviewing the problem-solving process just completed, the team can determine such things as what was done well, what could have been done better, and what could have been done differently. This reflection allows for team as well as individual growth. (After all, every problem attacked, whether solved or unsolved, is a learning experience that better prepares each individual to solve the next problem.)

In determining future action, the team establishes by whom and how often the process indicator needs to be checked to ensure that the preventive action is still working. The team also identifies any remaining problems and lays out a plan to solve them. If the remaining problems are outside of the team's direct control, then the team makes suggestions for their improvement. If there are no direct problems the team can begin working on, then the team is disbanded. Typically, the individual team members will quickly become involved in other teams since the overall company goal is continuous improvement.

OTHER CONSIDERATIONS

Time Frame

The QI story provides an organized framework to solve any problem. By having a structure to follow, a team can expect to reduce the amount of wasted time typically spent in the problem-solving process. It is important to note that the QI story framework is not a shortcut. The seven steps in the process help ensure that no aspect of the problem is ignored. Depending on the complexity of the problem, it will still take anywhere from several months to over a year to completely solve it.

Quality Improvement Story Requirements

The proper use of a QI story requires many things. First, it requires knowledge of the quality control tools. These tools could be the seven basic tools, they could include the seven management tools, or they could even more be advanced statistical tools like design of experiments. Whatever tools are required, it is important that the team members are properly trained in their use. Typically, everyone in the company is trained in the use of the seven basic tools. As the need arises, more advanced training is provided. For example, most managers need training in the seven management tools because they deal more with ideas than data. If a team feels that a certain tool could be effectively used and no team member is trained in its use, then support and guidance must be available to the team. Otherwise, the team will feel lost and may give up.

Second, the QI story requires that team members be able to effectively communicate their ideas. They must be able to communicate verbally in order to present their ideas. They must also be able to communicate their ideas in writing. This also requires training and support in the form of in-house short courses and guidance personnel.

Third, knowledge about the actual construction of a QI story is needed. This too requires training and support. The best training for this comes from experience. The more stories a team works on, the easier the process becomes. The best support comes from an experienced person acting as team leader. Every team should include one or more experienced people. This ensures that the team does not become discouraged and quit.

Finally, the QI story requires teamwork. Working as a team member does not come naturally to most people. Therefore, all employees need to be trained in how to work as a team. They need to know how to interact as a team, how to delegate work to other team members, and how to trust and respect everyone on the team. This may seem to be a trivial matter, but more teams have failed not because the problem was too difficult but because the team members themselves could not work together.

In summary, the QI story is a very effective means of addressing a problem. It requires the use of a variety of tools, the ability to communicate and persuade, and the ability to work as a team. With proper training and support, the goal of continuous improvement can be achieved through the use of QI stories. The following form will serve as a guide for teams that are interested in systematically developing a QI story line.

QI STORY LINE

Team Name: ________________________________

Team Mission: State why this team was established. [e.g., To improve the handling of clean claims; to reduce the average length of stay (LOS); etc.]

Team Vision: After your team has solved this problem, what would the improved situation look like? What indicators would you use to measure how successful your team has been? [e.g., To reduce the incidents of medication errors to 0 by June; To answer all incoming calls within three rings; To reduce turnaround time in claims processing by 50%; To reduce the time to become effective in the system by 25%, etc.] The initial vision statement must be provided by the Quality Council with input from members of the team.

Note: The team vision may be revised once the team decides to focus on the "vital few."

Constraints: State constraints that would limit the extent to which the team can achieve its objectives — for example, no hiring of additional staff, or physical space limitations.

Process Boundaries: What are the start and end points for this problem? [e.g., Problem starts when a customer returns damaged merchandise; and ends when the customer receives new merchandise.]

Current Status of the Problem: How bad is the problem today? [e.g., 65% of the ER patients wait for more than 3 hours before being seen by a doctor; Percent of merchandise returned; average time to fill a vacancy by HR, etc.]

Problem Justification and the Cost of Non-Conformance: Why is this problem important? Who is affected by this problem, and in what ways are they affected? What does it cost this organization to live with this problem? [lawsuits, member dissatisfaction, employee dissatisfaction, provider dissatisfaction, overtime costs, high risk, fatality, reputation, etc.]

TANGIBLE COSTS (provide $ amounts)	*INTANGIBLE COSTS*

DATA COLLECTION

What Data Will the Team Need?

Data Needed	*Who will collect them?*	*For what period?*	*How much data is needed?*	*Where will data be gathered?*	*How will data be gathered?*	*When will data be gathered?*

Data Gathering Training Requirements:

1. Is training required for the data collector? __Yes __No
2. If "Yes," provide the name of person to provide the training:

   ____

3. When will data collection training take place?

   ____

4. When will data collection begin?

   ____

5. Estimated completion date for data collection:

 ____

6. Date when analysis of data will begin:

 ____

7. Date when analysis of data will be completed:

 ____

8. Date when data will be presented to the team:

 ____

EXERCISES

18-1 What is a QI story?

18-2 Discuss several reasons for using a QI story.

18-3 List the basic steps of a QI story.

18-4 What is one of the best ways to demonstrate the importance of a problem to management?

18-5 What are the two types of action that can be taken? Give an example of each type.

18-6 What are the basic requirements of a QI story?

18-7 Within your own work environment, identify a problem that you (or your team) can work on. Use the QI story to address and document the problem-solving effort.

REFERENCES

FPL Quality Improvement Program, *QI Story and Techniques,* Miami: Florida Power and Light Company, 1987.

Gitlow, Howard S., Shelly J. Gitlow, Alan Oppenhelm, and Rosa Oppenhelm, "Telling the Quality Story," *Quality Progress,* Sep. 1990, pp. 41–46.

Kume, Hitoshi, *Statistical Methods for Quality Improvement,* Japan: The Association for Overseas Technical Scholarship, Tokyo, Japan, 1985.

Swift, J. A., *Introduction to Statistical Quality Control and Management,* Delray Beach, Fla.: St. Lucie Press, 1995.

19

QUALITY FUNCTION DEPLOYMENT

HISTORY

The first formal use of quality function deployment (QFD) can be traced to the Kobe Shipyard, Mitsubishi Heavy Industries, Ltd., Japan, in 1972. In 1977, Toyota began using QFD extensively. It wasn't until 1983 that QFD was introduced to U.S. companies. In 1983, Ford and several supplier companies went to Japan and had several meetings with Dr. Ishikawa and other member of the Union of Japanese Scientists and Engineers. It was at these meetings that the power of QFD was recognized. In 1984, Dr. D. Clausing, then of Xerox, introduced the operating mechanism of QFD to Ford and its supplier companies. Also in 1984, the American Supplier Institute organized three Japanese study missions for several U.S. supplier companies. The purpose of the study missions was to review Toyota supplier QFD case studies, which in turn would aid in the transfer of the technique to U.S. industries. The study missions took place in December 1984, June 1985, and April 1986. It was during the last study mission that two of the supplier companies, Budd Company and Kelsey Hayes, presented the first U.S. QFD case studies. In 1987, Ford and General Motors began QFD training in their plants. Since that time, QFD has been studied and implemented by companies all across the United States.

WHAT IS QUALITY FUNCTION DEPLOYMENT?

The overall objective of QFD is "to improve (reduce) the product development cycle while improving quality and delivering the product at lower costs."[1] QFD itself is a systematic and structured approach used to translate the voice of the customer into the appropriate technical requirements and actions for each stage of product or service development and production. In other words, it connects customer requirements to production or service requirements. QFD is driven by the voice of the customer, which is the customer's requirements expressed in the customer's own words.

Two major components make up the heart of QFD: product quality deployment and deployment of the quality function. Product quality deployment encompasses the activities associated with translating the voice of the customer into technical quality characteristics and features. These technical quality characteristics and features are known as the final product or service control characteristics. Deployment of the quality function encompasses the activities associated with ensuring that customer-required quality is actually achieved. Included in these activities is the assignment of specific quality responsibilities to specific groups or departments.

The QFD process (see Figure 19-1) is driven by the "voice of the customer." Therefore, the process begins by capturing the "voice of the customer" or the customer requirements. Because these requirements are usually stated in qualitative terms, they must be translated or converted into technical or company terminology. This translation results in the formulation of the product/service design characteristics. These characteristics need to be measurable because it will be the monitoring of these characteristics that determines whether the customer's requirements are

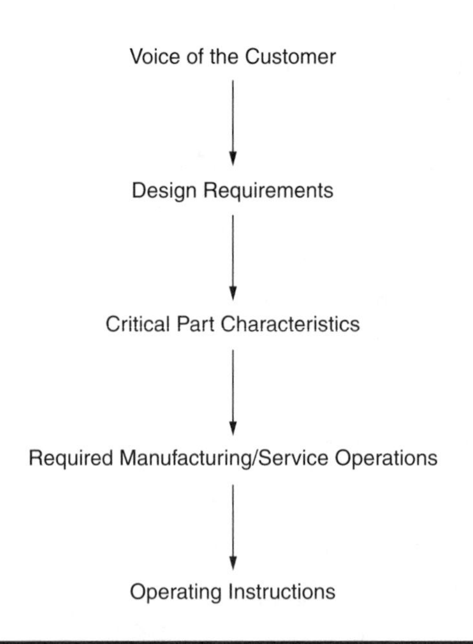

Figure 19-1 Quality Function Deployment Process

being satisfied. However, these product/service design characteristics need to be broken down further for proper implementation. Therefore, the design characteristics are translated into specific parts, with the critical characteristics of these parts identified. When the critical part characteristics have been identified, the required manufacturing/service operations are then determined. The critical process parameters are also identified so that checks can be made to ensure that the critical part characteristics are being met. Also at this stage, any constraints, such as new equipment needed or limited capital, are identified and dealt with. When the manufacturing/service operations have been specified, the operating instructions are developed. These instructions constitute the entire set of procedures and practices that will be used to consistently make products that satisfy customer requirements.

This process appears to be simple. However, the fundamental problem is that the initial customer requirements do not get properly translated into the final product. Another problem is that some customer requirements often conflict with one another. The concept of QFD is based on four key documents which aid in avoiding these problems:

1. Overall customer requirement planning matrix
2. Final product characteristic deployment matrix
3. Process plan and quality control charts
4. Operating instructions[2,3]

The planning matrix translates the "voice of the customer" into specific final product/service control characteristics. The deployment matrix takes the final product/service control characteristics and translates them into critical part characteristics, thus moving the customer requirements deeper into the design process. The process plan and quality control charts identify the critical product, service, and process parameters that are vital to meeting the critical part characteristics. They also identify checkpoints for each of the critical parameters. The operating instructions constitute the entire set of procedures and practices that will be performed by all personnel to ensure that the critical parameters are achieved. The main purpose of these documents is to translate and deploy the customer's requirements throughout the product/service design, development, and production process of an organization. The customer's requirements are ultimately addressed by the operational personnel who produce the product and deliver the services. The flow and relationship of these documents are illustrated in Figure 19-2.

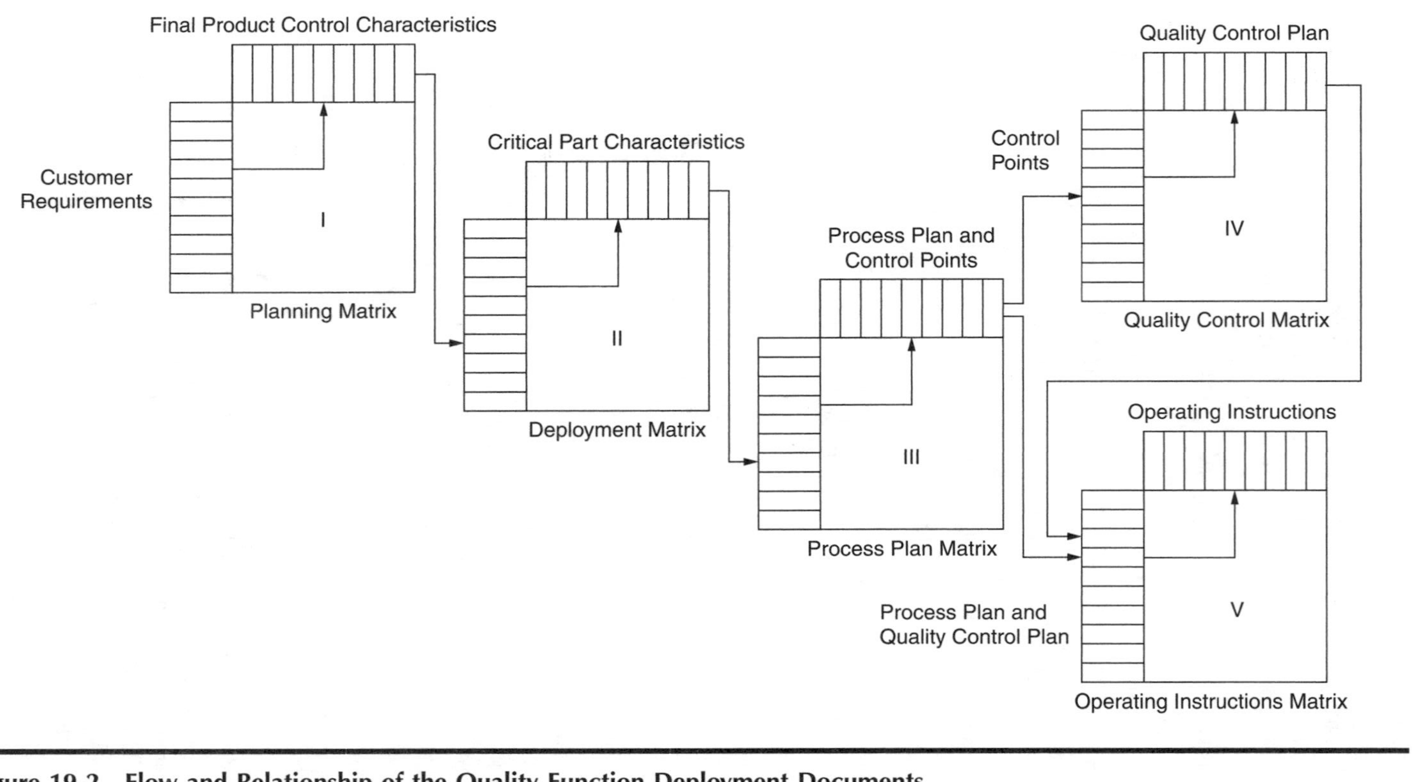

Figure 19-2 Flow and Relationship of the Quality Function Deployment Documents

Voice of the Customer

This part of the QFD matrix is often referred to as the "What" component. It is a catalogue of *what* the customers want and need. For example, a patient might express a need for "a nurse who knows what she is doing." This requirement, which is in the patient's own words, may then be translated to mean the following:

1. Qualification
   - Unlicensed nurse with bachelor's degree
   - Licensed nurse without bachelor's degree
   - Licensed nurse with bachelor's degree
   - Licensed nurse with bachelor's and master's degree
2. Experience
   - Number of years working in the nursing field
   - Number of years in hands-on nursing
3. Technical competence
   - Knows how to take vital signs
   - Can answer basic questions about patient's condition
   - Can efficiently draw and handle patient's blood
   - Appears confident and poised
   - Shows a willingness to find out what he/she doesn't know
   - Knows knowable information
4. Professionalism and quality of behavior
   - Empathy
   - Patience
   - Sensitivity
   - Listening skills
   - Respect for the patient
   - Linguistic skills
   - Appearance
   - Manner of speaking
   - Body language

A Case Study

The top management of a service organization wished to redesign its marketing brochure in response to the growing number of questions from its customers. A QFD team was established to accomplish this objective. Utilizing a number of focus groups, the team was able to obtain the necessary input for a QFD process. The customers identified four key requirements for a marketing brochure: They wanted a brochure that was easy to use, concise, clear, and contained accurate information. Figure 19-3 represents a complete QFD matrix using all

BROCHURE EXAMPLE		Font Size	Glossary	Use of Colors	Answers Ques	Illustrations	Graphics	Table Contents	Languages	Update Info	Importance Ratg	Orgn's Rating	Chief Competitor	PLAN	Rate Imprvmt	Absolute Wts.	Demanded Wts.
		1	2	3	4	5	6	7	8	9	1	2	3	4	5	6	7
Easy to Use	1	○	●	○	●	●	△	●	●	○	5	2	4	5	2.5	12.5	28
Conciseness	2	△	△	△	○	△		△		△	3	3	3	3	1	3	7
Clarity	3	●	○	○	○	●	●	●	○		4	1	4	4	4	16	36
Accurate Information	4		△		△					●	5	2	4	5	2.5	12.5	28
TOTAL		415	395	199	409	583	352	583	360	343							
Percent		12	11	5	11	16	10	16	10	9							
Company Now		10	10	0	5	1	1	7	1	1							
Chief Competitor		12	20	10	10	3	3	15	3	3							
PLAN		12	25	10	15	4	3	20	3	3							

Main Correlation

◎ 9 = strong correlation
○ 3 = some correlation
△ 1 = possible correlation

Figure 19-3 Brochure Example

the input from the customers and the organization. The remainder of this chapter will be devoted to the description of the various components of the QFD matrix, as well as the application of QFD to this organization's case example.

Customer Requirements

The customers of this service organization identified four key requirements, as follows:

- Ease of use
- Conciseness
- Clarity
- Accurate information

Technical Requirement

The technical requirement is the *how* of the QFD process. The technical requirement section of the QFD matrix represents *how* the organization will respond to each of the items stated in the customer requirements. These are the items that are controlled to assure that customer demands are met. Technical requirements are limited to only quantifiable items. The following technical requirements are pertinent to this case example:

- Adjustment of font size
- Expanding the section on glossary of terms
- Use of colors in the brochure
- Providing answers to often-asked questions
- Introducing some graphical illustrations
- Use of pictures
- Expanding the table of contents
- Presenting the brochure in more than one language
- Maintaining up-to-date information

Strength of Relationships

The intersection between the *what* and the *how* defines the strength of the relationship between the two components of the QFD matrix. The entries in this section of the QFD matrix reflect the correlations between the customer demands and the technical requirements. It is possible for a particular technical requirement to constitute a response to a number of customer requirements. Also, a particular customer requirement may be addressed by a number of technical requirements. The strength of the relationships between the customer requirements and the technical requirements are described with the aid of symbols. A solid black circle indicates a strong relationship; a circle signifies a moderate relationship; and a triangle signifies a weak relationship. A blank is used to indicate no correlation. The symbols provide a powerful visual impression of the strength of relationships between the customer requirements and the technical requirements. In this service organization's example, it is determined that the following responses by the company would have the strongest correlation with the customers' requirement of a brochure that has *ease of use*:

- Expanding the glossary of terms
- Provide answers to often asked questions
- Increase the use of graphic illustrations
- Expand the table of contents
- Provide the brochure in an additional language

Increasing the font size, use of colors, and update of information are deemed to have a moderate correlation with the customers' requirement of a brochure that has *ease of use.*

Adding or increasing the number of pictures is believed to have only a weak correlation to *ease of use.* Similarly, *conciseness* has a moderate correlation with *answers to question*, and only weak or no correlation with the other technical requirements. The other correlations are noted as shown in Figure 19-3.

The Vertical Entries

1. Importance Rating

This is a measure of the relative importance that customers assign to each of the stated requirements. The importance rating is usually stated on a numerical scale, as follows:

1 = Low importance
5 = High importance

For this example, two customer requirements (*ease of use* and *accurate information*) were assigned high importance rating of 5. A rating of 4 was given for *clarity*, and 3 for *conciseness.*

2. Surveying Company

Assuming this QFD table is being generated by your organization, the entries in this column represent how your organization's customers rate your performance with respect to their stated requirements. The rating is usually based on a numerical scale such as 1 = poor; and 5 = excellent. This organization received a performance rating of 2 for *ease of use*, 3 for *conciseness*, 1 for *clarity*, and 2 for *accurate information.*

3. Chief Competitor

The entries in this column represent how your customers rate your chief competitor with respect to their stated requirements. Again, as was the case with the "surveying company," the rating is usually based on a numerical scale such as 1 = poor; and 5 = excellent. The company in this example lagged behind its chief competitor in *ease of use*, *clarity*, and *accurate information* as perceived by its customers. It obtained a rating of 3 (same as its chief competitor) in terms of *conciseness* of the brochure.

4. *Plan*

The plan is an indication of where the company wishes to be with respect to each of the quality requirements stated by its customers. There is a plan for each customer requirement. The plan is determined by looking at where the company is today in relation to the competitor and in relation to the customer's rate of importance. The plan is also educated by the organization's strategic plan. After taking all things into account, the QFD team set a goal of achieving a performance rating of 5 for *ease of use,* 3 for *conciseness,* 4 for *clarity,* and 5 for *accurate information.* The team expects to achieve these levels of performance rating from this company's customers the next time they are surveyed.

5. *Rate of Improvement*

The rate of improvement is the ratio of two values — where the company plans to be and where the company is today. It is determined by dividing the value for *plan* by the value for *surveying company.* The rate of importance for *ease of use* is obtained by dividing the plan, 5, by where the conpany is today, 2. The resulting value for the rate of improvement is 2.5, approximated to 3. Similarly, the rate of improvement for *conciseness, clarity,* and *accurate information* are 1, 4, and 3 respectively.

6. *Absolute Weight*

The absolute quality weight is determined by multiplying the rate of importance by the rate of improvement. The absolute quality weight is an attempt to assign some weight to what the customer considers to be important and where the organization needs to be. In certain QFD applications, the absolute quality weight value is determined by multiplying by a third component known as a sales point. A customer requirement that is considered to represent a strong sales point is assigned a value of 1.5, and a lesser sales point, 1.2, and a 1.0 for an item that is not a sales point. Since sales point figures are not available, the absolute weight figures for *ease of use* is 5 × 2.5, which is equal to 13. The other values for the absolute weight are 3, 16, and 13 respectively. The sum of the absolute weight is 45.

7. *Demanded Weight*

The demanded quality weight is determined by converting each absolute weight to a percentage of the total absolute weight. First, the sum of the total absolute weight is determined. Next, each value of the absolute weight is divided by the sum of the absolute weight and multiplied by

100, to convert each entry to a percentage. For this example, the demanded weight for easy to use is 13/25 × 100%, which gives 28%. The highest demanded weight is 37% for clarity. On the basis of what is important to the customers, where the company is currently, its chief competitor's current superiority, and the plan, *clarity* appears to be the most important customer requirement to act on.

Competitive Technical Assessment

1. Total

The figures in this row represent the sum of the products of each column symbol value and the corresponding demanded weight. Using the sample calculation for font size, the figures for the remaining technical requirements are 398, 202, 412, 592, 361, 592, 363, and 343, respectively. The sum of all the entries in the total row is 3687. The two most important technical requirements are *graphic illustrations* and *table of contents.*

2. Percentage (%)

Each entry in this row is divided by the sum of all the entries in that row and multiplied by 100, to convert it into a percentage. Consequently, 424/3687 x 100 = 12%. Similarly, the remaining percentages are 11%, 5%, 11%, 16%, 10%, 16%, 10%, and 9% respectively.

3. Company Now

This row gives the values of the measurable technical requirements. Table 19-1 provides the performance indicators whose values are used in Figure 19-3.

Table 19-1 Measures for Technical Requirements

Technical Requirements	*Measures Now*	*Company*
Font Size	Actual size	10
Glossary of Terms	# of Terms	10
Use of Colors	# of Items with Colors	0
Answers to Questions	# of Questions Answered	5
Graphic Illustrations	# of Illustrations	1
Pictures	# of Pictures	1
Table of contents	# of Items	7
Languages	# of Languages	1
Update Information	# of Updates per Year	1

4. *Chief Competitor*

An analysis of the chief competitor's brochure and information gathered from outside sources revealed the values for Technical Requirement for the chief competitors, as shown in Figure 19-3. The competitor outperforms the company in this example in all aspects of the technical requirements.

5. *Plan*

The most aggressive plans were targeted at the two technical requirements with the highest totals — graphic illustrations and table of contents. The plan represents the target of the team's effort for the next year as it seeks to redesign the brochure.

BENEFITS

By using QFD, many benefits are realized:

1. QFD makes product quality a function of product design. Quality is built in, and product quality is no longer the result of quality control efforts.
2. Total product development time is reduced. (The actual time to define the product typically increases, but the total design cycle is reduced.)
3. Products are produced at lower cost with higher quality (product design is better).
4. The number of start-up problems is reduced.
5. Documentation and communication between groups and departments are improved, which results in an improved working environment.
6. Any conflicting design requirements are usually identified early. Also, any omissions that are typically the result of oversight are avoided.
7. The critical quality characteristics that need to be controlled are identified.
8. Customer requirements are identified and translated directly into product characteristics, which leads to increased customer satisfaction. It also reduces misinterpretation of customer requirements during subsequent stages.
9. The technology and job functions required to carry out the design are identified and assigned to specific individuals or groups.
10. Specific tools and techniques that will provide the greatest payoff are identified.

11. The organization's means of satisfying customer requirements is documented.
12. A historical database is established. This is a very valuable resource for future design and process improvements.

QFD has a major downside in that it requires extensive training and can be very laborious. However, there is a hidden benefit to the extensive training. Once training is completed, all of the functional areas are using the same approach and criteria. Hence, they are speaking the same language.

CONCLUSION

Several points need to be kept in mind when using QFD. First, it is applicable to more than new product design. It can be used effectively for existing design or process improvements. Second, QFD can be applied to any process within an organization, especially service processes. Third, it is not a tool restricted to the quality department. It is a valuable tool for everyone.

EXERCISES

19-1 Where and when was QFD first used?
19-2 What is the objective of QFD?
19-3 Discuss the two major components of QFD.
19-4 What are the four key documents of QFD? What is the purpose of each document?
19-5 What are the benefits of using QFD?
19-6 What is the major downside to QFD? Is there a hidden benefit to this downside? If so, what is it?

ENDNOTES

1. See articles written by Lawrence P. Sullivan and Norman E. Morrell published in *Quality Function Deployment: A Collection of Presentations and QFD Case Studies,* Dearborn, Mich.: American Supplier Institute, 1987.
2. Lawrence P. Sullivan, "Quality Function Deployment (QFD): The Beginning, the End and the Problem In-Between," in *Quality Function Deployment: A Collection of Presentations and QFD Case Studies*. Dearborn, Mich.: American Supplier Institute, 1987.
3. Lawrence P. Sullivan, "Quality Function Deployment," *Quality Progress,* June 1986, pp. 40–41.

REFERENCES

Conti, Tito, "Process Management and Quality Function Deployment," *Quality Progress,* Dec. 1989, pp. 45–48.

Day, R. G., *Quality Function Deployment,* Milwaukee, WI: ASQC Quality Press, 1993.

Denton, Keith D., "Enhance Competition and Customer Satisfaction — Here's One Approach," *Industrial Engineering,* May 1990, pp. 24–30.

DeVera, Dennis, Tom Glennon, Andrew A. Kenny, Mohammad A. H. Khan, and Mike Mayer, "An Automotive Case Study," *Quality Progress,* June 1988, pp. 35–38.

Einspruch, E., Omachonu, V. K., and Einspruch, N. G. "Quality Function Deployment (QFD): Application to Rehabilitation Services," *International journal of Health Care Quality Assurance,* Vol. 9, Number 3, 1996, p. 42.

Fortuna, Ronald M., "Beyond Quality: Taking SPC Upstream," *Quality Progress,* June 1988, pp. 23–28.

Gopalakrishnan, K. N., Barry E. McIntyre, and James C. Sprague, "Implementing Internal Quality Improvement with the House of Quality," *Quality Progress,* Sep. 1992, pp. 57–60.

Hauser, John R. and Don Clausing, "The House of Quality," *Harvard Business Review,* May–June 1988, pp. 63–73.

King, B., *Better Design in Half the Time,* Methuen, MA: GOAL/QPC Publisher, 1987.

Kogure, Maseo and Yoji Akao, "Quality Function Deployment and CWQC in Japan," *Quality Progress,* Oct. 1983, pp. 25–29.

McElroy, John, "QFD: Building the House of Quality," *Automotive Industries,* Jan. 1989, pp. 30–32.

Morrell, Norman E., "Quality Function Deployment," in *Quality Function Deployment: A Collection of Presentations and QFD Case Studies,* Dearborn, Mich.: American Supplier Institute, 1987.

Omachonu, V. K. *Health Care Performance Improvement,* Norcross, GA: Engineering and Management Press (IIE), 1999.

Quality Function Deployment: Executive Briefing, Dearborn, Mich.: American Supplier Institute, 1987.

Sullivan, Lawrence P., "Quality Function Deployment," *Quality Progress,* June 1986, pp. 39–50.

Sullivan, Lawrence P., "Quality Function Deployment (QFD): The Beginning, the End and the Problem In-Between," in *Quality Function Deployment: A Collection of Presentations and QFD Case Studies,* Dearborn, Mich.: American Supplier Institute, 1987.

Sullivan, Lawrence P., "Policy Management through Quality Function Deployment," *Quality Progress,* June 1988, pp. 18–20.

Swift, J. A. *Introduction to Statistical Quality Control and Management,* Delray Beach, Fla.: St. Lucie Press, 1995.

III

CRITERIA FOR QUALITY PROGRAMS

In Part III, the commonly accepted standards for measuring the effectiveness of an organization's quality program are presented. Chapter 20 discusses the most widespread standard in Europe, ISO 9000. Chapter 21 discusses the Malcolm Baldrige Award. In conjunction with ISO 9000, Chapter 22 discusses the European Union Directives. Chapter 23 discusses the QS-9000 standards and Chapter 24 discusses the ISO 14000 standards.

20

ISO 9000

"Simply put, ISO 9000 has come to be the price of admission for doing business in Europe," says Robert Caine, president of the American Society for Quality Control (ASQC). "Ask any business person who has given up trying to gain entry into the European market what stopped him, and he's likely to answer in code: ISO 9000," concludes Kymberly Hockman of Du Pont's Quality Management and Technology Center. These are among the many experts who are urging U.S. firms to take the ISO Series standards seriously.

Even if a firm does not do business in Europe or does not plan to do so, it should not ignore this accelerating movement to international standards. As will be discussed, the movement is expanding into other areas of the world and into many areas of the U.S. public and private sectors as well.

ISO 9000 is a set of five worldwide standards that establish requirements for the management of quality. Unlike *product* standards, these standards are for *quality management systems*. They are being used by the nations of the European Union to provide a universal framework for quality assurance — primarily through a system of internal and external audits. The purpose is to ensure that a certified company has a quality system in place that will enable it to meet its published quality standards. The ISO standards are generic, in that they apply to all functions and all industries, from banking to chemical manufacturing. They have been described as the "one size fits all" standards.

ISO AROUND THE WORLD

The European Union (EU) consists of 15 member nations: Belgium, Denmark, France, Germany, Greece, Ireland, Italy, Luxembourg, the Netherlands, Portugal, Spain, the United Kingdom, Austria, Finland, and

Sweden. The goal of the EU is to create a single internal market, free of all barriers to trade. For products and services to be traded freely, there must be assurance that those product meet certain standards, whether they are produced in one of the EU nations or in a non-EU nation, such as the U.S.[1] The EU is using the standards to provide a universal framework for quality assurance and to ensure the quality of goods and services across borders.

The International Organization for Standardization (ISO) is the specialized international agency for standardization and at present comprises the national standards bodies of 91 countries. The American National Standards Institute (ANSI) is the member body representing the U.S. ISO is made up of approximately 180 technical committees. Each technical committee is responsible for one of many areas of specialization, ranging from asbestos to zinc. The purpose of ISO is to promote the development of standardization and related world activities in order to facilitate the international exchange of goods and services and to develop cooperation in intellectual, scientific, technological, and economic activities. The results of ISO technical work are published as international standards, and the ISO 9000 Series is a result of this process.

In 1987 (the same year the ISO 9000 Series was published), the U.S. adopted the ISO 9000 Series verbatim as the ANSI/ASQC Q-90 Series. Thus, the use of either of these series is equivalent to the use of the other.[2] The ISO standards are being adopted by a varying number of companies in over 50 countries around the world that have endorsed them.[3]

By 1992 more than 20,000 facilities in Britain had adopted the standards and became certified.[4] The Japanese not only have adopted the standards but also have mounted a major national effort to get their companies registered.[5]

The EU adopted ISO 9000 in 1989 to integrate the various technical norms and specifications of its member states. By 1991, ISO compliance became part of hundreds of product safety laws all over Europe, regulating everything from medical devices to telecommunications gear. Such products accounted for only about 15% of EU trade at that time, but the list of products is growing. Entire industries are encouraging the adoption of the standards.

One example of the impact is reflected in the requirements of Siemens, the huge German electronics firm. The company requires ISO compliance in 50% of its contracts and is pressing all other suppliers to conform. A major justification for this action is that it eliminates the need to test parts, which saves time and money and establishes common requirements for all markets.

Even for companies whose products are unregulated, ISO standards are becoming a *de facto* market requirement for doing business with other EU companies. If two suppliers are competing for a contract or an order, the one that has registered its quality systems under ISO 9000 has a clear edge.

The impact of these standards is reflected by the widespread distribution of the ISO 9000 Series, which has become the best-seller in the history of ISO, under whose auspices they were developed. ISO 9000 even outsold the universal and long-standing international weights and measurement standards. However, it is worth repeating that ISO 9000 is not standards for products but standards for operation of a *quality management system.*

ISO 9000 IN THE U.S.

U.S. companies have been slow to adopt these international standards, despite the fact that 30% of the country's exports go to Europe. Moreover, to the extent that the standards are adopted elsewhere in the world, additional exports will be affected as well. Additional markets both within and outside the U.S. may be closed to those firms that ignore the requirement or fail to become certified. Du Pont, now a leader in adopting the standards, began its ISO drive in 1989 only after losing a large European order for polyester film to an ISO-certified British firm.

Some people perceive ISO 9000 as a barrier to competition and even a plot to keep U.S. firms out of Europe. This view, of course, is not the case, but a barrier can exist unless the standards are clearly understood.

Additional evidence of growing acceptance lies in the fact that the standards are being integrated into the requirements for manufacturers that make products under contract for several U.S. government agencies, including NASA, the Department of Defense, the Federal Aviation Administration, and the Food and Drug Administration.[6] To date, ISO 9000 registration is required of suppliers to the governments of Canada, Australia, and the United Kingdom.

Du Pont, Eastman Kodak, and other U.S. pioneers adopted ISO 9000 in the late 1980s to ensure that they were not locked out of European markets. They then found that the standards also helped to improve their quality. Now, Baldrige winners such as Motorola, Xerox, IBM, and others are making suppliers adopt ISO. As the movement catches on and as suppliers to suppliers are required to come on board, there may be a geometric leverage effect in the number of companies adopting the standards. This effect may give additional meaning to the often-repeated description of the market as *global* in dimension.

Despite the weight of the evidence that suggests the need to adopt ISO 9000, it appears that many U.S. firms have not done so, nor do they plan to do so.

The good news is that for those firms planning to become ISO 9000 certified, the process is not all that difficult, especially if a company already has a quality effort underway. Indeed, those companies using total quality management (TQM) are more than halfway there. For Baldrige winners, certification would be a relatively simple process.

What is the impact of ISO 9000 for service industries and for those manufacturing firms whose products fall outside the *regulated* product areas? The answer is provided by ASQC:[7]

> Outside of regulated product areas, the importance of ISO 9000 registration as a competitive market tool varies from sector to sector. For instance, in some sectors, European companies may require suppliers to attest that they have an approved quality system in place as a condition for purchase. This could be specified in any business contract. ISO 9000 registration may also serve as a means of differentiating "classes" of suppliers, particularly in high-tech areas, where high product reliability is crucial. In other words, if two suppliers are competing for the same contract, the one with ISO 9000 registration may have a competitive edge with some buyers. Sector and product areas where purchasers are more likely to generate pressure for ISO 9000 registration include aerospace, autos, electronic components, measuring and testing instruments, and so on. ISO 9000 registration may also be a competitive factor in product areas where safety or liability are concerns.

Some U.S. manufacturers have criticized the EU's adoption of ISO 9000, suggesting that the standards are inferior to those used in the U.S. Moreover, it is suggested that requiring U.S. companies to conform to the standards will force them to incur larger production costs.[8]

The counter arguments are that the standards will eliminate the hodgepodge of standards that now exist around the world, and production costs will be more than offset by other savings and the increase in productivity and quality.

Criticisms and ignorance of ISO 9000 notwithstanding, there is evidence of a growing acceptance of the standards among U.S. firms. One source reports an increase in registration of 500% between 1992 and 1993. Of course, this increase is computed on a somewhat smaller 1992 base.[9] It is interesting to note that the Japanese experience is similar to that in the

U.S.: Initial resistance was largely overcome by pressure to conform to the requirements of the international marketplace.[10]

Involvement of professional and trade associations appears to be growing as firms within a particular industry band together to research how best to meet ISO requirements. The chemical industry has been a leader in this movement. Professional engineers, public utilities, software vendors, and manufacturers of information technology are among the groups with organized efforts.[11] Some have formed a network of support groups.[12]

ISO 9000

In 1979, ISO established Technical Committee 176 to develop a generic set of quality system management standards. The original committee had 20 participating members and 14 observing members.* This committee relied heavily on the U.K. standard BSI 5750 as a guide to developing the ISO 9000 series of standards. The first ISO 9000 series of standards was published in 1987. A revised version was published in 1994.

The ISO 9000 series of standards has been translated into various languages and is known by different names in different countries. A list of some of the different names by which the ISO 9000 series is known is provided in Table 20-1. Note that most of the national versions bear some code number that includes 9000 or 90. Also note that the European Community has adopted its own version of the standard series, EN 29000.

Table 20-1 National Equivalents of ISO 9000 Series

Country	*Standard*
Australia	AS3900
Brazil	NB 9000
Denmark	DS/ISO 9000
France	NF-EN 29000
Germany	DIN ISO 9000
Japan	JIS Z 9900
Portugal	FM 29000
Spain	UNE 66 900
United Kingdom	BS 5750
United States	ANSI/ASQC Q90
European Union	EN 29000

* ISO/TC 176 now has 15 participating members and 21 observing members.

Components of ISO 9000 Standard Series

The standards in the ISO 9000 Series are intended to provide a generic core of quality system standards applicable to a broad range of industry and economic sectors. They are not standards for products. Instead, they are standards for governing quality management systems. Therefore, products do not meet ISO 9000 standards; organizations do.

Historical Perspectives

There were five parts to the ISO 9000 series:

Part 1: ISO 9000:1994 (E)	**Quality Management Standards**
ISO 9000-1	Part 1: Guidelines For Selection and Use of ISO 9001, ISO 9002, and ISO 9003
ISO 9000-2	Part 2: Generic Guidelines for the Application of ISO 9001, ISO 9002, and ISO 9003
ISO 9000-3	Part 3: Guidelines for the Application of ISO 9001 to the Development, Supply, and Maintenance of Software
ISO 9000-4	Part 4: Guide to Dependability Program Management
Part 2: ISO 9001:1994(E)	**Quality Systems — Model for Quality Assurance in Design, Development, Production, Installation, and Servicing**
Part 3: ISO 9002:1994(E)	**Quality Systems — Model for Quality Assurance in Production, Installation, and Servicing**
Part 4: ISO 9003:1994(E)	**Quality Systems — Model for Quality Assurance in Final Inspection and Test**
Part 5: ISO 9004:1994(E)	**Quality Management and Quality System Elements**
ISO 9004-1	Part 1: Guidelines
ISO 9004-2	Part 2: Guidelines for Services
ISO 9004-3	Part 3: Guidelines for Processed Materials
ISO 9004-4	Part 4: Guidelines for Quality Improvement
ISO 9004-5	Part 5: Guidelines for Quality Plans
ISO 9004-6	Part 6: Guidelines on Quality Assurance for Project Management
ISO 9004-7	Part 7: Guidelines for Configuration Management
ISO 9004-8	Part 8: Guidelines on Quality Principles and Their Application to Management Practices

Today the old ISO 9001, 1994 standard has ben replaced by ISO 9001, 2000. Similarly, the old ISO 9002, 1994 and ISO 9003, 1994 quality standards have been discontinued.

Briefly, ISO 9000 is a guideline for selecting at which level (9001, 9002, or 9003) to be certified. ISO 9001, ISO 9002, and ISO 9003 are the guidelines for each specific level of certification. ISO 9004 is a management model.

ISO 9000 certification is done on a site basis. In other words, a company cannot get an ISO 9000 certification that covers all sites and facilities of that company. The company must have each individual site and facility independently certified. Certification can be obtained at one of three different levels: 9001, 9002, or 9003. ISO 9001 certification is the most comprehensive level of certification in the series. Certification at this level requires conformance to all 20 functional areas of the standard. ISO 9002 certification requires conformance to 19 of the 20 functional areas. ISO 9003 requires conformance to 16 elements. The 20 functional areas of standards and which elements are required for each level of certification are listed in Table 20-2.

Management Responsibility

The commitment and involvement of top management are requirements for the success of any significant cultural or operational change. So it is with both the Baldrige and ISO 9000. The concern of ISO with management responsibility is reflected in the following series excerpts:[13]

Quality policy. The supplier's management shall define and document its policy and objectives for, and commitment to, quality. The supplier shall ensure that this policy is understood, implemented, and maintained at all levels in the organization.

Management review. The quality system adopted to satisfy the requirement of the standard shall be reviewed at appropriate intervals by the supplier's management to ensure its continuing suitability and effectiveness. Records of such reviews shall be maintained.

Internal quality audits. The supplier shall carry out a comprehensive system of planned and documented internal quality audits to verify whether quality activities comply with planned arrangements and to determine the effectiveness of the quality system. Audits shall be scheduled on the basis of the status and importance of the activity. The audits and follow-up actions shall be carried out in accordance with documented procedures.

The results of the audits shall be documented and brought to the attention of the personnel having responsibility in the area audited. The management personnel responsible for the area shall take timely corrective action on the deficiencies found by the audit.

Table 20-2 Functional Areas Required by Each Level of Certification

Functional area	*ISO 9001*	*ISO 9002*	*ISO 9003*
1. Management Responsibility	♦	♦	♦
2. Quality System	♦	♦	♦
3. Contract Review	♦	♦	♦
4. Design Control	♦	♦	
5. Document and Data Control	♦	♦	♦
6. Purchasing	♦	♦	
7. Control of Customer-Supplied Product	♦	♦	♦
8. Product Identification and Traceability	♦	♦	♦
9. Process Control	♦	♦	
10. Inspection and Testing	♦	♦	♦
11. Control of Inspection, Measuring, and Test Equipment	♦	♦	♦
12. Inspection and Test Status	♦	♦	♦
13. Control of Nonconforming Product	♦	♦	♦
14. Corrective and Preventive Action	♦	♦	♦
15. Handling, Storage, Packaging, Preservation, and Delivery	♦	♦	♦
16. Control of Quality Records	♦	♦	♦
17. Internal Quality Audits	♦	♦	♦
18. Training	♦	♦	♦
19. Servicing	♦	♦	
20. Statistical Techniques	♦	♦	♦

Corrective action. The supplier shall establish, document, and maintain procedures for:

- Investigating the cause of nonconforming product and the corrective action needed to prevent recurrence
- Analyzing all processes, work operations, concessions, quality records, service reports, and customer complaints to detect and eliminate potential causes of nonconforming product
- Initiating preventive actions to deal with problems to a level corresponding to the risks encountered
- Applying controls to ensure that corrective actions are taken and that they are effective
- Implementing and recording changes in procedures resulting from corrective action

Functional Standards

ISO 9000 standards also require documentation and follow-up performance for all functions affecting quality. Functional requirements are illustrated by the following examples:[14]

- ***Design*** — Sets a planned approach for meeting product or service specifications
- ***Process control*** — Provides concise instructions for manufacturing or service functions
- ***Purchasing*** — Details methods for approving suppliers and placing orders
- ***Service*** — Details instructions for carrying out after-sales service
- ***Inspection and testing*** — Compels workers and managers to verify all production steps
- ***Training*** — Specifies methods to identify training needs and keeping records

BENEFITS OF ISO 9000 CERTIFICATION

The benefits to an organization gained by improving quality in products and services were outlined in Chapter 1. To repeat:

1. Greater customer loyalty
2. Improvements in market share
3. Higher stock prices
4. Reduced service calls
5. Higher prices
6. Greater productivity and cost reduction

These same benefits would be achieved by ISO 9000 certification, to the extent that actions leading to certification result in a quality management system. Moreover, certification provides the additional benefit of acceptance by EC customers and others whose criteria of acceptance include ISO 9000 certification.

Experience tends to confirm that companies do achieve these benefits. Consider the following examples:

- A British government survey revealed that 89% of ISO 9000 registered companies reported greater operational efficiency: 48% reported increased profitability, 76% reported improvements in marketing, and 26% reported increased export sales.[15]

- The British Standards Institution, a leading British registrar, estimates that registered firms reduce operating costs by 10% on average.[16]
- Du Pont attributes the following results to the adoption of ISO standards in its plants:
  - On-time delivery at one plant increased to 90% from 70%
  - Cycle time at one plant went from 15 days to 1.5 days
  - First-pass yield at one plant went from 72% to 92%
  - Test procedures were reduced from 3000 to 1100
- A number of U.S. firms have reported benefits ranging from increased sales to improved communications.[17]

GETTING CERTIFIED: THE THIRD-PARTY AUDIT

Many managers perceive an audit of any kind as a necessary bureaucratic activity that has a very low priority. This negative perception may increase when it is learned that preparation for ISO 9000 certification may take from six to twelve months and that the failure rate the first time around can be as high as two out of three. Nevertheless, a third-party audit is a prerequisite to certification. Speaking of certification, Deming noted, "You don't have to do this — survival is not compulsory!"

The traditional two-party quality audit system relies on the buyer–seller relationship, where the buyer (customer) "audits" the supplier. This puts a burden on both parties. Imagine a supplier with a hundred or more customers, each with its own specific requirements. From a customer's point of view, it would be beneficial if all suppliers could be judged by a single set of criteria.

The third-party audit places great importance on quality systems, a critical factor in the EC. The independent third-party *registrar* certifies that the quality system meets the requirements of ISO 9000.

What is the rationale for a third-party audit? Financial results are measured by financial statements, while product and service outputs are measured by quality. If the impartial third-party audit is required for financial systems, why not a similar check on quality systems? This is particularly important in helping to guarantee quality across international borders.

DOCUMENTATION

There are three basic steps to the registration process:

1. Appraisal of the organization's quality manual
2. Evaluation of conformance to documented procedures
3. Presentation of findings, with recommendations for corrective action

A great deal of *documentation* is required. The justification is reflected in a management axiom: "If you haven't written it out, you haven't thought it out." Moreover, as people come and go, change jobs, and forget a procedure, documentation ensures that a record is maintained for continuity. The simple rule is that if all personnel involved in a given system or procedure were replaced, the new people could continue making the product at the same quality level.

The amount of documentation depends on the nature and complexity of the business. A hierarchical approach involving three levels is generally acceptable:

- ***Level 1*** — An overview type of quality manual consisting of policies that meet the requirements of the ISO standard for which certification is sought
- ***Level 2*** — Functional or departmental operating procedures in terms of "who does what"
- ***Level 3*** — Work instructions that explain how each task is to be accomplished

The criteria for approval are simple: "Can you say what you do and do what you say?" Questions such as the following may be asked: Is the process control system adequate for your needs? Is it understood by those who run the process? Are they properly trained to operate the process? Is the documentation up to date? Do you have an internal audit system that regularly assesses whether the control system is functioning as it should be?

POST-CERTIFICATION

The third-party audit and subsequent certification, if achieved, should be viewed as a means, not an end to be achieved. The importance of preparation for certification lies not so much in the certification itself but in the quality system that results from the effort leading to it.

The customer is the ultimate beneficiary of the quality system, and any effort to obtain ISO 9000 certification without customer communication can be a waste of time and a compromise of any system that may result.

Certification is a beginning, not an end. Continuous evaluation, feedback, and fine-tuning are suggested. Who will perform this internal and continuing "audit" following certification? The responsibility, of course, is top management's. The role of the internal auditor, if any, is not clear. Should the role include getting ready for certification or maintaining post-certification requirements, or both?[18] The role is not clearly assigned and may represent an opportunity for internal auditors.

CHOOSING AN ACCREDITED REGISTRATION SERVICE

Quality managers who decide to implement an ISO 9000 system are confronted by two related issues: how best to implement the new system and how to ensure that certification will be recognized by customers. This latter issue will normally be settled if certification is recognized by legitimate accreditation bodies.

U.S. firms located in Europe normally utilize one of the many accrediting bodies in those countries. Many are government sanctioned, such as Raad voor de Certificatie (RvC) in the Netherlands and the National Accreditation Council for Certification Bodies (NACCB) in the United Kingdom. IBM's Application Business Systems Division was the first U.S.-based firm to be certified in all of its business lines. Certification was gained after an audit by Bureau Veritas Quality International.[19]

No single firmly established registrar-accredited authority is recognized in the U.S., and confusion exists as to which auditors are accredited by whom. Two non-governmental groups — the Registrar Accreditation Board (RAB) (an offshoot of the ASQC) and ANSI — have carried out a joint effort to develop accreditation requirements for ISO 9000 auditing companies operating in the U.S.[20] The ANSI/RAB accreditation program is the best source of credible U.S.-based registrars.

A number of criteria should affect the decision on the choice of a registrar, including the registrar's knowledge in a specific industry and in the auditing of quality systems, how many similar firms it has registered, its turnaround time for audit results, its re-audit schedule (which should complement the business cycle of the firm), and, most importantly, accreditation.

As a general rule, it is probably not wise to shop around for the lowest price, because the cost of an audit is small compared to the overall cost of the registration effort.

ISO 9000 AND SERVICES

The standards apply not only to the manufacturing process but to after-sale service and to service departments, such as design, within the manufacturing firm as well. Additionally, the standards translate to the service sector: They specifically address quality systems for service as well as production. Indeed, ISO 9000-2, a separate guideline, was issued to explain ISO criteria in terms of selected service industries.

In the United Kingdom, standards are being used by educational institutions, banks, legal and architectural firms, and even trash collectors. At London's Heathrow Airport, British Airways PLC adopted ISO standards to reduce complaints of lost cargo and damaged goods. In the U.S., a growing number of transportation companies will not transport hazardous material unless the shipper is ISO certified.

There is some evidence that the ISO 9000 Series is receiving more interest from service organizations in the U.S. than in Europe. Service firms in consulting, purchasing, and materials management are expressing interest. It is believed by some that the greater interest by U.S. service firms is based on strategic considerations, as ISO 9000 is perceived as a "market differentiator."[21]

THE COST OF CERTIFICATION

A frequently asked question is: "How much does certification cost?" This is a legitimate concern, although the question may be accompanied by another one: "What is the payoff?"

There is no set answer to how much it costs and how long it takes. Each company is different. The answer depends on such factors as company size, product line, how far along a company's existing systems are on the quality continuum, whether consultants are used, and the implementation strategy adopted. It can cost a small company $2000 to $25,000 in consulting fees for advice on developing a quality system.[22] Employee time in creating the system is additional and can be the largest cost.

The major determinant is a firm's starting position. If a company has just won a Baldrige Award, registration of a plant or business might take just a few days. However, if the system must be created from the ground up, it can take a year and cost $100,000 or more.[23]

ISO 9000 VS. THE BALDRIGE AWARD

The ASQC reports that one of the most frequently asked questions regarding the ISO Series is: "Aren't the Baldrige Award, the Deming Prize, etc., equivalent or better 'standards' than the ISO Series?" The answer, replies ASQC, is quite simple: "You can't hope to meet the expectations of any of these programs if you aren't already implementing the ISO 9000 (ANSI/ASQC Q-90) standards in your company. These standards provide the foundation on which you can build your quality management and quality assurance systems so you may ultimately achieve a high level of success. Moreover, the ISO 9000 Series is the only system accepted internationally."

The Baldrige is a much more comprehensive program than ISO 9000. It is truly a TQM system, whereas the ISO Series is much more limited in scope. It is a basic standard, a minimal requirement, and can be worth about 200 to 300 points in the Baldrige program. For example, it does not address the human resource dimension, as does the Baldrige. On the other hand, a company implementing the Baldrige criteria is in a much better position to implement the ISO standards.

The Baldrige criteria are much more specific. The guidelines spell out what is expected in detailed language. In contrast, the ISO Series is designed to be inclusive, not exclusive. It does not mandate that one approach be used over another. As long as you can say what you do and do what you say, you can get your system registered. This generic nature of the standards can be a source of frustration as well as liberation.

For those companies whose quality systems are on the low end of a TQM continuum, ISO may be a starting place on the road to eventually achieving a TQM system. Certification also has the advantage of putting an organization on a level playing field with the competition worldwide.

IMPLEMENTING THE SYSTEM

Although the series provides guidance on the required attributes of the quality system, the standards do not spell out the means of implementation. Once a decision is made to adopt the standards and seek certification, the following major steps will facilitate successful change:

1. Recognize the need for change and get the commitment of top management.
2. Incorporate quality in the strategic plan as the linchpin of differentiation.
3. Formulate and adopt a holistic quality policy statement adapted to ISO requirements. Get support and commitment from all managers.
4. Determine the scope of the business to be certified. Will it be a particular process, related facilities, a geographical site, or the whole company?
5. Determine the status of the current quality system through an internal audit. Define the *gap* between where you are and what it will take to close the gap.
6. Estimate the cost in time and money, and implement the plan by organizing the necessary action steps.

FINAL COMMENTS

The question becomes, "Do we implement TQM or register for ISO 9000?" These two are not independent goals. They are intertwined and complement one another. TQM is a philosophy, and ISO 9000 can be used as a structural framework for implementing a company's TQM philosophy. ISO 9000 needs TQM. ISO certification does not guarantee high-quality products and services, but integration with TQM does. The real benefit of the emergence of ISO 9000 as the international standard is that in order to compete in the international marketplace, top management now must be committed to quality.

EXERCISES

20-1 Why is it important for U.S. firms to comply with ISO 9000?

20-2 Compare the standards of ISO 9000 with those of the Baldrige Award.

20-3 Does ISO 9000 contain product standards or standards for operation of a quality management system? Explain the difference.

20-4 Answer the criticism that meeting ISO standards will add to production costs.

20-5 What are the five sets of standards? Summarize each.

20-6 What are the benefits of ISO 9000 certification?

ENDNOTES

1. Gary Spizizen, "The ISO 9000 Standards: Creating a Level Playing Field for International Quality," *National Productivity Review,* Summer 1992, p. 332. This is an excellent summary of the provisions of ISO 9000.
2. The ANSI/ASQC Series is available from ASQC headquarters through the customer service department (phone 800-248-1946). The ISO 9000 Series is available from ANSI (phone 212-642-4900). Keep in mind that the ANSI/ASQC Q-90 Series is *identical* to the ISO 9000 Series.
3. Suzan L. Jackson, "What You Should Know about ISO 9000," *Training,* May 1992, p. 48. This is a good primer on ISO standards. ISO 9000 was adopted by the EC in 1990 as a global standard of quality. Its stringent requirements ensure that products manufactured along ISO 9000 are world class. See Jack Cella, "ISO 9000 Is the Key to International Business," *Journal of Commerce and Commercial,* Jan. 25, 1993, p. 88. Even China has moved toward adoption of the standards. The Shanghai-Foxboro Company Ltd., an affiliate of the U.S.-based Foxboro Company, became the first company in China to attain ISO 9000 certification.
4. "Want EC Business? You Have Two Choices," *Business Week,* Oct. 19, 1992, p. 58. In the U.K., where the standards have become most widely embraced, over 80% of large employers with payrolls over 1000 are registered. See Kymberly K. Hockman and David A. Erdman, "Gearing Up for ISO 9000 Registration," *Chemical Engineering,* April 1993, p. 128.
5. Donald W. Marquardt, "ISO 9000: A Universal Standard of Quality," *Management Review,* Jan. 1992, p. 50.
6. "U.S. Firms Lag in Meeting Global Quality Standards," *Marketing News,* Feb. 15, 1993.

7. American Society for Quality Control, "ISO 9000," a brochure prepared by the Standards Development Department of ASQC, P.O. Box 3005, Milwaukee, WI 53201 (phone 414-272-8575).
8. Milton G. Allimadi, "New Quality Standards Draw Fire from US Group," *Journal of Commerce and Commercial,* Jan. 4, 1993, p. 4.
9. Mark Morrow, "International Agreements Increase Clout of ISO 9000," *Chemical Week,* April 7, 1993, p. 32. This article attributes the success of ISO 9000 to its brevity (20 pages) and its simplicity. Since its inception, over 30,000 companies have registered.
10. Marjorie Coeyman, "ISO 9000 Gaining Ground in Asia/Pacific," *Chemical Week,* April 28, 1993, p. 54. In some parts of the Asia/Pacific region, the ISO 9000 quality standards have almost become domestic standards.
11. In a 1993 conference of the National Society of Professional Engineers, the topic of compliance with the EC's ISO 9000 quality control standards was discussed. See Jane C. Edmunds, "Engineers Want Quality," *ENR,* Feb. 8, 1993, p. 15. For Power Transmission Distributors (The Association), see Beate Halligan, "ISO Standards Prepare You to Compete," *Industrial Distribution,* May 1992, p. 100. The concern of the public utilities industry is reported in Greg Hutchins, "ISO Offers a Global Mark of Excellence," *Public Utilities Fortnightly,* April 15, 1993, p. 35. The computer industry's concern is reflected in Gary H. Anthes, "ISO Standard Attracts U.S. Interest," *Computerworld,* April 26, 1993, p. 109.
12. "Support Group Formed for Companies Seeking ISO 9000," *Industrial Engineering,* March 1993, p. 8. The National ISO 9000 Support Group will provide information, support, advice, and training at low cost to any U.S. company interested in the ISO 9000 process. The goal of the group is to allow the free exchange of information and questions between companies seeking ISO registration.
13. Adapted from Kymberly K. Hockman and David A. Erdman, "Gearing Up for ISO 9000 Registration," *Chemical Engineering,* April 1993, p. 129.
14. Adapted from *Business Week,* Oct. 19, 1992 and ASQC *ANSI/ASQC Q-90-1987 — Quality Management and Quality Assurance Standards.*
15. Gary H. Anthes, "ISO Standard Attracts U.S. Interest," *Computerworld,* April 26, 1993, p. 109.
16. Donald W. Marquardt, "ISO 9000: A Universal Standard of Quality," *Management Review,* Jan. 1992, p. 52.
17. See Elisabeth Kirschner, "Nalco: Registration in Context," *Chemical Week,* April 28, 1993, p. 71. See also Marjorie Coeyman, "FMC: The Benefits of Documentation," *Chemical Week,* April 28, 1993, p. 69.
18. See Giovanni Grossi, "Quality Certifications," *Internal Auditor,* Oct. 1992, pp. 33–35 and Gary M. Stern, "Sailing to Europe: Can Auditing Play a Role in the New International Quality Standards?" *Internal Auditor,* Oct. 1992, pp. 29–33. Both of these authors, who are members of the internal auditing profession, argue for an expanded role for internal auditors in the certification and follow-on process.
19. "IBM, Help/Systems Receive ISO Certification," *Systems 3X-400,* Feb. 1993, p. 16. ABS won the Baldrige Quality Award in 1991.

20. Emily S. Plisher, "Seeking Recognition: U.S. Auditors Build Their Base," *Chemical Week,* Nov. 11, 1992, pp. 30–33. See also a special report entitled "Confusion Persists on Issue of Registrar Accreditation," *Chemical Week,* April 28, 1993, p. 42. As of April 1993, ANSI/RAB had accredited 27 quality system registrars. The "unofficial" list is contained in this endnote citation.
21. Gary Spizizen, "The ISO 9000 Standards: Creating a Level Playing Field for International Quality," *National Productivity Review,* Summer 1992, p. 335.
22. This is the estimate of OTS Registrars of Houston, an ISO 9000 registrar. "Small Companies Are Finding It Pays to Think Global; Firms Win New Business by Adopting International Quality Standards," *Wall Street Journal,* Nov. 19, 1992, Section B, p. 2.
23. Donald W. Marquardt, "ISO 9000: A Universal Standard of Quality," *Management Review,* Jan. 1992, p. 51. See also Ian Hendry, "ISO Standardizes Quality Efforts," *Pulp & Paper,* Jan. 1993, p. S4. Several of these firms report a cost of certification of about $112,000. Also, General Chemical's Green River plant achieved certification at an estimated cost of $150,000. Rick Mullin, "General Registers Green River Site: First ISO 9002 for Natural Soda Ash," *Chemical Week,* April 28, 1993, p. 59.

REFERENCES

Clements, R. B., *Quality Manager's Complete Guide to ISO 9000,* Englewood Cliffs, N.J.: Prentice-Hall, 1993.

Rabbitt, J. T. and P.A. Bergh, *The ISO 9000 Book,* New York: Quality Resources, 1993.

Reimann, Curt W. and Harry S. Hertz, "The Baldrige Award and ISO 9000 Registration Compared," *Journal for Quality and Participation,* Jan./Feb. 1996.

Scotto, Michael J., "Seven Ways to Make Money From ISO 9000," *Quality Progress,* June 1996, pp. 39–41.

Struebing, Laura, "9000 Standards?" *Quality Progress,* Jan. 1996, pp. 23–28.

Vloeberghs, Daniel and Jan Bellens, "Implementing the ISO 9000 Standards," *Quality Progress,* June 1996, pp. 43–48.

Wilson, Lawrence A., "Eight-Step Process to Successful ISO 9000 Implementation: A Quality Management System Approach," *Quality Progress,* Jan. 1996, pp. 37–40.

Zuckerman, Amy, "ISO 9000 Skepticism," *Industry Week.* Vol. 243 No. 13, July 4, 1994, pp. 43–44.

21

THE BALDRIGE AWARD

Quality has come a long way since renewed interest began in the middle and late 1970s. Prior to that time, many people considered the emphasis on quality as just one more passing phase in a string of business fads — value analysis...management by objective...Theory X and Y...portfolio management...and so on. The impact of the Baldrige Award has laid to rest any notion that quality is not here to stay.

There is little doubt that quality will continue to be the major competitive issue in industry and beyond. Increasing global competition and customer sensitivity have given quality increasing visibility. An additional impetus was provided when Congress established the Baldrige Award in 1987 as a result of Public Law 100-107. Background information on the law mentions foreign competition as the major rationale. No other business prize or development in management theory can match its impact. As evidence of this impact, over 20 states are working to develop regional quality programs.[1]

The award has set a national standard for quality, and hundreds of major corporations use the criteria in the application form as a basic management guide for quality improvement programs. Although the award has its detractors,[2] it has effectively created a new set of standards — a benchmark for quality in U.S. industry.

Applicants must address seven specific categories. These categories of examination items and their respective point values are listed in Table 21-1. The Baldrige Award framework and the dynamic relationships among the criteria are shown in Figure 21-1.

Meeting the criteria is not an easy matter. A perfect score is 1000. The distribution of scores for the 203 applicants during the first three years (1988, 1989, 1990) is shown in Table 21-2. Of the 1203 applicants, only 9 were selected for the award.

Table 21-1 1997 Examination Items and Point Values[3]

Examination categories/items			*Point values*
1.0 Leadership			**110**
Senior leaders' personal leadership involvement in creating and sustaining values, company directions, performance expectations, customer focus, and a leadership system that promotes performance excellence. How the values and expectations are integrated into the company's leadership system, including how the company continuously learns and improves, and addresses its societal responsibilities and community involvement.			
1.1	Leadership System	80	
1.2	Company Responsibility and Citizenship	30	
2.0 Strategic Planning			80
How the company sets strategic directions, and how it determines key action plans. How the plans are translated into an effective performance management system.			
2.1	Strategy Development Process	40	
2.2	Company Strategy	40	
3.0 Customer and Market Focus			80
How the company determines requirements and expectations of customers and markets. How the company enhances relationships with customers and determines their satisfaction.			
3.1	Customer and Market Knowledge	40	
3.2	Customer Satisfaction and Relationship Enhancement	40	
4.0 Information and Analysis			**80**
The management and effectiveness of the use of data and information to support key company processes and the company's performance management system.			
4.1	Selection and Use of Information and Data	25	
4.2	Selection and Use of Comparative Information and Data	15	
4.3	Analysis and Review of Company Performance	40	
5.0 Human Resource Development and Management			**100**
How the work force is enabled to develop and utilize its full potential, aligned with the company's objectives. The company's efforts to build and maintain an environment conducive to performance excellence, full participation, and personal and organizational growth.			
5.1	Work Systems	40	
5.2	Employee Education, Training, and Development	30	
5.3	Employee Well-Being and Satisfaction	30	
6.0 Process Management			**100**
The key aspects of process management, including customer-focused design, product and service delivery processes, support processes, and supplier and partnering processes involving all work units. How key processes are designed, effectively managed, and improved to achieve better performance.			
6.1	Management of Product and Service Processes	60	
6.2	Management of Support Processes	20	
6.3	Management of Supplier and Partnering Processes	20	

(continued)

Table 21-1 1997 Examination Items and Point Values (continued)

Examination categories/items			*Point values*
7.0 Business Results			**450**
The company's performance and improvement in key business areas — customer satisfaction, financial and marketplace performance, human resources, supplier and partner performance, and operational performance. Also examined are performance levels relative to competition.			
7.1	Customer Satisfaction Results	130	
7.2	Financial and Market Results	130	
7.3	Human Resource Results	35	
7.4	Supplier and Partner Results	25	
7.5	Company-Specific Results	130	
Total Points			**1000**

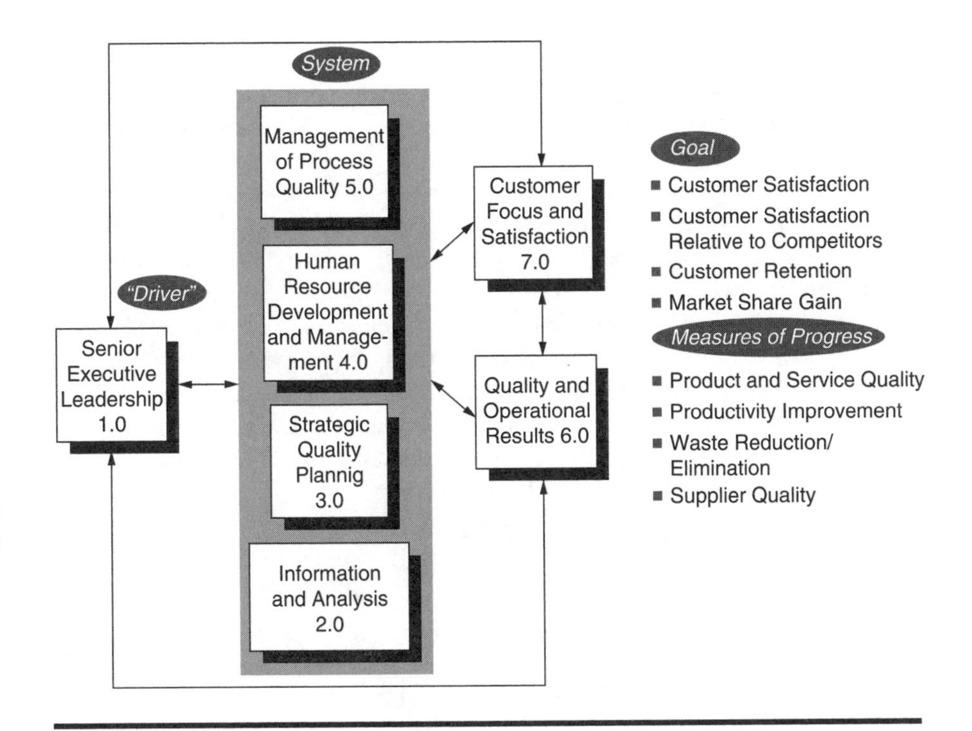

Figure 21-1 Baldrige Award Criteria Framework: Dynamic Relationships

An indication of the interest in the Baldrige is the number of application guidelines (167,000 in 1990) requested. In the first three years, 203 companies applied and 9 won: 6 manufacturers, 2 small companies, and 1 service company (Federal Express). Winners of the award are required to share their successful strategies with other companies. IBM's Rochester,

Table 21-2 Distribution of Scores

Scoring range	*Number of applications* *1988*	*1989*	*1990*
0–125	0	0	0
126–250	0	1	7
251–400	1	8	18
401–600	31	15	51
601–750	23	12	19
751–875	11	4	2
876–1000	0	0	0
Total	66	40	97

Minnesota site, home of the Applications System/400 and a 1990 winner, attributes the success of the division to the way in which it appropriated the ideas of Motorola, Xerox, and Milliken, winners in prior years. This sharing of ideas is a central purpose of the National Institute of Standards and Technology (NIST), the administering agency.[4] The sharing policy by winners ensures a multiplier effect.

Another indication of the award's leverage is the stringent criteria related to quality assurance for products and services purchased by external providers (suppliers) of goods and services. It is clear that suppliers are a critical link in the chain of processes that constitute total quality management. As a result, many companies require their suppliers to apply for the Baldrige. For example, Motorola and Westinghouse, two winners, will not do business with a supplier that has not applied for the award and does not use its criteria. Another winner, Globe Metallurgical, is certified as a supplier by Ford. Globe in turn requires certification by its suppliers. Thus, the number of firms using the Baldrige criteria may grow geometrically as first-tier suppliers certify second-tier suppliers and so on.

Hewlett-Packard, IBM, Motorola, Westinghouse, and 3M are among the many companies that use the application as a guide for managers and a checklist for internal quality standards:

> But basically, the Baldrige criteria will be the way we judge our own operations from now on. The reason is simple: The Baldrige Award process is a basic blueprint on how to do the quality process.[5]
>
> *3M*

> Competing for the award motivated people to a level I didn't think possible.[6]
>
> **General Manager**
> *GM Cadillac Division*

> The National Quality Award process enabled the company to look at itself through the eyes of the customer, and every aspect of the business came under scrutiny.[7]
>
> *Xerox*

> Managers of IBM's Santa Teresa, California lab are required to score their operations every 90 days using the criteria.

The winners for the five-year period since the beginning of the award in 1988 are shown in Table 21-3.[8]

Table 21-3 Baldrige Award Winners: 1988 to 1996

Year	*Award winner*
1988	Motorola
	Westinghouse Commercial Nuclear Fuel Division
	Globe Metellurgical, Inc.
1989	Milliken & Company
	Xerox Business Products and Systems
1990	Cadillac Motor Car Company
	IBM Rochester
1991	Selectron Corporation
	Zytec Corporation
	Marlow Industries
1992	AT&T Network Systems Group-Transmission Systems Business Unit
	Texas Instruments, Inc. Defense Systems & Electronics Group
	AT&T Universal Card Services
	The Ritz-Carlton Hotel Company
	Granite Rock Company
1993	Ames Rubber Company
	Eastman Chemical Company
	Carolina Eastman
1994	AT&T Communication Services
	GTE Directories Corporation
	Wainwright Industries
1995	Armstrong World Industries Building Products Operations
	Corning Incorporated Telecommunications Products Division
1996	Adac Laboratories
	Custom Research
	Dana Commercial Credit
	Trident Precision Manufacturing

ENDNOTES

1. Curt W. Reimann, "America Unites Behind the Baldrige Quality Crusade," *Electronic Business,* Oct. 15, 1990, p. 63. Reimann was director of the Malcolm Baldrige National Quality Award and associate director for quality programs at the National Institute of Standards and Technology, the agency that administers the Baldrige Award program. A good summary of what it takes to compete for the Baldrige is contained in Curt Reimann, "Winning Strategies for Quality Improvement," *Business America,* March 25, 1991, pp. 8–11. See also "A Standard for All Seasons," *Executive Excellence,* March 1991, p. 9 and "The Baldrige Award: Leading the Way to Quality," *Quality Progress,* July 1989, pp. 35–39.
2. See Jeremy Main, "Is the Baldrige Overblown?" *Fortune,* July 1, 1991, pp. 62–65. Philip Crosby, of *Quality Is Free* fame, scorns the paperwork, thinks that customers rather than the company applying should do the nominating, and deplores the lack of financial measures. Tom Peters, co-author of *In Search of Excellence,* complains that the criteria are "strangely silent on the subject of bureaucracy." There was also a bit of sour grapes when Cadillac won the award in 1990.
3. A more detailed description is contained in the 1993 Award Criteria and the 1993 Application Forms and Instructions. These two documents can be obtained from Malcolm Baldrige National Quality Award, National Institute of Standards and Technology, Route 270 and Quince Orchard Road, Administration Building, Room A537, Gaithersburg, MD 20899 (phone 301-975-2036).
4. Michael Fitzgerald, "Quality: Take It to the Limit," *Computerworld,* Feb. 11, 1991, pp. 71–78.
5. Remarks of A. G. Jacobson in a presentation to the Conference Board Quality Conference, April 2, 1990.
6. Jeremy Main, "Is the Baldrige Overblown?" *Fortune,* July 1, 1991, p. 63.
7. Company brochure entitled "The Xerox Quest for Quality and the National Quality Award."
8. Texas Instruments is one of the several Baldrige winners that attribute their turnaround to the adoption of the principles of TQM. See *Fortune,* Nov. 30, 1992, pp. 80–83. In October 1992, the Ritz-Carlton Hotel Company became the first hotel company to win the Baldrige Award. Its approach to quality relies on traditional TQM principles. Edward Watkins, "How Ritz-Carlton Won the Baldrige Award," *Lodging Hospitality,* Nov. 1992, pp. 22–24. In 1990, AT&T chairman and CEO Robert Allen created the Chairman's Quality Award, the criteria and examination process for which were taken from the Malcolm Baldrige Award. See Rick Whiting, "AT&T Started a Quality Bonfire to Learn How to Put it Out," *Electronic Business,* Oct. 1992, pp. 95–103.

22

QS-9000

HISTORICAL PERSPECTIVE

Until the early 1990s, suppliers to the automotive industry were required to adhere to an excess of customer-specific standards and requirements when dealing with the "Big Three" automakers (Chrysler, Ford, and General Motors). Each automaker had its own detailed set of standards and requirements, which increased the cost of making parts and made the suppliers less efficient and therefore less competitive. To deal with this problem, the Supplier Quality Requirements Task Force was formed. In September 1994, the task force released the Quality System Requirements: QS-9000 and its companion document, Quality System Assessment (QSA).

BASIC QS-9000

QS-9000 is a harmonization of DaimlerChrysler's *Supplier Quality Assurance Manual,* Ford's *Q-101 Quality System Standard,* and General Motors' NAO *Targets for Excellence,* with input from the following truck manufacturers: Freightliner Mack Trucks, Navistar International, PACCAR, and Volvo GM Heavy Truck. QS-9000 was created to improve product quality and company productivity by effectively reducing waste, variation, and defects in the automotive industry's products, just as the ISO 9000 series of standards did for general industrial operations. In fact, QS-9000 is based upon the 1994 revision of ISO 9001 and includes the text of the ISO 9001 standard. Over 100 additional auditable requirements were added to make the types of applications required by the Big Three more specific. The goal of QS-9000, as stated in its introduction, is "the development of fundamental quality systems that provide for continuous improvement, emphasizing defect prevention and the reduction of variation and waste in the supply chain."

STRUCTURE OF QS-9000

The document consists of three sections:

- Section 1: ISO 9000-Based Requirements
- Section 2: Sector-Specific Requirements
- Section 3: Customer-Specific Requirements

Section 1 contains the main part of QS-9000. This section includes all 20 elements of the old ISO 9001:1994 Section 4 and three supplemental automotive requirements. Each element is exactly the same as in the original ISO 9001 document (ISO 9001 elements are italicized). Within each italicized section, additional interpretation information and supplemental quality system requirements that have been harmonized by the Big Three are printed in non-italic type.

Section 2 contains requirements that go beyond ISO 9000 and address three areas within the automotive and truck industry:

- Production part approval process
- Continuous improvement
- Manufacturing capabilities

These programs are already in place within the automotive and truck industry.

Section 3 contains miscellaneous and detailed requirements that Ford, General Motors, and DaimlerChrysler have stipulated for their suppliers. These requirements cater only to unique, specific product needs of the Big Three. For example, GM has specific procedures for prototype material that Ford and DaimlerChrysler do not.

Compliance with QS-9000 is mandatory for tier-one suppliers to the Big Three and other OEM (original equipment manufacturer) customers subscribing to QS-9000. It should be noted that QS-9000 applies only to suppliers of production materials, production and service parts, heat treating, painting and plating, and other finishing services. In other words, all worldwide suppliers of these services, whether external or internal, must comply with QS-9000, but all tier-one suppliers are not bound by QS-9000.

DOCUMENT CONTROL AND REGISTRATION

As part of the ISO 9001 requirements embedded in QS-9000, a quality manual is required from the supplier. This manual must outline the structure of the documentation used. Also, the intent of defining policies, responsibilities, methods, and record requirements must be satisfied. Furthermore, QS-9000 calls for improved contract reviews and requires a system to review drawings before accepting purchase orders. Production part approval process control is also stressed, since parts must be approved before shipping.

The registration process consists of several steps. First, the supplier must conduct a self-assessment of its own system. Then, a second-party audit is performed. In the case of GM and Chrysler, third-party registration by a qualified registrar that belongs to a recognized accreditation body is required. Current GM suppliers are required to obtain QS-9000 registration by December 31, 1997. New suppliers to GM were required to obtain registration by January 1, 1996. All current and new suppliers to DaimlerChrysler need third-party registration. Ford suppliers are required to demonstrate compliance with QS-9000 and be registered by a third party.

SUMMARY

QS-9000 requires a systematic managerial strategy and is structured to enhance a company's quality system. It also requires that a company maintain both short- and long-term business plans, perform feasibility studies, maintain control plans from part prototype all the way through to final part production, and observe strict process control requirements.

EXERCISES

22-1 Who make up the Big Three?
22-2 How did QS-9000 come about?
22-3 What is the goal of QS-9000?
22-4 What is the QS-9000 registration process?

REFERENCES

Avery, Susan, "ISO 9000: The Auto Impact," *Purchasing,* Jan. 12, 1995. pp. 68–72.

Bandyopadhyay, Jayanta K., *QS-9000 Handbook: A Guide to Registration and Audit,* Delray Beach, Fla.: St. Lucie Press, 1996.

Harrison, Gerald, "A Report from the Automotive Trenches," *Quality Progress,* June 1996, pp. 35–37.

Hughey, Dennis and Marek Piatkowski, "What Is QS-9000?" *Plant Engineering and Maintenance,* Vol. 19 No. 2, April 1996, pp. 19–21.

Larson, Melissa, "Document Control Will Get You Certified," *Quality,* Nov. 1995.

Larson, Melissa, "QS-9000: Not Firing on All Cylinders," *Quality,* Sep. 1996.

Lovitt, Mike, "Continuous Improvement Through the QS-9000 Road Map," *Quality Progress,* Feb. 1996, pp. 39–43.

Marquedant, Stephen, "QS-9000: Quality Regs for Suppliers," *Quality,* March 1995, p. 26.
"QS-9000: Chrysler, Ford & General Motors' New Quality Standard — An Executive Overview," Southfield, Mich.: Perry Johnson, Inc., 1995.
Quality System Requirements: QS-9000, Chrysler Corp., Ford Motor Co., and General Motors Corp., Aug. 1994.
Scrimshire, David, "QS-9000 from the Big Three," *Modern Casting,* Oct. 1994, pp. 38–39.
Sorge, Marjorie, "Big Three Find Common Bond: QS-9000 Marries Automaker Quality Standards," *Ward's Auto World,* May 1995, pp. 44–46.
Stamatis, D. H., *Integrating QS-9000 with Your Automotive Quality System,* Milwaukee: ASQC Quality Press, 1995.
Struebing, Laura, "QS-9000 Standards?" *Quality Progress,* Jan. 1996. pp. 23–29.
Zuckerman, Amy, "QS-9000 Basics," *Quality Progress,* May 1996, p. 22.
Zuckerman, Amy, "Registration Changes Loom in QS-9000's Future," *Quality Progress,* May 1996, p. 21.

23

ISO 14000

In 1993, the International Organization for Standardization (ISO) established Technical Committee 207 to develop new standards for environmental management. The standards being developed are known as the ISO 14000 series and are patterned after the British Standard BS7750. This series is a collection of voluntary standards that have been developed to assist organizations in achieving environmental and economic gains through the implementation of effective environmental management systems. The standards provide a means of documenting an organization's ability to manage its environmental affairs. The overall goal of ISO 14000 is to establish an objective and verifiable system of environmental management.

COMPONENTS OF ISO 14000

The ISO 14000 series covers a wide variety of environmental disciplines, ranging from basic management systems to auditing, labeling, and product standards. There are many standards in the series. All but one (ISO 14001) are guidance documents. ISO 14001 is the standard against which companies will be certified. The standards are:

ISO 14000 Environmental Management Systems. General guidelines on principles, systems, and supporting techniques

ISO 14001 Environmental Management Systems. Specification with guidance for use

ISO 14004 Environmental Management Systems. General guidelines on principles, systems, and supporting techniques

ISO 14010 Guidelines for Environmental Auditing. General principles of environmental auditing

ISO 14011/1 Guidelines for Environmental Auditing. Audit procedures — Part 1: Auditing of Environmental Management Systems

ISO 14012 Guidelines for Environmental Auditing. Qualification criteria for environmental auditors

ISO 14013 Guidelines for Environmental Auditing. Management of environmental management system audit programs

ISO 14014 Guidelines for Initial Environmental Reviews

ISO 14015 Guidelines for Environmental Site Assessments

ISO 14020 Environmental Labeling. Principles of all environmental labeling

ISO 14021 Environmental Labeling. Self-declaration, environmental claims — terms and definitions

ISO 14022 Environmental Labeling. Symbols

ISO 14023 Environmental Labeling. Testing and verification methodologies

ISO 14024 Environmental Labeling. Practitioner programs, guiding principles, practices, and certification procedures of multiple criteria

ISO 14030 Environmental Performance Evaluation

ISO 14031 Evaluation of the environmental performance of the management system and its relationship to the environment

ISO 14040 Environmental Management — life cycle assessment — principles and guidelines

ISO 14041 Environmental Management — life cycle assessment — goal definition/scope and inventory analysis

ISO 14042 Environmental Management — life cycle assessment. Impact assessment

ISO 14043 Environmental Management — life cycle assessment. Improvement assessment (or evaluation and interpretation)

ISO 14050 Terms and Definitions

ISO 14060 Guide for the inclusion of environmental aspects in product standards

The ISO 14000 series is composed of five major components:

1. Environmental Management Systems
2. Environmental Auditing
3. Environmental Performance Evaluation
4. Environmental Labeling
5. Life Cycle Assessment

The Environmental Management System (documents 14001 and 14004) was published in 1996. This standard provides the "core" requirements for developing and implementing an environmental management system that can be certified or registered by a third party. The Environmental Auditing standards (documents 14010 through 14015) provide requirements for general principles of environmental auditing, guidelines for auditing environmental management systems, and qualification criteria for environmental auditors. The Environmental Labeling standards (documents 14020 through 14024) provide harmonization of the criteria to determine which products will be able to use the three types of ecolabels. The first type of label is a "seal of approval" for products that meet specified requirements within a product class. The second type of label is a single-claim label for such things as recycled content, energy efficiency, etc. The third type is an "environmental report card" that uses a life cycle approach and allows comparison of the environmental effects of the manufacturing and use of products. The Environmental Performance Evaluation (documents 14030 and 14031) is scheduled for publication in 1997. These evaluations are a means to measure, analyze, assess, and describe an organization's environmental performance against agreed-upon criteria for management purposes. The publication date for the Life Cycle Assessment guidance (documents 14040 through 14043) has not been set. Life Cycle Assessment is a tool for evaluating the environmental attributes associated with a product, process, or service.

ISO 14001

As previously stated, ISO 14001 is the actual document that is used for implementation and registration. The structure of ISO 14001 is very similar to the structure of ISO 9001. There are five major elements of the ISO 14001 standard:

4.1 Environmental Policy
4.2 Planning
  2.1 Environmental aspects
  2.2 Legal and other requirements
  2.3 Objectives and targets
  2.4 Environmental management program(s)
4.3 Implementation and Operation
  3.1 Structure and responsibility
  3.2 Training, awareness, and competence
  3.3 Communication
  3.4 Environmental management system documentation
  3.5 Document control
  3.6 Operational control
  3.7 Emergency preparedness and response
4.4 Checking and Corrective Action
  4.1 Monitoring and measurement
  4.2 Non-conformance and corrective and preventive action

4.3 Records
4.4 Environmental management system audit
4.5 Management Review

Like ISO 9001, ISO 14001 views the document system as a condition to achieve the main objective of the standard, which is an effectively functioning environmental management system.

REGISTRATION

In order to comply with ISO 14000, an organization will have to:

1. Create an environmental management system
2. Demonstrate that its procedures comply with relevant regulations and laws
3. Demonstrate a commitment to continuous improvement and pollution prevention

There are several general steps that can be followed to prepare for eventual ISO 14000 certification:

1. Establish management commitment
2. Gather environmental impact data
3. Develop an environmental policy
4. Evaluate the organization's existing environmental management system against ISO 14000
5. Develop and implement an action plan
6. Perform internal audits against ISO 14001 requirements
7. Correct any deficiencies
8. Pursue ISO 14001 certification

These are general guidelines that can be adapted to most organizations.

BENEFITS

Implementation of ISO 14000 can produce many benefits. The main benefit for an organization and the environment as a whole is the reduction of pollution and the increased efficiency of resources. Other benefits include:

- Reduced exposure to liability
- Improved compliance with regulatory requirements
- Improved public and community relations
- Better management of resources such as electricity, water, and gas
- Reduced insurance premiums

It is important to note that the overall goal of ISO 14000 is to establish an objective and verifiable system of environmental management. Therefore, ISO 14000 is only a management system. It does not replace current environmental performance regulations, codes, etc. What it does do is provide a means for tracking, managing, and improving performance in reference to those regulations and codes.

EXERCISES

23-1 What does the ISO 14000 series represent?

23-2 What are the main components of the ISO 14000 series?

23-3 How does a company prepare for ISO 14000 certification?

23-4 Discuss the advantages and disadvantages of a company being ISO 14000 certified.

REFERENCES

Alexander, Forsyth, "ISO 14000: What Does it Mean for IE's?" *IIE Solutions,* Jan. 1996.

Aspan, Howard N. "Environmental Performance Evaluation: The ISO 14000 Scorecard," *Total Quality Environmental Management,* Vol. 5 No. 2, Winter 1995/96, pp. 101–106.

Diamond, Craig P., "Voluntary Environmental Management System Standards: Case Studies in Implementation," *Total Quality Environmental Management,* Vol. 5 Issue 2, Winter 1995/96, pp. 9–23..

"ISO 14001: A Critical View," *Environmental Manager,* Vol. 7 No. 10, May 1996, pp. 13–15.

Jump, Rodger A., "Implementing ISO 14000: Overcoming Barriers to Registration," *Total Quality Environmental Management,* Vol. 5 No. 1, Autumn 1995, pp. 9–14.

Pouliot, Chuck, "ISO 14000: Beyond Compliance to Competitiveness," *Manufacturing Engineering,* Vol. 116 No. 5, May 1996, pp. 51–56.

Sayre, Don, *Inside ISO 14000: The Competitive Advantage of Environmental Management,* Delray Beach, Fla.: St. Lucie Press, 1996.

Van Houten, Gerry, "The ISO Document Tidal Wave," *Records Management Quarterly,* April 1996.

"What Is ISO 14000," available at http://theinformedoutlook.com/main/iso14000/, Sep. 9, 1996.

Zuckerman, Amy, "Don't Rush into ISO 14000," *Machine Design,* Jan. 11, 1996.

24

ISO 9000: A PRACTICAL STEP-BY-STEP APPROACH*

Many publications address the philosophy of, and reasons and benefits for, becoming ISO 9000 registered. The reasons for obtaining registration are numerous, and there are tangible benefits. But what does an organization do once the scope of certification has been determined and the decision to proceed has been made?

Most companies that have been through the process recognize that obtaining ISO 9000 registration is a major accomplishment, and they are more than willing to share their experiences. A division of W.R. Grace & Co., Grace Specialty Polymers (GSP) is one of those companies, and the details of its registration efforts might inspire other companies to achieve ISO 9000 registration.

PREPARING FOR ISO 9000 REGISTRATION

In early 1993, GSP management established a company goal to become ISO 9000 registered by the end of 1994. It also decided to pursue a multisite certification of four separate locations to ISO 9001-1987: headquarters, the research and development (R&D) facility, and two manufacturing locations. All four were within 30 miles of one another, so travel barriers were minimal.

The company had several points in its favor when it began the registration process:

---

* By Roger S. Benson and Richard W. Sherman, *Quality Progress,* Oct. 1995, pp. 75–78. ©October 1995 American Society for Quality Control. Reprinted with permission.

- The vice president/general manager was a visionary who had recognized the benefits of total quality improvement (TQI) many years earlier.
- A strong TQI program was in place, providing a solid base for team activities and training.
- The company had a background in military standards (MIL-I-45208A) and had achieved automotive quality (Ford Q1) requirements.

ISO 9000 registration seemed to be a natural fit and an extension of existing activities. To begin, an executive steering committee was chartered, consisting of the general manager and employees who reported directly to him. As top-level management, they provided the commitment, direction, and resources that are an absolute must in achieving registration.

Next, an ISO Implementation Team (IIT) was chartered. This seven-member cross-functional team was made up of mostly department managers or supervisors from four local sites and three major departments, including R&D, manufacturing, and quality.

The TQI director was designated by the steering committee as the team leader. Since he was also a member of the steering committee, he acted as a liaison between the committee and the IIT, reporting progress and relaying important resource needs and decisions. Each member was directly responsible for coordinating ISO 9000 activities at his or her site.

An in-house lead assessor training program was arranged by pooling resources and requirements with other W.R. Grace & Co. business units. Each team member attended the program and received a certificate for successful completion.

Immediately after the completion of lead assessor training, the team began meeting on a regular basis. Weekly meetings were scheduled well in advance so team members could arrange their schedules to prevent conflicts.

The team leader prepared a road map to ISO 9000 registration that was based on former DuPont's stairstep approach and outlined the procedures GSP would follow to achieve registration by the end of 1994 (see Figure 24-1). This map was reviewed and approved by the IIT at the first meeting.

While team members were in lead assessor training, the steering committee attended a one-day management awareness training session that discussed basic ISO 9000 requirements and explained management responsibilities. Additionally, companywide awareness training was conducted at each site, using commercially available videotape presentations and customized information pertinent to that site. Each functional manager instructed his or her staff. This showed upper management's commitment and helped ensure buy-in to the program at all levels.

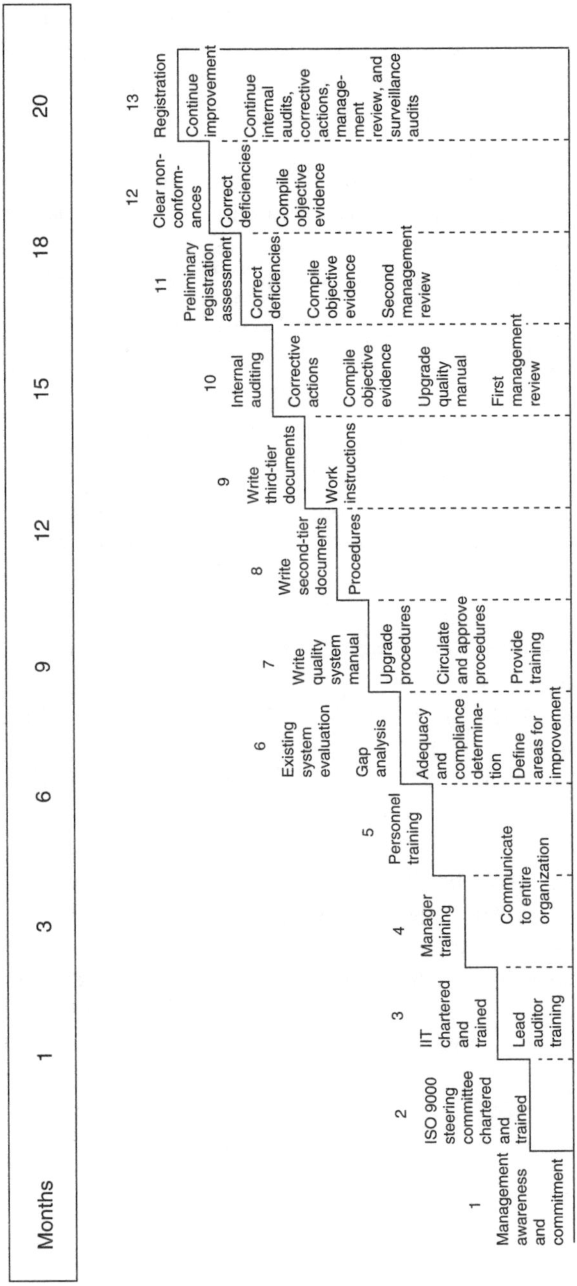

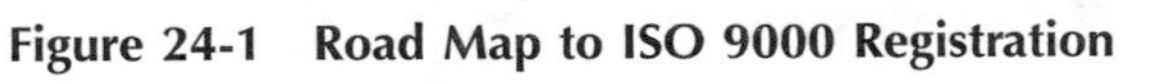

Figure 24-1 Road Map to ISO 9000 Registration

PRE-AUDIT CONDUCTED

Discussions with other registered companies indicated a lack of highly qualified consultants at that time. It wasn't uncommon for the client to know as much, or more, about the standard than the consultant. Based on this information, the cost, and GSP's collective quality systems experience, they decided to forgo consulting services and conduct pre-audit or gap analysis within the company.

The IIT was divided into three audit teams that systematically audited the sites to all sections of ISO 9001-1987. Noncompliance reports (NCRs) were written for all noncompliances, with the designation "major" or "minor." Audit teams presented their NCRs at weekly IIT meetings where team members had a chance to question and critique them. This helped ensure that all team members agreed with the noncompliance and understood its importance. Typically, the team also discussed possible solutions that could be used for corrective action.

The NCRs were used to drive corrective actions and push the overall quality system toward ISO 9000 compliance. For many of the minor NCRs, the responsibility for corrective action was assigned to the team member who had the greatest knowledge in that area. These NCRs usually required modifying existing procedures either to comply with the standard or reflect current practices.

After reviewing the NCRs that were assigned to them, the team members estimated time frames and the resources needed for completion. All of these estimates were then put into a master time line that was used to measure progress and estimate total project completion time.

Major noncompliances consisted of missing ISO 9000 elements and corrective action requirements that were beyond the ability of any single team member. Many of these cases had been documented but did not sufficiently meet ISO 9000 requirements. These corrective actions, which required a change in basic business practice or philosophy, had to be handled in a completely different manner.

For most major noncompliances, the team created subcommittees of two to four members who met outside the normal scheduled meetings. The subcommittees researched potential solutions and drew up a proposal listing possible solutions, projected implementation time, costs, benefits, advantages, and disadvantages. These proposals were then submitted to the appropriate functional manager.

The functional manager reviewed the various proposals and made a final decision based on the company's philosophy and availability of resources. Once a decision was made, it was reported to the IIT and resources were assigned. The IIT leader kept the steering committee informed of all proposals submitted.

From the beginning, benchmarking was used extensively to provide a baseline or reference point from which to begin the registration process and assist in setting direction. Other business units within W.R. Grace & Co. and suppliers and customers that had completed the registration process were contacted. The information obtained was very valuable and, in many cases, served as a model for documentation format and content.

DOCUMENTING CONTROL PROCEDURES AND WORK ACTIVITIES

Developing document control procedures was a major hurdle. Like most other companies, GSP settled on the three-tier approach (see Figure 24-2).

And, like most other companies, GSP struggled with the question of how many work activities needed to be documented, especially at level 3. Because GSP had only 225 employees and already had an existing quality manual, the team decided to centralize the control of all documents in the quality assurance department. One exception was made for design control. Product design was entirely under the auspices of R&D; therefore, product design created and controlled its own procedures, which eventually evolved into the new-product development manual.

The existing quality assurance manual served as a foundation from which a new manual consistent with ISO 9000 requirements was built. A working copy of the manual was issued to the IIT members. Team members reviewed, commented on, and further revised these procedures. The procedure review process was conducted during and outside regular team meetings. Ultimately, this process of writing, reviewing, and modifying procedures transformed the existing quality assurance manual into the new ISO 9000 manual.

R&D used monthly staff meetings and ad hoc meetings facilitated by the R&D ISO coordinator to develop and refine the design control process, as well as the supporting procedures.

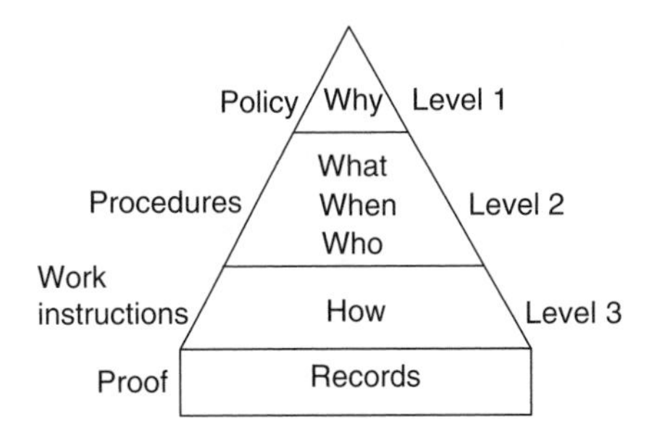

Figure 24-2 Quality System Document Structure

IMPLEMENTING A POLICIES AND PROCEDURES TRAINING PROGRAM

Upon completion of the new manual, a training program in policies and procedures — not a direct requirement of the standard — was created to ensure smooth implementation. This effort was facilitated by the creation of a training program development responsibility matrix that assigned generic training modules to IIT members. The modules were developed by the IIT members and distributed to training administrators as required. In addition, a representative from each site developed a training plan that designated who needed training and what level of detail was required. IIT members were prepared to administer the training modules to the employees in group meetings.

In addition to procedures training, training guidelines for each employee's job function were developed. Unlike the policy and procedures training, this training is a requirement of the standard under section 4.18. Resident site training coordinators and administrators were designated. Because training records had not been formally maintained in the past, a training system was created that included a training procedure, a training requirement matrix for each employee, request forms for training, and training certificates. These documents were used to establish and record minimum job requirements, request training, certify that training had been provided, or certify that the employee already met the minimum job requirements.

Based on benchmarking of several companies, employees who had been in the same job function at GSP for five years or more were considered to have met the minimum requirements of the job. Employees who had not been in the same job function at GSP for at least five years required evidence of certification, evidence that training was provided, or, at minimum, that a personal training plan was in place. A member of the steering committee administered a quality policy training module for all employees because the standard requires that all employees be aware of and understand their company's quality policy.

THE FINAL ASSESSMENT

The IIT leader was responsible for determining which registrar to use for the assessment. This was done by benchmarking other organizations and contacting different registrars directly.

The original road map had planned for a pre-assessment with additional time for corrective action before the final assessment. The company felt comfortable that its systems were strong, so it decided to forgo the pre-assessment and proceed directly to the final assessment. This approach had several advantages. It saved an appreciable amount of time in meeting the company's goal. It saved money; a pre-assessment by a third party is expensive. Further, the company knew that if it didn't pass the actual assessment, it would be given a certain amount of time (usually 40 days) to correct noncompliances. During this time the company could focus all its efforts on these issues since they would be the only areas reaudited to confirm compliance. The assessment took place over three days: one-half day each at R&D and headquarters, and one day at each of the two manufacturing sites. Two auditors conducted the assessment; one of the auditors was designated as lead auditor. Two guides — IIT members who were selected for their experience and knowledge of the areas being audited — were assigned to the auditors at each site to help facilitate the audit process. The quality assurance manager, who was one of the guides, accompanied the auditors throughout the entire assessment.

Having two guides accompany the auditors was efficient because it gave them the opportunity to immediately address noncompliances as they arose without interrupting the assessment. This approach proved effective at R&D, where one guide addressed design control issues while the other guide addressed calibration issues, correcting "observations" as they arose. (An observation is not considered a noncompliance, but it is an issue that could lead to a noncompliance and should be corrected in a timely manner.)

The IIT leader was one of the guides at headquarters. During the assessment of the other sites, he remained at headquarters, but attended the daily closing meetings to review the day's events at each site. From headquarters he was able to track the assessment's progress and relay important information ahead to the sites not yet audited. Additionally, he was able to plan and coordinate corrective action activities, addressing noncompliances as they were uncovered. Using this method, the company was able to solve 7 of 11 minor noncompliances before the end of the assessment.

After the assessment of the final site, GSP had no major noncompliances and only four minor noncompliances. As a result, on September 16, 1994, GSP was recommended for registration — a full three months ahead of schedule.

GSP'S RECOMMENDATIONS FOR ACHIEVING ISO 9000

Although there were not many ISO 9000 guidelines in place when GSP began the registration process, careful preparation, execution, documentation, and training enabled it to reach its goal. Other companies may wish to consider the following recommendations:

- Gain full support and commitment from the highest level of management. It must cascade down through the entire organization for successful implementation to take place.
- Form teams and committees to drive the ISO 9000 implementation process. Such teams include an executive steering committee to support and guide the process from the highest level; an implementation team to plan, assess, and execute actions as required; subcommittees to assess and provide recommendations to management; and corrective action teams to resolve specific noncompliances uncovered during internal assessments.
- Set a goal with milestones, conduct status reports, and pay close attention to timing.
- Enroll key individuals in a lead assessor's training program. This brings the knowledge and skills for ISO 9000 certification in-house, which may be more cost effective than hiring an outside consultant.
- Engage in benchmarking and networking activities. Typically, those who have been through the certification process are willing to share their experiences and might provide their policies and procedures for perusal.
- Attempt to capture what is being done today when writing procedures for an initial assessment. Make minor modifications and enhancements only where feasible and necessary.
- Structure quality manuals to follow the ISO 9000 format. A quality manual that is arranged according to standard format looks well planned and effective and is less confusing for the auditors. This helps the audit progress faster and more efficiently.
- Consider forgoing a formal pre-assessment. Pre-assessments, especially by a third party, can be costly and time consuming. Many minor noncompliances can be cleared as the regular assessment progresses, and efforts can be focused on areas that need attention.

Becoming ISO 9000 registered doesn't have to mean flying blind. These practical steps can make the process easier and can assist your company in reaching its ISO 9000 goals.

EXERCISES

24-1 Discuss the advantages and disadvantages of forgoing the preassessment audit.

24-2 Discuss the importance of having a commitment from the highest level of management.

24-3 Study Figure 24-1. Can you adopt this roadmap for your company? If yes, plan it out. If no, discuss difficulties.

IV

SPECIAL TOPICS IN QUALITY

In Part IV special topics are presented, dealing with the subjects of process capability, reliability, Six Sigma, and Health Care Service Excellence. Chapter 25 presents the concept of process capability. It examines how to measure process capabilities for both attribute and variable data. In Chapter 26, the reader will learn about the basics of reliability. Six Sigma is discussed in Chapter 27. Chapter 28 discusses how to achieve service excellence in the healthcare industry.

25

PROCESS CAPABILITY

INTRODUCTION

Thus far, we have focused our attention on determining whether a process is in statistical control or predictable. In this chapter, we deal with another important question — Is the in-control process capable of producing products that meet the specifications of the customers? After a process has been stabilized, the behavior of the process defines its capability. It is important to understand that a process in statistical control will not necessarily produce units that meet the specifications established by the customer. Any attempt to improve a process must be preceded by a process capability study. The following scenarios highlight the significance of process capability studies:

- Suppose a hospital that belongs to a major healthcare system receives a mandate from the corporate office to reduce the waiting time in the emergency room from its current level of 4 hours (for acuity level 3) to 2 hours. If the process is statistically in control, is the emergency room capable of "producing" to that specification?
- Suppose the customers of a major manufacturing company have entered into a contract with the manufacturer regarding the production of some specialty parts. This time, the customers have asked for part tolerances so fine that the company's machines may not be capable of producing to that level of precision.
- Suppose the management of a managed care company has decided to set higher standards with respect to telephone call abandonment. It currently has an abandonment rate of 17%, and the new mandate calls for 5%. It needs to determine whether it has the capability to meet the new specifications.

Determining the capability of a process aids companies in responding to customer requirements. Process capability studies can also help reduce variations in product characteristics, and therefore improve predictability.

Process capability is defined as a statistical measure of the inherent process variability for a given characteristics. In other words, process capability refers to the ability of a process to produce products that meet the specifications set by the customer or design engineer. There are two types of process capability studies — attribute process capability studies and variable process capability studies.

ATTRIBUTE PROCESS CAPABILITY

Attribute control charts remain the main approach for determining the process capability for attribute data. The following table shows the measures of process capability for attribute data:

Attribute control chart type	*Measure of process capability*	*Statistic for process capability*
P Chart	The average proportion defective produced by the process when it is operating in statistical control	$\bar{p}$
Np Chart	The average number defective units produced by a process for a given subgroup size when it is operating in statistical control	n $\bar{p}$
C Chart	The average number of defects per unit produced by a process (when it is operating in statistical control) when the area of opportunity is constant	$\bar{c}$
U Chart	The average number of defects per unit produced by a process (when it is operating in statistical control) when the area of opportunity changes	$\bar{u}$

For example, if $\bar{P}$ = 0.04; It means that (on average) 96% of the products produced by this process (when it is operating in statistical control) are acceptable.

VARIABLE PROCESS CAPABILITY

The main approach for studying process capability for variable data is variable control charts. The process capability is defined in terms of the process output or a specific product characteristic. The capability of a process is based on the performance of individual products or services against specifications. Organizations rely on sampling rather than measure every product they produce. The data gathered through sampling are then used to understand the behavior of the individual products generated by the process. It is important to understand the relationship between the individual product values and their subgroup averages. It is noteworthy that individual values spread out more widely than their averages — refer to Figure 25-1.

For the sake of computational convenience, assume that the process has a normal distribution; the standard deviation can then be estimated from either the standard deviation associated with the sample standard deviation (s) or the range (R):

$$\hat{\sigma} = \frac{\bar{s}}{c_4} \quad \text{or} \quad \hat{\sigma} = \frac{\bar{R}}{d_2}$$

Where

$\hat{\sigma}$ = estimate of population standard deviation

$\bar{s}$ = sample standard deviation calculated from process

$\bar{R}$ = average range of subgroups

c_4 = as found in Appendix A

d_2 = as found in Appendix A

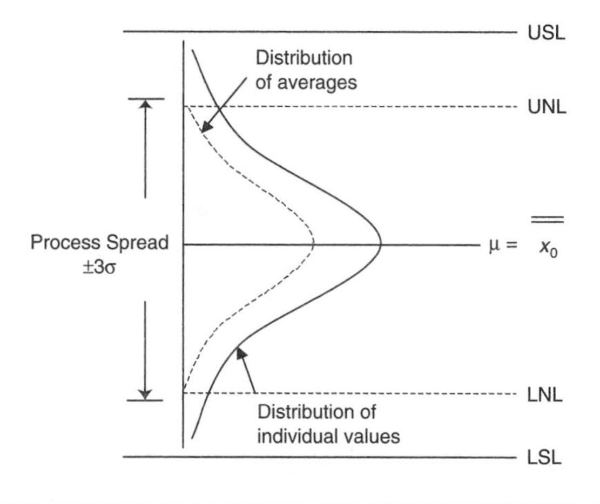

Figure 25-1 Spread of Individual Product versus Averages

As a result of the use of the estimators c_4 and d_2, these two formulas will produce similar but not identical values for $\hat{\sigma}$.

TOLERANCE AND SPECIFICATIONS

Specifications when applied to individual units of a product describe the boundaries that apply to some characteristics of the product. An individual unit of a product is said to conform to specification if it falls within the boundaries for that particular characteristic. The specifications for individual units consist of a *nominal value* and a *tolerance.* The nominal value is the desired value for process performance, as specified by the customer. The nominal value represents the ideal value of the quality characteristic, such that the product will perform optimally during its life. A tolerance is an acceptable departure from the nominal value, as established by the design engineer. This departure creates a band around the nominal value that would still allow the product to perform adequately during its life. Tolerances are added to and subtracted from the nominal value. A *specification limit* represents the boundaries created by adding/or subtracting tolerances from a nominal value. A two-sided specification limit would consist of the following:

Upper Specification Limit (USL) = Nominal + Tolerance
Lower Specification Limit (LSL) = Nominal – Tolerance

A one-sided specification limit consists of either the USL or LSL.

An example: Suppose the individual units of a product would be said to conform if the diameter were 5.0 mm ± 1.5 mm.

The nominal value in this specification	5.0 mm
The two-sided tolerance	1.5 mm
LSL (5.0 mm – 1.5 mm)	3.5 mm
USL (5.0 mm + 1.5 mm)	6.5 mm

Each individual unit will conform if it falls between 3.5 mm and 6.5 mm.

When the tolerance is established without consideration for the spread of the process, the consequences can be grave. The process spread will be referred to as process capability, and is equal to 6σ (refer to Figure 25-2).

The upper and lower "natural (tolerance) limits" of the process fall at $\mu + 3\sigma$ and $\mu - 3\sigma$, respectively:

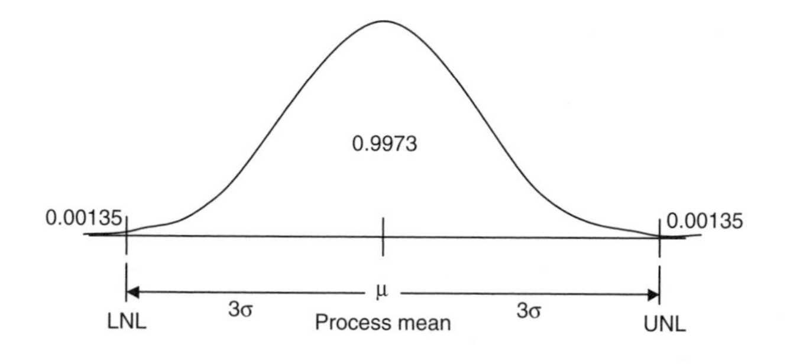

Figure 25-2 Upper and Lower Natural Tolerance Limits in the Normal Distribution

$$\text{UNL} = \mu + 3\sigma$$
$$\text{LNL} = \mu - 3\sigma$$

For a normal distribution, the natural tolerance limits include 99.73% of the variable. In other words, only 0.27% of the process output will fall outside the natural tolerance limits. The following two points are important:

- Although 0.27% outside the natural tolerance limits seems insignificant, this number corresponds to 2700 nonconforming parts per million.
- If the distribution of process output is nonnormal, then the percentage of output falling outside $\mu \pm 3\sigma$ may differ significantly from 0.27%.

Three different situations can occur when the process spread and the specifications are compared: (1) The process spread is less than the spread of the specification limits; (2) the process spread is equal to the spread of the specification limits; or (3) the process spread is greater than the spread of the specification limits.

Case I: $6\sigma < \text{USL} - \text{LSL}$. Process capability is less than the spread of the specification limits.

This is the most desirable case. Figure 25-3 illustrates this relationship, which allows the process to produce units that meet the specifications even when there is a shift in the process average. This process is in control at (a). Since the spread of the specification limits is considerably greater than the process capability, there is no difficulty even when the process average shifts, as shown in (b). Even though this shift creates an

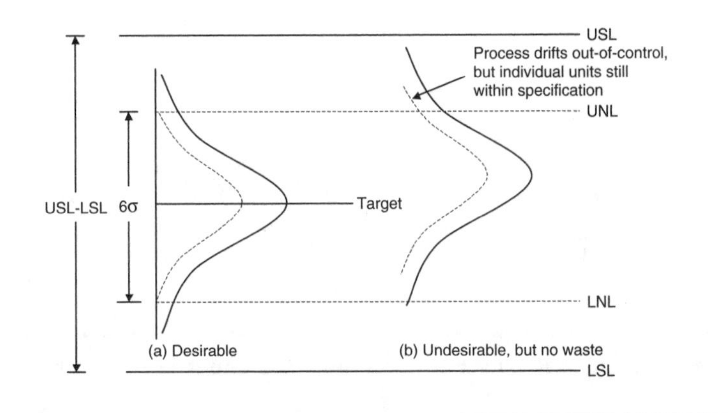

Figure 25-3 Case I: 6σ < USL – LSL

out-of-control situation, no waste is generated, because the distribution of the individual values is still less than the Upper Specification Limit.

Case II: 6σ = USL – LSL. Process capability is less than the spread of the specification limits.

As long as the process stays in statistical control and centered, with no change in the process variation, the units produced will be within specification (Figure 25-4a). However, a shift in the process average (Figure 25-4b) will result in the production of units that are nonconforming.

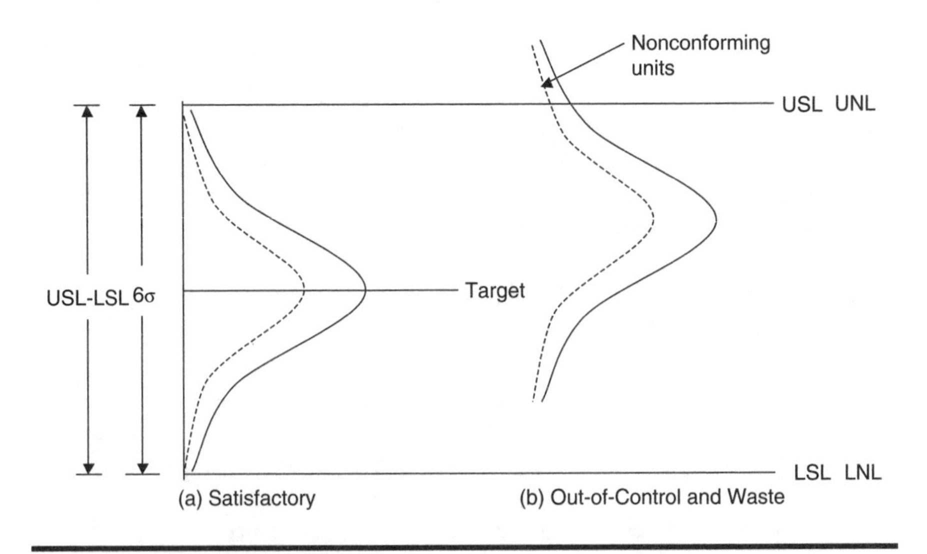

Figure 25-4 Case II: 6σ = USL – LSL.

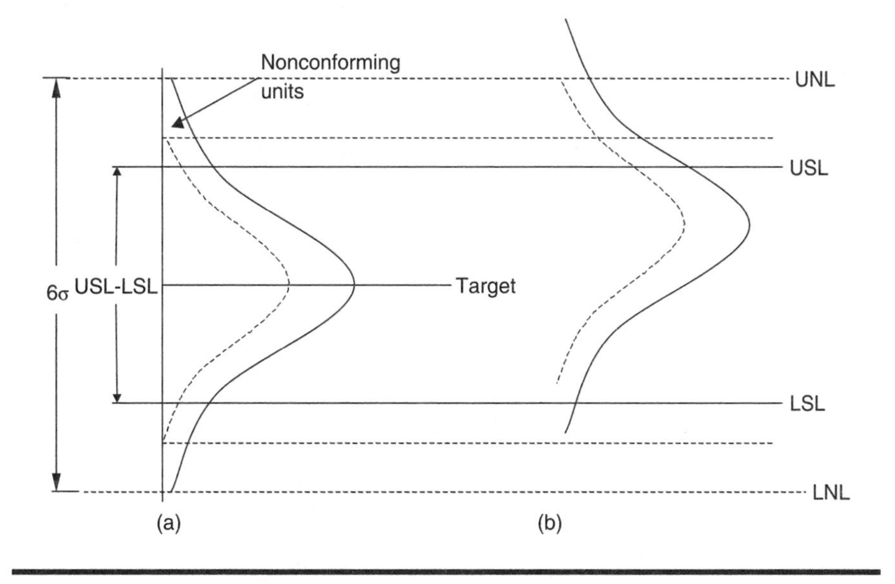

Figure 25-5 Case III: 6σ > USL – LSL.

Case III: 6σ > USL – LSL. Process capability is less than the spread of the specification limits

Whenever the process capability (6σ) is greater than the tolerance spread, an undesirable situation occurs (Figure 25-5a). Even though the process is exhibiting natural patterns of variation, it is not capable of producing units that meet the customers' specifications.

There are essentially three conditions that explain why a stable process may not be capable.[1] The first condition is observed when the process exhibits too much unit-to-unit variation, which causes output to exceed specification limits. The second condition occurs when the process mean is not centered on nominal, which causes output to exceed specification limits. The third condition occurs as a result of any combination of the first two conditions.

CAPABILITY INDEXES

It is often convenient to have a simple, quantitative way to express process capability. One approach to doing so is through the process capability ratio, C_p. The existence of condition 1 can be determined by calculating the capability ratio, C_p:

$$C_p = \frac{USL - LSL}{6\sigma}$$

where USL and LSL are the upper and lower specification limits, respectively. If C_p is greater than or equal to 1.0, the process would produce conforming parts. A C_p value of less than 1.0 means the process would produce some nonconforming output. The greater this value, the better.

Shortcomings of C_p

C_p can't be used without both Upper and Lower Specifications Limits. C_p does not account for process centering. If the process average is not exactly centered on nominal, the C_p index will give misleading results. If the process is not centered, a better measure of actual capability is C_{pk}:

$$C_{pk} = \text{Min}[\frac{USL - \mu}{3\hat{\sigma}}, \frac{\mu - LSL}{3\hat{\sigma}}]$$

In effect, C_{pk} is a one-sided process capability ratio that is calculated relative to the specification limit nearest to the process mean. The estimate of the process capability ratio, C_{pk}, is

$$C_{pk} = \text{Min}[\frac{USL - \bar{\bar{X}}}{3\hat{\sigma}}, \frac{\bar{\bar{X}} - LSL}{3\hat{\sigma}}]$$

Many U.S. companies use $C_p = 1.33$ as a minimum acceptable target and = 1.66 as a minimum target for strength, safety, or critical characteristics. Some companies require that internal processes and those at suppliers achieve a $C_{pk} = 2.0$. A process with $C_{pk} = 2.0$ is referred to as a Six Sigma process because the distance from the process mean to the nearest specification is six standard deviations.[2] In a Six Sigma process, if the process mean shifts off-center by 1.5 standard deviations, the C_{pk} decreases to $4.5\sigma/3\sigma = 1.5$. Assuming a normally distributed process, the fallout of the shifted process is 3.4 parts per million. Consequently, even when the mean of a Six Sigma process shifts by 1.5 standard deviations from the center of the specification, it can still maintain a fallout of 3.4 parts per million opportunities.

Comments Concerning C_p and C_{pk}

1. C_p gives misleading results when the process is not centered.
2. $C_p = C_{pk}$ when the process is centered.
3. C_{pk} is always C_p.
4. A C_p value of 1.0 indicates that the process is producing units that meet specifications.
5. A C_{pk} value less than 1.0 indicates that the process is producing units that do not meet specifications.
6. A C_p value less than 1.0 indicates that the process is not capable.
7. A C_{pk} value of zero indicates the average is equal to one of the specification limits.
8. A negative C_{pk} value indicates that the average is outside the specifications.

Example 25-1

The following represents the diameter of a C14 piston in millimeters. Five observations are collected for each subgroup, with a total of 25 subgroups. Assuming that the data came from a stable process, determine the process capability based on range and standard deviation.

Subgroup	X_1	X_2	X_3	X_4	X_5	$\overline{X}$	R	s
1	12.05	11.96	11.88	11.78	12.00	11.93	0.27	0.11
2	11.89	12.00	12.07	11.96	12.02	11.99	0.18	0.07
3	12.02	11.99	12.08	11.88	11.97	11.99	0.20	0.07
4	11.99	12.12	11.99	12.07	11.95	12.02	0.17	0.07
5	11.96	12.10	12.02	12.08	12.03	12.04	0.14	0.05
6	12.01	12.02	12.04	11.99	11.78	11.97	0.26	0.11
7	12.00	11.97	11.95	12.02	11.96	11.98	0.07	0.03
8	12.02	11.93	11.98	12.04	11.88	11.97	0.16	0.07
9	11.97	12.13	12.06	11.95	12.07	12.04	0.18	0.07
10	11.95	11.98	11.97	11.98	12.08	11.99	0.13	0.05
11	12.03	11.99	11.98	12.06	11.99	12.01	0.08	0.03
12	11.78	12.09	11.78	11.97	12.02	11.93	0.31	0.14
13	11.96	11.89	11.96	11.98	12.04	11.97	0.15	0.05
14	11.88	12.02	11.88	12.04	11.95	11.95	0.16	0.08
15	12.07	11.99	12.07	12.09	11.98	12.04	0.11	0.05
16	12.08	11.96	12.08	11.89	12.06	12.01	0.19	0.09
17	11.99	12.01	11.99	12.02	11.97	12.00	0.05	0.02
18	12.02	12.00	12.02	11.99	11.98	12.00	0.04	0.02
19	12.04	12.02	12.04	11.96	12.04	12.02	0.08	0.03
20	11.95	11.97	11.95	12.01	12.09	11.99	0.14	0.06
21	11.98	11.95	11.98	12.00	11.89	11.96	0.11	0.04
22	12.06	12.03	12.06	12.02	12.02	12.04	0.04	0.02
23	11.97	11.78	11.97	11.97	11.99	11.94	0.21	0.09
24	11.98	11.96	11.98	11.95	11.96	11.97	0.03	0.01
25	12.04	11.88	12.04	12.03	12.01	12.00	0.16	0.07
SUM						**299.74**	**3.62**	**1.50**

Process Capability Based on Range (R) Values

$$\bar{R} = \frac{\sum R}{n} = \frac{3.62}{25} = 0.145$$

$$\hat{\sigma} = \frac{\bar{R}}{d_2} = \frac{0.145}{2.326} = 0.062$$

$$\text{Process Capability} = 6\hat{\sigma} = 6(0.062) = 0.374$$

Process Capability Based on Standard Deviation (s) Values

$$\bar{s} = \frac{\sum s}{n} = \frac{1.50}{25} = 0.06$$

$$\hat{\sigma} = \frac{\bar{s}}{c_4} = \frac{0.06}{0.9400} = 0.064$$

$$\text{Process Capability} = 6\hat{\sigma} = 6(0.064) = 0.383$$

Example 25-2

Using the data from Exercise 25-1, suppose the USL and the LSL are given as 12.40 mm. and 11.82 mm., respectively. Determine the value for C_p, using $\hat{\sigma} = 0.064$. What conclusion can you draw from your answer?

$$C_p = \frac{USL - LSL}{6\hat{\sigma}} = \frac{12.40 - 11.82}{6(0.064)} = 1.51$$

Since the value for is greater than 1.0, the process is capable.

Example 25-3

Determine C_{pk} for Example 25-2 (USL = 12.40, LSL = 11.82, and $\hat{\sigma} = 0.064$).

$$\bar{\bar{X}} = \frac{\sum \bar{X}}{n} = \frac{299.74}{25} = 11.99$$

$$C_{pk} = Min[\frac{(USL - \bar{\bar{X}})}{3\hat{\sigma}}, \frac{(\bar{\bar{X}} - LSL)}{3\hat{\sigma}}]$$

$$C_{pk} = Min[\frac{(12.40 - 11.99)}{3(0.064)}, \frac{11.99 - 11.82}{3(0.064)}] = Min[2.135, 0.885]$$

$$= 0.885$$

Since C_{pk} provides a more accurate picture regarding the capability of the process, we conclude that the process is not capable.

COLONY FASTENERS

Colony Fasteners, Inc. (CFI) utilizes process design, monitoring, and control as a basic strategy for the business. The company has determined that increased quality and productivity with minimized costs are the results of such actions.

CFI closely follows the Motorola concept of Six Sigma (6σ), using the capability index of $C_p = 2$ and C_{pk} of 1.5, considering a 1.5 sigma variation shift of the mean. To meet the process capability of $C_p = 2$, three sigma limits of the process are 0.5 of the 6σ specification limits of the particular process step. This concept is carried throughout the organization in products and non-product service support functions and with suppliers.

A C_{pk} of 1.5 results in 3.4 parts per million (ppm) outside the specification limits. Figure 25-6 shows this relationship and the calculations of C_p and C_{pk}. This figure is from a Motorola publication, "The Nature of Six Sigma Quality," by Mikel Harry, Ph.D, 1994.

The in-control process of ±3σ is shown as B. The specification limits of ±6σ are shown as A. The capability index $C_p = 2$ represents A/B = 6/3 = 2.

If the mean of the process moves 1.5σ, the mean would be 4.5σ from a limit and is shown as C. For an individual item, $C_{pk} = C/(0.5\ B) = 4.5/3 = 1.5$.

If the mean of the in-control process shifts 1.5σ, the results are that 3.4 ppm will be outside the specification limit. If the mean shifts only 1σ, the result will be 0.39 ppm, and with no shift, 0.002 ppm.

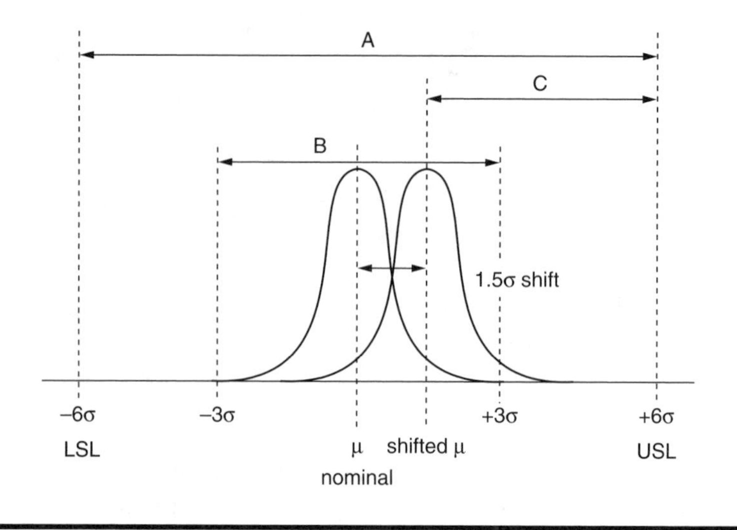

Figure 25-6 C_p and C_{pk} Relationships

The aims of the processes are to hold the mean at the center, but it may shift 1.5σ and still have only 3.4 ppm outside the specification limit for the individual process step.

CFI has used this approach successfully, and the customers and suppliers understand the concept and participate in the fruits of the results. As described in this Category, process teams continually reduce process variability. Many processes have been subjected to several years of improvement.

As the individual process variations become less, the possibility for mean shifts becomes greater. Although the operators and responsible teams strive to hold the means, some movement usually results.

Another aspect of the approach used is that the customers, both internal and external, through the many contact opportunities, have understood the need for moving out specification limits. The limits are expanded not to deliver products and services over a wider range, but to allow process capabilities of at least $C_p = 2$.

Industry specifications are generally wider than customers desire, and it is customary for producers to deliver to tighter tolerances, often as tight as one-half to one-quarter of industry standards. CFI has led discussion in the industry and with customers to utilize the process capability approach rather than tighter limits. With wider tolerances, more processes operate with a C_{pk} of 1.5 or more.

DESIGN AND INTRODUCTION OF PRODUCTS AND SERVICES

How Products and Delivery Processes are Designed

Translation of Customer Requirements

The design process is handled in a slightly different manner for products in each of the four sectors, due to different product requirements. However, both internal as well as external customer satisfaction is always considered as the prime driver for new or improved processes. Another driver is manufacturability, as described in a program called Design for Manufacturability (DFM).

Inputs from the concerned functions convert customer requirements into product and service design requirements. This ensures customer needs are recognized as the basic reason for any changes, improvements, or new concepts.

A very important input to the design process is through the field sales engineer (FSE) who brings detailed customer knowledge to the design process. The FSE has spent a lot of time in the customer facilities, has talked to customer employees, and has firsthand information about their needs and expectations. The FSEs and marketing representatives make recommendations in conjunction with suppliers for the selection of appropriate materials, so that the application results in reduction of waste material. The results of effective material selection reduce in material wastage, as shown in Figure 25-7.

In the Consumer Product Sector (CPS), industry standards are utilized for configuration, and CFI processes are used. Commercial and Automotive Sector (CAS) parts are a combination of both standards and user specifications. DAS products are designed to military specifications with CFI processes.

A brief, high-level summary of the product design cycle at CFI is summarized below and in Figure 25-8. The first activity in the cycle focuses on a new concept and its feasibility of design and production. Customers provide detail designs which are reviewed, evaluated, and concurred with by CFI sector product teams. After agreement by the customer and CFI, the order is accepted. The first review is a Preliminary Design Review (PDR) where dimensions, materials, and suppliers are screened for capabilities, proper design margins, and safety allowances. At this PDR, preliminary processes are presented for customer concurrence, and PTP (Pass Through Partnerships) relationships are established between CFI, the customer, and associated suppliers.

During the engineering and validation test activity phase (EVT), the new product is fully engineered and tested according to product

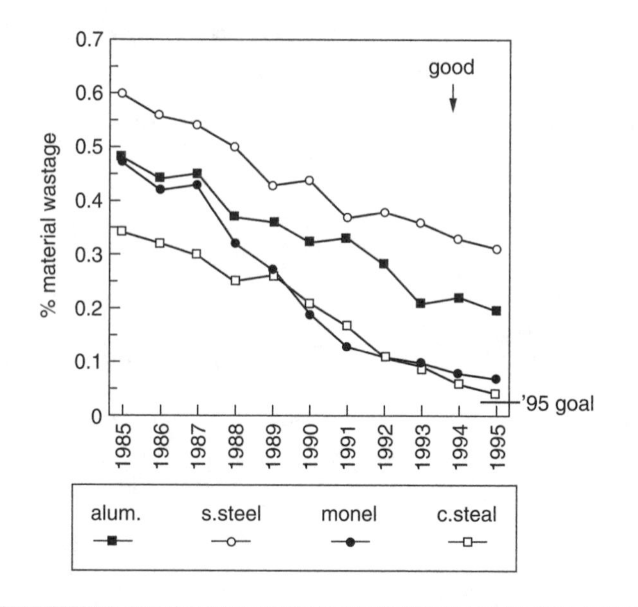

Figure 25-7 Wastage Reduction

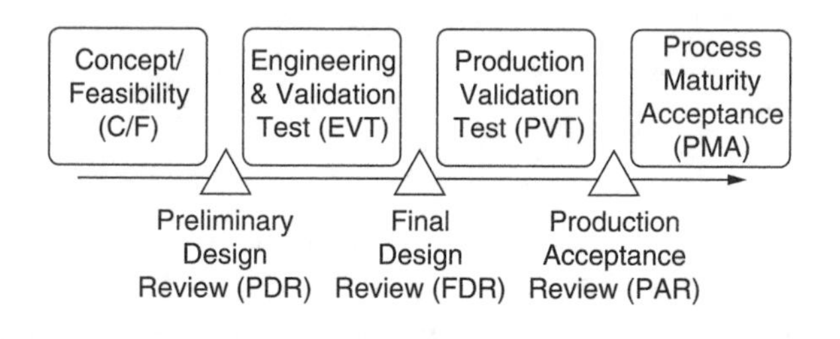

Figure 25-8 Product Design Cycle

specifications. Design capability margins and reliability are ensured. The final results of the design and its testing are reviewed at the Final Design Review (FDR). The initial production processes have been designed and checked for capability and verification that a process capability of C_p = 2.0 is attainable. The process itself and the limit specifications provided by knowledgeable customers are considered.

The production validation testing (PVT) activities follow the FDR. In this set of activities, the initial processes are put into place and early production begins. Processes are constantly reviewed, analyzed, and tuned to ensure that proper capabilities are attained. The results of these activities are reviewed during the product acceptance review (PAR). This is the final review before a new product is put into full-scale production. Once in full production, the product goes into the phase focused on process maturity acceptance (PMA). This is where process capabilities are constantly reviewed, production output is ramped up, and product maturity begins.

In each sector, customer requirements are fulfilled, suppliers are consulted, and, if applicable, customers approve the design.

In the Technical Data System (TDS), statistical techniques are maintained in the Engineering Information Architecture (EIA). All designed products are subjected to a DOE analysis for assurance that the design is robust and that applied environmental conditions will not adversely affect the reliability or the performance of the product in the end-use conditions. The DOE is contained in the Design for Quality and Reliability (DFQR) section of the EIA.

Each sector maintains its own design capability, and all procedures and processes are documented and maintained. All sectors use the two-design review process steps noted in Figure 25-8, but they modify and adjust them according to the particular design program requirements.

Within each sector, designs are documented in a single file that ensures coordination for the many designs utilized when they are formalized and placed under configuration control. These files are summarized for the review of other sectors for the possibility of adaptation or overlap. The complete files are always available to other sectors if desired.

An important link between customers and CFI during new product design, as well as other times where close communication is required, is the Electronic Data Interchange (EDI). Data are transferred between both suppliers and customers of CFI. EDI is also utilized between the engineering organizations and the factories to facilitate transfer of information at a high degree of accuracy, which is particularly important with the factories spread worldwide. The manual transfer of data would take considerably more time and be subject to many errors.

Another important communications tool with outside customers and suppliers is the voice mail hook-up, in which the outside organizations are treated as an arm of CFI.

Requirements Translated into Processes

As stated earlier, CFI maintains a strategy that all processes must maintain a C_p of 2.0 or greater.

Production process changes for new products are usually modifications of those used for similar products. However, before FDRs can be approved, objective data must be presented demonstrating the distribution pattern of the material in the specific process steps. It must reference the measured sigma compared with the tolerance limits determined for that particular step. Usually the variation is well within the capability index of two and often has a margin of three or more. The design review team determines what characteristics will be measured and when the measurements will be accomplished.

Each of the process steps is documented with the sampling plan that is used to ensure that process variability is monitored and maintained. The sample size, frequency of sampling, and parameters to be checked are listed. The sampling proves the distribution to be in control, and the sigma are automatically calculated by the Operations Data System (ODS).

How All Requirements Are Addressed Early in the Design Cycle

Design reviews vary in scope depending on the complexity of the product and whether the changes are minor or major. In all cases, the reviews are accomplished with all concerned organizations.

Typically, the functions represented in the design review are Engineering, Manufacturing, Quality, Logistics, Marketing, Field Service, and Purchasing, as well as supplier representatives and customers. At the PDR, all customer and CFI requirements are reviewed. All design reviews are documented, and the results maintained in the TDS.

The Computer Aided Fastener Design (CAFD) process embedded in the EIA is utilized for consistency and completeness.

Suppliers are included in the early design phase. CFI realizes that the suppliers provide the basic raw materials crucial to product and process consistency. This includes the chemicals used for cleaning, plating, and other process steps. Suppliers are often able to recommend better materials to meet customer requirements, such as expected environments, strength or other special needs, and the workability in the drawing, upsetting, and threading processes.

This careful selection of materials is part of the data used by the customer for consideration of allowing broadened specifications.

How Processes Are Reviewed Prior to Launch

The FDR provides data that show that the materials or services subjected to the specific processes meet the criteria of a distribution by attaining at least one-half the specification limit for the specific process step.

Many designs are modifications to existing products, and the established processes have been subjected to a continuing improvement process over several years. The process teams have the experience and knowledge, along with the data systems described in Category 2.0, to fully understand where the processes might need possible additional modification. The DFQR process establishes assurance that production processes have been fully considered prior to a production launch.

Inputs to the design reviews come from functional organizations and are used to determine the necessary actions by the design team. These inputs are furnished by the functional representative on the design team, who also provides two-way communications into the functional organizations. This ensures that all levels of employees have reviewed the designs so that no details have been overlooked.

How Processes Are Evaluated and Improved

Within each department of each sector, processes are maintained by teams assigned the responsibility of designated processes. This same approach is used in the support service departments as well as the production departments. These teams continually review the status of processes to ensure that the relationship between the element being acted on and the limits assigned are within a C_{pk} of 1.5.

In the few cases when the $C_p = 2.0$ is not attainable initially, teams continue to work on improvements to the process steps in various ways, including benchmarking, R&D, reengineering, and examining similar processes within CFI. Customers are also contacted to relate to process capability indices rather than the use of expanded specification limits, if the design permits. Special summary reports are structured for those processes with a C_p of less than 2.0 and reported in the departmental reviews.

Every quarter, a review of the design process in each sector is accomplished. Representatives from the functional areas in the design process, as well as corporate representatives, review past data for results in quality and cycle time.

To ensure continuous improvement, goals are set for improvement and reviewed for attainment at the next meeting. If improvements have not been attained, assignments are established with specific dates set for accomplishment.

One success of this approach is shown in Figure 25-9. Design cycle time improvement results have occurred in all sectors, with NSS still showing the greatest improvement opportunity, primarily due to the extreme customer and regulatory requirements.

At monthly departmental reviews for action teams, various characteristics are compared to goals set, and process teams report, on a rotating basis, the results of their activities. They report on how many of the processes have a C_{pk} of 1.5 and how many exceed that figure. Team recognition is awarded to the top five teams each month in each department.

Teams are continually examining processes in their areas of responsibility and looking for better ways to accomplish the task. Benchmark

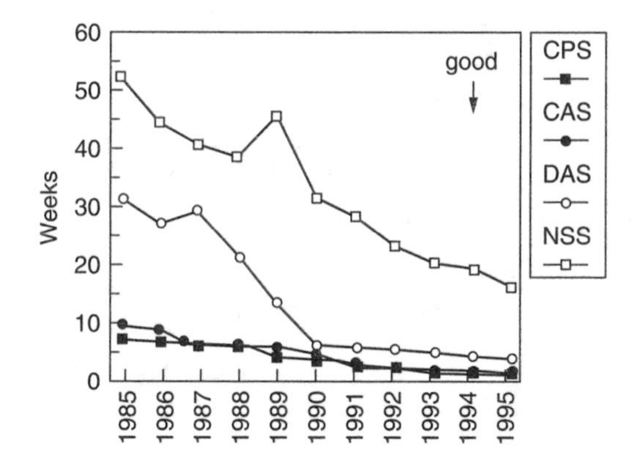

Figure 25-9 Product Design Cycle Time

activities are extensively used for comparison with other organizations that use alternative techniques that result in less variability, shorter cycle time, or lower cost.

Although each sector has different products, many processes are similar. This gives the teams during monthly reviews opportunities to compare process steps and results. In the case of support services, the differences in the needs of the organizations are less, and usually the modified processes can be applied directly.

PROCESS MANAGEMENT: PRODUCT AND SERVICE PRODUCTION AND DELIVERY

How the Company Maintains Performance Production

Key Processes

In the production of fasteners, all production processes are considered important, because each process contributes to the quality, cycle time, and cost of the products. CFI has maintained the concept that all production processes will be at least a C_{pk} of 1.5.

The key processes that produce the products and services supplied to the customers are shown in Figure 25-10. The principal requirements for the key processes are as follows:

Receiving Material Check

The receiving material check varies with the material, end customer requirements, and status of the supplier of that material. If the supplier is a key supplier that has been certified through the PTP process, the

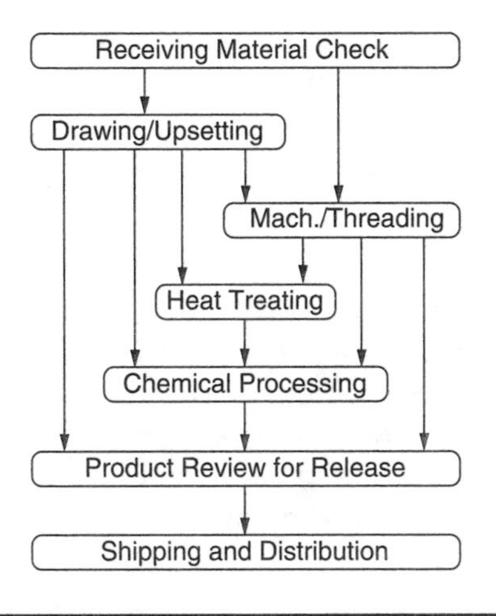

Figure 25-10 Key Process Flowchart

checking process only identifies the material and any certifications required by the end customers, as is usually required in the Nuclear and Specialty Sector. The PTP partner continually feeds statistical data through the EDI communication capabilities.

Other materials from non-certified suppliers will have chemical analysis or checks of physical properties. These checks are on a statistical sampling basis unless otherwise demanded by the individual customer. It is the primary goal of all sectors to reduce this effort as suppliers are certified through training and proven performance.

In handling incoming materials, bar codes are utilized to reduce cycle time and improve accuracy. Certified suppliers are provided with special bar code strips that identify the material, lot number, and required statistical process results information.

Drawing and Upsetting

The drawing process typically changes the materials that are received in spools, such as wire, to the proper diameter for the particular fastener. The controls on this process vary with the end requirements of size tolerance, strength, material coatings, or other special customer needs.

The upsetting process forms the heads on the end of the fastener, as with bolts or rivets. Some upsetting, as in the drawing process, requires precise temperature and environmental controls. Each situation is clearly defined during the design cycle, and the process is controlled to maintain the C_{pk} that was qualified during the process design drawing reviews.

These processes are dependent on lubricants used during the metal forming operations. Productivity gains have resulted from changes in lubricants as a result of supplier inputs and benchmarking visits.

Machining and Threading

Most of the fasteners produced use machining and thread cutting or rolling. Differences lie in the size, materials, and end use of the product.

The processes involved include the type of machine, usually with high levels of automation, and the type of cutting tools used with the requirement of adequate sharpness, and required coolants. Threading is accomplished in a variety of ways depending on the physical characteristics and the customer end use.

During the machining operations, the operators are empowered to monitor the process and are expected to halt the process if excessive variations occur. The operators use optical comparators on a regular basis. On a periodic basis, technicians from the labs select random samples for a complete analysis.

Process steps are handled during the process design with the sampling and assurance of the proper distributions to ensure the C_{pk} as determined in the design of the process.

Heat Treating

Heat treating is considered a key process because of the critical nature of the end use, where the safety of people is usually involved.

The heat treating process includes chemicals, temperature controls, and various quenching techniques. The heat treating process is highly automated, with little intervention by operators and, therefore, with less chance for process variability. The processes are certified during the design cycle, as explained in "How Products and Delivery Processes are Designed," to ensure an index of at least $C_{pk} = 1.5$.

Chemical Processing

Chemical processing is vital to customers who use fasteners in environments hostile to the materials used. The amount and type of protection vary by the sector and the specific customer.

The process requirements may be plating, which includes the chemicals used and the configuration of the baths with the times, temperatures, and agitation required. The process might be a dipped coating or a coating applied by an electrostatic process.

Processes have individual steps, each of which contributes to the end result. Each individual process step was defined during the design process, and the process steps were documented to result in a C_{pk} = 1.5. During the process, performance sampling procedures are followed and documented in the ODS. Periodic audits of the production areas and data system ensure ongoing compliance of the chemical processing system.

Product Review for Release

This key process is for assurance that all processes have been applied to the proper materials for the proper customers. With the facilities distributed around the world delivering to customers around the world, and with the large number of products, overall reviews are necessary.

Production control of all products is centrally controlled in the ODS. Bar codes are used extensively to result in rapid, accurate records. Problems are seldom discovered. However, to ensure customer satisfaction, this process is retained.

Shipping and Distribution

The distribution system is a key process for CFI due to the necessity of delivery to customers throughout the world from the 16 manufacturing facilities in various countries.

Customers have become increasingly demanding of just-in-time deliveries. CFI has learned to balance off-shore manufacturing to stage products in warehouse sites at strategic locations close enough to major customers to meet just-in-time requirements.

To accomplish these actions, a multi-layer distribution system has been developed. This system provides bulk shipments from manufacturing facilities directly to large customers as well as distributed warehouses. In turn, warehouses ship directly to customers, as well as secondary facilities where materials from all sectors are accumulated. This is particularly important for the CPS.

As a result, the distribution system requirements of accuracy of records and protection of parts are essential. The concept of C_{pk} = 1.5 is maintained, with processes being determined with sampling used to ensure minimized variability. Automatic counting utilizes sensitive scales and bar codes with the automatic stocking process being computer controlled.

Measurement Plan

During the design cycle, every process is designed and proven before the design is released. The process certification consists of demonstrated success that the process has a capability of at least a C_{pk} of 1.5. As products are processed, distributions are automatically calculated via the ODS to ensure that the sigma is remaining as was initially determined to be necessary.

When it is discovered that the distribution is out-of-control or the variability has increased, the process is immediately stopped by the operator, who is empowered and expected to halt the process. The team that has the responsibility for the specific process is immediately convened, and it requests any technical capabilities that are needed to rectify the process problem. Design engineers, suppliers, quality engineers, data analysts, or whoever can contribute to the solution are utilized by the process team leader. The Plan, Do, Check, Act (PDCA) process is exercised at this time. After the root cause has been determined, the solution is installed, the process restarted, and the data taken once again.

The sample frequency is tripled, and no more deviations can be experienced for the next five shifts in order for the sampling frequency to go back to normal. The process team reviews the data records to determine any similar processes throughout the company. If any are found, the responsible teams are immediately notified in the other three sectors.

Another key service deliverable is the service of FSEs who perform in close liaison with customers solving problems and providing a vital link back to the various CFI businesses.

How Processes Are Improved

Process Analysis and Research

The assigned process teams are always on the alert to improve processes in order to reduce variability and increase capability. Benchmarking and communicating with other functions in the company have proven to be a fertile area for improvement ideas.

Employee operators are aware of distributions that begin to spread. They alert the appropriate team if the capability number drops although it may still be over the C_{pk} of 1.5.

For common processes, a section of the common R&D division researches processes for both improvements as well as totally new approaches. This information is available to the process teams in all sectors through the data system.

As a result of the data system providing information worldwide, any time a process shows improvement, the information is available to the other plants regardless of the sector affiliation. At the annual Recognition Celebrations, the process teams are recognized for the increases in capability indexes that they have accomplished.

Benchmarking

As has been pointed out earlier, benchmarking has proven to be one of the better methods for improving process characteristics, including simplification, reduced variation, and reduced cycle time.

When a benchmark partner in the PTP process is selected for a particular review, the process team members make the partners fully aware of their own process, with the specific measurements used to measure the process. During the visit, the partner's process is examined in detail with careful observance to incremental improvement data. Often, only a part of the process may be the superior part, and only that particular segment may be utilized.

Use of Alternative Technology

Technology has resulted in many improvements. Design engineers and process teams are always aware of opportunities to apply new techniques, equipment, and approaches to process control.

As an example of applying a new technique, a recent improvement in the DAS was the result of better temperature control of bolts that had a heat-treating requirement. One of the process teams learned of a new electronic method that used very high frequency radiation to heat bolts in a protective container. This new method reproduced the temperature variation to within ±0.3 degree Celsius, where past equipment would only reproduce to within ±1.0 degree. This order of magnitude improvement resulted in a new C_p of 2.1 against the earlier C_p of 1.8.

With this change, the process capability was improved to the point that heat treating testing was eliminated. Only the regular sampling required to determine that the process is in control is used. The new technology was developed for a drug manufacturing process, but an alert process team member visualized the application to heating individual bolts.

Information from Customers

CFI maintains close customer relationships. Some customers also manufacture fasteners, although none has the broad range of products of CFI. These customers have provided benchmark partnerships, PTPs, that have resulted in many process improvements.

Other customers have helped in service areas such as stocking and delivery techniques. Customers want just-in-time deliveries, and CFI is dedicated to provide the service. In several instances, customers have suggested solutions and improvements to CFI's service deliveries by describing and demonstrating how cycle times could be reduced, and in many cases have provided better protection for the product. This has been most prevalent in the CAS, where customers are more mature in ways of handling products and delivering just-in-time.

With many teams in action throughout the company in all sectors, internal customer feedback has been the source of many applications of improved processes and reduced cycle time.

Stock handlers in CPS provided an idea to handle large skids of product by using air to lift the skid for better maneuverability. A base with air outlets on the bottom connecting to an air line permitted a lone stock handler to move skids weighing greater than one ton around the shipping floor.

PROCESS MANAGEMENT: SUPPORT SERVICES

Here several examples are discussed that are representative of the total company. The process concept of the company, as described in "Design and Introduction of Products and Services" and "Process Management: Product and Service Production and Delivery," encompasses the support service functions as well. Formal procedures are utilized in the support service areas in a similar fashion to those used in the product and service portions of CFI.

The support functions now use measurements in parts per million (ppm) rather than percent. They are tied to the production facilities with EDI, and they design and monitor their processes to the goals of $C_{pk} \geq 1.5$. The support functions are also connected to customers, both internal and external, as well as suppliers, with voice mail capabilities. This improvement in communications reduces cycle times, improves accuracy, and results in lower costs.

How Support Service Processes Are Designed

How Key Requirements Are Determined

Each support service determines the key processes needed for support of its delivered products. The department determines the mission of its function as a result of the strategic planning process, and the departmental plans stem from the strategic plan and from discussions with employees in the function.

Planning teams are formed during the data gathering phase for the strategic plan, and priorities are established in the order of importance to the business aims of the sector. The cognizant team lists all of the requirements for specific actions, with the agreement of the team as to the key requirements, in order to establish priorities.

How Requirements Are Translated into Processes

Process teams are utilized in the support service functions as they are in the product and service producing functions. These teams are also trained in process mapping, problem solving, and benchmarking. Processes are flowcharted, compared to similar processes, and fully documented.

Processes in the support functions are structured with specific limits for each step, and the process is measured to determine the sigma for the controlled process. These processes are also expected to attain a capability index of $C_p \geq 2$. Dialog with internal customers results in limits that are more easily defined. The process steps are documented, and the necessary measurements are included. This includes the sampling plan with the specific parameters to be checked.

As an example, in the CAS, the Accounting Department designed an accounts payable process to be more responsive to small suppliers, who require regular cash flow to maintain their continuous improvement processes.

An Accounting Department team met with suppliers and Purchasing to determine the best method for submitting invoices. It also met with the material receiving organization to determine the fastest, most accurate methods of verifying material acceptance. The results are shown in Figure 25-11.

A new process was designed and installed, selecting limits in cycle time for each step in the paying process. The resulting sigma was determined to ensure that the process operated at $C_p \geq 2.0$. Sampling plans were installed to ensure that the process would remain “in-control.”

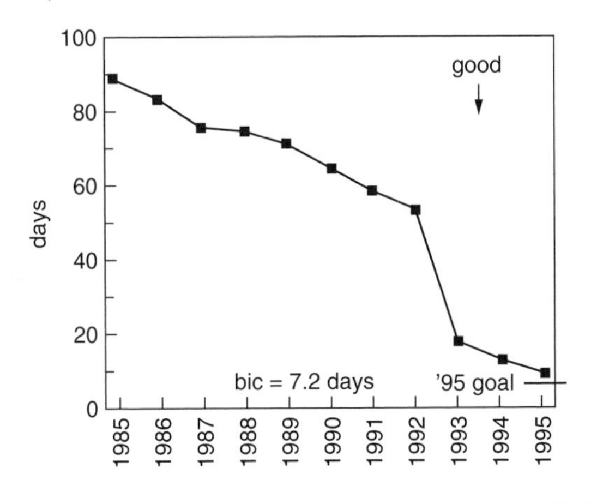

Figure 25-11 Time to Pay Invoices

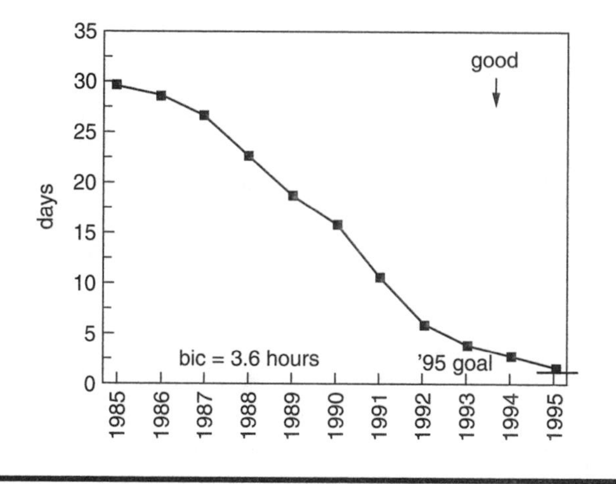

Figure 25-12 Days to Close the Books

Another process that has been significantly improved is the reduction in time to close the books at the end of each month. This required the efforts of a cross-functional team composed of members from Accounting, Operations, IS, Marketing, and Human Relations. The team met weekly and continues to meet to maintain the rate of improvement. The results of this activity are shown in Figure 25-12.

Requirement Addressed Early in the Design

Most of the processes in the functional support organizations are established and changed, using the continuous improvement methodology, by the assigned process teams.

New processes usually evolve from new technology and new techniques to go with additional requirements. An example of this has been the upgrading of the IS process. As the company has grown and expanded around the world, significant additional demands were placed on the system. At the same time, both hardware and software with additional capabilities had become available.

To take advantage of the new capabilities and to meet growing requirements, a significant amount of coordination with all the sectors, suppliers, and customers was needed. Facilities around the world were consulted to ensure adequate inputs for requirements and agreement of acceptable internal data system cross-communications and available data outputs.

The new IS system was completely designed before going on-line in 1989. The system was installed with measurements concerning response time and availability for data inputs. Limits were established, and the resulting sigma operated within the required process capability considerations.

The system went on-line with few problems, a real tribute to the many teams that worked together to structure the system. It could also be a tribute to the work of designing the process to the requirements of the internal customer and determining the capability of the process to operate within the set limits.

Figure 25-13 shows the success of the system for availability and improved response time. This has been accomplished during a period of sales growth and increased dependence on the data system.

How the Company Maintains Performance

Key Processes and Requirements

Some key processes in support areas are contained in Table 25-1. These are the highlights of a wealth of additional processes.

Each of these representative key processes has principal requirements that are determined by assigned teams in the areas of the process. For example, the requirements of the Financial Data System (FDS) are concerned with accuracy, time to generate reports, containment of all costs and revenues, and timely output reports that are clear and understandable by those receiving the reports.

The planning teams use an established methodology for reviewing processes. Each key process has established measurements with goals of continuous improvement assigned. Processes are reviewed through process mapping, using questions such as the following:

- Is the process step needed?
- Is it accomplishing the requirement?
- Can it be done better by modification of equipment or operator training?

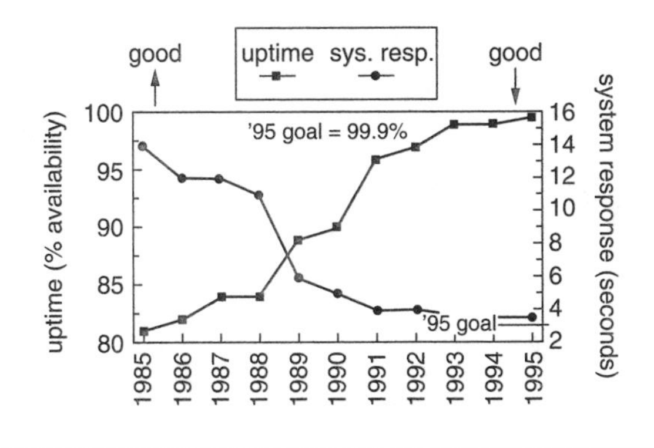

Figure 25-13 Information Systems Availability

Table 25-1 Key Company Processes

Area	*Processes*	*Figure*
Accounting	Time to Pay Invoices	5.3.1
	Sales Volume	6.2.5
	Profit Margin	6.2.7
	Days to Close Books	5.3.2
Marketing	Market Share	6.2.6
	Customer Complaint Calls	7.4.6
	Lead Time to Order Fulfillment	6.1.7
R&D	Design Cycle Time	6.2.8
	Product Setup Cycle Time	6.2.9
Human Resources	Employees Engaged in Teams	6.2.21
	Training Hours per Employee	6.2.18
	Team and Individual Awards	6.2.19
Administration	Corporate Citizenship Engagements	6.2.22
	Information Systems	6.2.10
Operations	Preventive Maintenance	5.3.4
	Operational Productivity, Non-Products	6.2.2

This same methodology is repeated in all functions, by all teams, to ensure continuous improvement.

As an example, preventive maintenance (PM) is a function that has a significant effect on overall operations. When the maintenance is performed on time, the production equipment performs better. Figure 25-14 shows the results of improved maintenance over the last ten years with machine downtime approaching 1.0%.

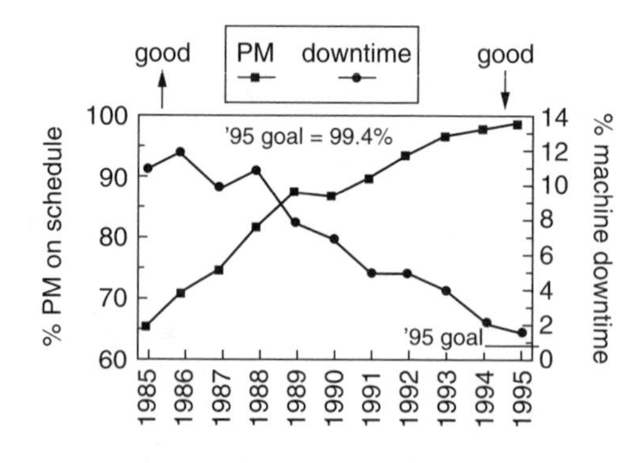

Figure 25-14 Preventive Maintenance

The measurement plan in the service and support areas is very similar to those in the product areas. All process steps are designed with a documented process. The steps are continually assessed to be in control with a calculated sigma to ensure a C_{pk} of 1.5 or better.

As in production, the process steps have specific plans for the frequency of sampling as well as the parameters to be measured. In support areas, usually the measurement is for cycle time, with accuracy and costs also measured. Examples of the reduction in cycle times are the invoice payment cycle and "days to close the books" indicators shown in Figures 25-11 and 25-12.

The Measurement Plan

Support functions require continual training of employees involved in statistical techniques, in order to ensure their comfort and competence with these types of measurements. The process control applications in support functions have normally been heavily weighted toward measuring and controlling paper handling and administrative type processes. This has proven to be a significant driver in increased organizational performance, resulting in less cycle time and lower operating costs.

How Processes Are Improved

Process Analysis and Research

Support processes are analyzed to define the steps for measurement improvement. Research is accomplished primarily when new techniques or equipment are utilized, such as the data system redesign with new computer capabilities.

Process teams in all functions are continually investigating new capabilities that often require additional training. If required, the teams arrange for and sometimes perform the training.

Benchmarking

Benchmarking activities have been mentioned in several areas. CFI has discovered the advantages of benchmarking in the goal-setting mode and the more important advantage of process review capability and improvements.

Benchmarking has been utilized to a high degree in support areas. A real advantage is that new processes have applicability in most sector businesses. The major difference in support processes from product processes is that we generally go outside our industry to find high performance processes.

A new process, such as preventive maintenance (which was described earlier), can be used in the same manner for many facilities in the company. As was also stated earlier, the NIH factor has been overcome, and processes learned in a benchmark study are frequently used "as-is" in many support organizations.

Use of Alternative Technology

Alternative technology has been addressed in many areas in the field of personal computers. As new equipment with significantly greater capability and many new software programs become available, teams are continually investigating and evaluating their applicability. Due to the rapid changes in costs and capabilities, comparisons are ongoing.

Voice mail is used extensively in the support functions. Communications have been vastly improved, and actual paper use has almost been eliminated. Bar codes are used in the support areas for routing of reports and documents.

Process teams in the support functions are continually researching and reviewing ideas for the application of alternative technologies. Inputs for ideas and actions derive from benchmarking visits of organizations outside the fastener industry, visits to trade shows, internal visits to other sectors, and combing business periodicals. Monthly meetings are utilized to measure progress and emphasize the need for improvements.

Information from Customers

CFI is focused on outside customers and has always maintained a good dialog with them. Although much of the information transferred relates to products for both today and in the future, many times support processes are covered in customer discussions. Usually this concerns an interface condition, such as billing or credits, and sometimes occurs as a result of a visit where an observation can lead to improvements.

Internal customers are also solicited for information, and actions are often taken as a result of their observations and suggestions.

Most of the sectors have utilized the PTP technique of having internal suppliers and customers meet at a regular time (such as Friday afternoons at 3:00 PM) to discuss mutual needs and services. These meetings allow better understanding between the parties and result in continually improved, more effective services.

All sectors of the company are encouraged to take advantage of customer inputs. The NSS recently learned of a better process to tabulate and maintain records of audits required by the Nuclear Regulatory Commission. The process utilized a portable device that transferred records by radio directly from the auditor to the records retention area. The device also produced a bar code strip that would be attached to the material.

This was faster and more accurate, with less overall cost, and was an application of new technology provided by an outside customer. In turn, this process was picked up by the CPS as a method of inventory counting where counts from the end of the production line are fed into the data system immediately. Upon receipt in the stockroom, a verification count would either accept the count or immediately alert the material handlers of possible misplaced material.

MANAGEMENT OF SUPPLIER PERFORMANCE

CFI recognizes the importance of suppliers to the success of the business in all sectors and has established a process called Pass Through Partnerships (PTP) to share lessons learned to improve quality, reduce cycle time, and pass on technology advances.

This process establishes special relationships with suppliers. It fosters sharing of benchmark information, provides training in process control concepts, and periodically holds seminars in "lessons learned." When suppliers qualify, with their processes attaining $C_{pk} > 1.5$ and passing other requirements to show that capabilities will be maintained, they become "Certified" and their products are not subject to receiving inspection at any of the CFI facilities. Figure 25-15 shows a high-level view of the supplier certification process.

Certified Suppliers receive special bar code strips to attach to products delivered. This simplifies the incoming process and results in the suppliers receiving payment earlier. Certified Suppliers are connected with EDI, which provides them with product specifications and production requirements information as soon as CFI makes needs determinations. The

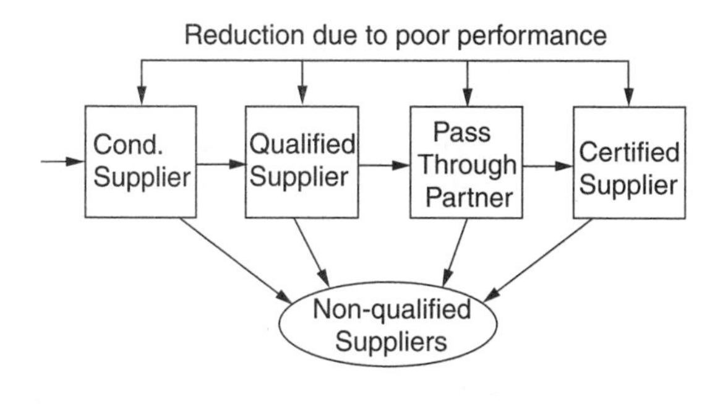

Figure 25-15 Supplier Certification Process

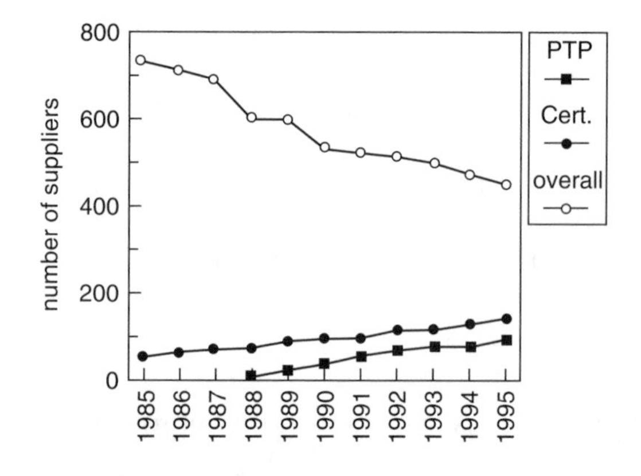

Figure 25-16 Supplier Base Transitions

supplier certification program is shared with all sectors to minimize costs of developing suppliers. Figure 25-16 shows the growth in PTP partners and the increase in Certified Suppliers.

SUMMARY OF COMPANY REQUIREMENTS

Principal Requirements for Key Suppliers

CFI presently has 420 suppliers for both raw materials and for products and materials integrated into salable products. Key materials are metals such as aluminum, carbon steel, stainless steel, monel, and brass. Chemicals for processing and cleaning are also considered to be key materials.

Quality, delivery, and price are all very important for the following reasons:

- ***Quality*** — The foundation of CFI has been built on quality and continues to be a driving force in the competitive markets. Due to the nature of the products, the raw materials establish the foundation for output product quality.
- ***Delivery*** — A percentage of customers are purchasing proprietary parts and depend on CFI to deliver on time. If delivery were to be interrupted, customers' deliveries would be quickly affected as most customers are utilizing just-in-time concepts with little material in stock.
- ***Price*** — Price is always important, as competitors are emerging on a daily basis. Although customers understand quality and dependable on-time delivery, they are continually concerned about price. Fasteners are usually considered a "C" item and expected to be a very low price.

Utilizing the process control philosophy with process capability measurements, CFI passes the same requirements on to suppliers. Suppliers are expected to have their processes measured and to have a C_{pk} of 1.5 or greater. This includes the support services at the suppliers.

CFI has a business strategy to be the technology leader in the introduction of revolutionary fasteners and has established a supplier base with similar goals. These knowledgeable suppliers have simplified the transfer of understanding of process control capability techniques.

How the Company Determines Performance

Through the PTP and supplier certifications with process capability techniques, in which many suppliers participate, incoming product quality is ensured. Many suppliers are part of large companies and have participated in similar programs with other customers.

In some cases, suppliers are not certified and alternative suppliers have not yet been developed to replace them, although an active program is in process. For suppliers not certified, product inspection is performed on all incoming materials. These inspections consist of samples of chemical analysis and physical properties. The measurements to be sampled are determined during the design process to maintain material capabilities that are utilized in the production process. In addition to quality requirements, price and delivery are also important and are consistently monitored by the purchasing group.

How Performance Is Fed Back

The PTP program is structured with a series of meetings with involved suppliers that occur at least quarterly. These meetings are held with all concerned sectors represented to minimize the number of meetings of the company.

The number of suppliers in the PTP program is shown in Figure 25-16. Also shown is the number of Certified Suppliers.

For suppliers in the PTP program, all have the EDI link to receive real-time feedback on their performance. In addition, most have electronic mail connectivity for regular communications with CFI.

For suppliers not in the formal PTP program, information is mailed to them from the Purchasing organization once per quarter in the form of a printout from the ODS. This Supplier Rating System (SRS) information is automatically compiled, and rejection rates are presented along with supplier receipt information.

If any lot rejection rate exceeds 250 ppm, the supplier is flagged with a Supplier Corrective Action Request (SCAR). A specific site visit is made to resolve the problem by determining the root cause and developing a Corrective Action Plan (CAP).

How the Company Improves Supplier Performance

Improve Suppliers' Abilities

As described earlier in the application, CFI utilizes several programs to improve suppliers' capabilities. The PTP program for participating suppliers has many ways to help.

A Cost of Doing Business (CODB) factor has been established for all suppliers, which calculates the costs to CFI due to suppliers' failure to meet requirements, such as the costs to return lots which fail to meet specification. The CODB factor is used in subsequent procurements to give preference to high-performing suppliers.

Annually, a planning meeting is held with the top-40 volume suppliers to develop plans for improvements. In this planning meeting, action plans with targets are developed for both CFI and the supplier for the upcoming year. These action plans and targets are fed into the company's annual planning process.

CFI has provided supplier training and recognition as incentives to improve. The data systems discussed in Category 2.0 explain how data are available, timely, and accurate. If the incentives and help are not enough for improvement to occur, the supplier may well become one of those in the supplier base reduction plan.

Improve Procurement

The PTP program provides a two-way dialog with individual suppliers. This gives them the opportunity to point out situations where the purchasing process can be improved.

Last year, CFI started a Preferred Customer Certification Program (PCCP). Principal suppliers have helped develop a set of criteria for rating whether CFI is the customer of choice. These criteria include: timeliness, quality, clarity of communication, and supplier satisfaction. Quarterly Survey Inc., an independent contractor, sends all CFI suppliers a survey. Analysis is conducted on the survey, and ratings are established similar to the PTP program. Action plans are established by a multi-functional internal team to improve the procurement and supplier management process. Results of the survey and action plans are fed to all suppliers.

In 1994, one of the suppliers that had not attained certification, Salter Inc., explained in a memo that it was not provided a detailed set of reasons why its product, a special washer, was being rejected. The investigation by the process team determined that the washer specification had been modified for use in another sector. Further analysis proved that a second supplier of the same washer had agreed to the change and had been shipping to a tighter specification. The purchasing procedures were modified, Salter was able to make the changes, and the rejections stopped.

Minimize Costs of Inspection

The process control capability concept is directed toward reduced costs in receiving inspection. In the case where suppliers have processes greater than a C_{pk} of 1.5, they also can cease final inspection. Thus, the thrust of capability studies is directed toward lower costs and reduced lead-times.

The suppliers who have achieved certification send process performance data through the EDI system to CFI. When these data show that the supplier is attaining a C_{pk} of at least 1.5, then all inspections and audits are suspended for that supplier. As long as the data continue to show the processes are under control, no audits or inspections are conducted. Further, if the supplier takes prompt actions for processes which are drifting out of control, inspections and audits are not resumed. Our PCCP survey results indicate that all suppliers find this proactive approach very helpful and meaningful.

EXERCISES

25-1 Give a scenario that highlights the significance of process capability studies in each of the following:

a. An airline
b. A commercial bank
c. A university
d. The mass transit system
e. A hotel
f. A hospital

25-2 The individual units of a product would be said to conform if the diameter is 6.0 mm ± 0.08 mm. Determine the two-sided tolerance, LSL, and USL. What conclusions can you draw from your answers?

25-3 The data presented below represent the length in cm of a rubber liner used in protecting the delicate side of a medical device. On the basis of range and standard deviation, determine the process capability. Using both estimates of the standard deviation values, calculate C_p and C_{pk}. What conclusions can you draw from your results?

Subgroup	X_1	X_2	X_3	X_4
1	11.05	10.96	10.88	10.78
2	10.89	11.00	12.07	11.06
3	11.02	11.99	12.08	11.88
4	10.99	11.12	11.99	11.07
5	10.96	11.10	12.02	12.08
6	12.01	11.02	12.04	11.99
7	12.00	11.97	11.95	11.02
8	12.02	11.93	11.98	12.04
9	11.97	12.13	12.06	11.95
10	11.95	11.98	11.97	11.98
11	12.03	11.99	11.98	12.06
12	11.78	11.09	11.78	11.97
13	11.96	11.89	10.96	11.98
14	10.88	11.02	11.88	12.04
15	12.07	11.99	12.07	12.09
16	12.08	11.96	11.08	11.89
17	11.99	12.01	11.99	10.02
18	12.02	10.00	12.02	11.99
19	11.04	12.02	11.04	11.96
20	11.95	11.97	11.95	12.01
21	11.98	11.95	11.98	10.00
22	12.06	12.03	12.06	12.02
23	11.97	11.78	11.97	11.97
24	11.98	11.96	11.98	11.95
25	11.04	11.68	11.04	11.03

Process Management at Colony Fasteners

Compare each of the following criteria to Colony Fasteners and indicate whether the situation in the company is a strength (S) or needs improvement (I). Justify your answers.

25-4 Design and Introduction of Products and Services

- Customer satisfaction and manufacturability are considered as prime drivers for new and improved processes.
- Suppliers are included in the early design phase to ensure that raw materials they supply can be controlled for process consistency.
- A Product Design Cycle (Figure 25-8) shows the flow of a systematic process, including customer requirements and product and service design requirements.
- Product and service design requirements are clearly translated into effective production/delivery processes.

25-5 Process Management: Product and Service Production and Delivery

- There is a systematic benchmarking approach for improving process management.

- During the design cycle, every process is designed and proven before the design is released.
- There is a systematic approach for identifying production and delivery processes that are falling short of customer requirements.
- Specifically assigned process teams focus on improving processes and reducing variability.

25-6 Process Management Support Services

- Support service requirements are determined based on strategic plan requirements and input from employees.
- Requirements determination is driven by the team process of each individual team and is systematic or repeatable within a division or across divisions.
- There are systematic approaches for using process analysis, benchmarking, and customer input to evaluate and improve support service processes, including cycle time.
- Processes in the support functions are defined with corresponding measures and indicators for each step.

25-7 Management of Supplier Performance

- The company's defect level goal of Six Sigma is matched by suppliers that also have the same defect level goal.
- Certified suppliers receive special bar codes to attach to their products to show that inspection upon receipt is not necessary since they have already gone through an in-depth certification process.
- A Cost of Doing Business (CODB) factor has been established for all suppliers.
- Supplier feedback is used to improve the procurement process.

ENDNOTES

1. Gitlow, H, Oppenheim, A. and Oppenheim, R. *Quality Management,* 2nd ed., Chicago, IL: Irwin, 1995, pp. 354–358.
2. Montgomery, D. C. and Runger, G. C. *Applied Statistics and Probability for Engineers,* 3rd ed., New York, NY: John Wiley & Sons, 2003.

REFERENCES

Besterfield, D. H. *Quality Control,* 5th ed., Upper Saddle River, NJ: Prentice Hall, 1998.

Goetsch, D. L. and Davis, S. B. *Quality Management,* 4th ed., Upper Saddle River, NJ: Prentice Hall, 2000.

Montgomery, D. C. *Statistical Quality Control,* 2nd ed., New York, NY: John Wiley & Sons, 1991.

Omachonu, V. K. *Healthcare Performance Improvement,* Norcross, GA: Engineering and Management Press, 1999.

Smith, G. M. *Statistical Process Control and Quality Improvement,* 5th ed., Upper Saddle River, NJ: Prentice Hall, 2004.

Summers, D. C. *Quality,* 2nd ed., Upper Saddle River, NJ: Prentice Hall, 2000.

Swift, J. A. *Modern Statistical Quality Control and Management,* Delray Beach, FL: St. Lucie Press, 1995.

26

INTRODUCTION TO RELIABILITY

INTRODUCTION

In the previous chapters, we discussed the concept of quality management during the manufacture of products or the delivery of services. We used the concepts of subgroup statistics and control charts to measure or monitor quality at a specific moment in time. The question we were unable to answer was, "Will the product continue to perform its intended function as prescribed over the course of its life?"

Reliability is a measure of the extent to which the product will retain its quality over time. Rather than look at a snapshot of time, we consider the concept of quality over the long run. Reliable products are products you can depend on to function the way they are supposed to. Everything an organization does (from raw materials to design, manufacture, packaging, and shipping) can have an important effect on the reliability of the product over time. The concept of reliability has some implications in the service sector as well. Everything a travel agent does (from creating the vacation package, understanding the needs and expectations, negotiating rates, checking the quality of accommodation, food, etc.) can have a profound impact on the quality of the service experienced by all the members of a family during their vacation. When the product does not perform optimally during its life, there is a tremendous cost to the customer, and warranty costs to the manufacturer.

A hospital has hundreds of pieces of diagnostic equipment, devices, and tools that are put to everyday use in the course of providing care. Today, a patient entering a hospital will most likely come in contact with or use one of these: patient monitor, respirator, electroencephalograph (EEG), heart pacemaker, electrocardiograph (ECG), high voltage radio therapy equipment, x-ray equipment, defibrillator, anesthesia machine, heart pump, dialysis machine, suction pump, hyperthermia apparatus, or heart-lung machine. There is a steady growth in the number, variety, and complexity of these devices. Healthcare professionals depend on these devices for accurate diagnosis, monitoring, treatment, and examination. Have you ever contemplated what happens if they fail? What happens if your blood chemistry produces false or inaccurate test results? What happens if an important diagnostic device stops working altogether? Hospitals depend on these high-tech devices to provide accurate diagnosis so that the proper treatment can begin in a timely manner. In most cases, hospitals become concerned about product reliability issues only when the device fails at a particularly critical time, or when they become immersed in a litigation battle.

RELIABILITY

Reliability is the probability that a product will perform its intended function satisfactorily for a prescribed period of time when it is used under the specified environmental conditions. This definition presents four important considerations in the study of reliability.[1]

1. **Numerical Value** — Reliability is a numerical value ranging between 0 and 1. For example, if the reliability of a mechanical pencil is 0.94, it means that the probability is 94% that the pencil will perform its intended function satisfactorily during its specified life under certain stated conditions. It also means that 94 of 100 pencils will perform, while 6 will not.
2. **Intended Function** — Most products are designed for particular applications. For example, a kitchen knife is not designed for opening canned products, nor is a screwdriver designed for opening paint cans. Kitchen knives are designed for cutting meat, vegetables, etc., and screwdrivers are designed for turning screws.
3. **Intended Life of the Product** — The intended life of a product is a function of usage, time, or both. For example, the intended life of automobile tires can be specified as 36 months or 36,000 miles. The intended life of a light bulb is stated as 2,000 hours.
4. **Environmental Conditions** — Certain products are designed for indoor use, outdoor use, or both. Certain types of ping pong tables are designed for both indoor and outdoor use, while others are designed for only indoor use. Certain medicines need to be stored at room temperature in order to preserve their potency.

Why the Emphasis on Reliability?

The Consumer Protection Act of 1972 marked a critical turning point in product safety. In 1972, Congress determined that "an unacceptable number of consumer products" presented unreasonable risks of injury to the consumer. This act, along with other government legislative actions, has raised the level of awareness of both manufacturers and consumers in matters related to product performance.

Products as we know them have become increasingly complicated over the years. Manufacturers are racing to add more features and more components. The probability of product failure increases as the number of features increases. Manufacturers are increasingly aware of the challenge of increasing the reliability of their products to match the additional features. In the age of automation, manufacturers grapple with how to design automated products that are still operable manually even when the automatic component fails.

Increased competition has created a barrage of competing products. Product reliability has become a competitive weapon in the battle for market share. Today, many manufacturers have made the goal of increased product reliability part of their strategic plan. History has shown us that when a company consistently produces unreliable products, it would not survive.

Product Life Cycle Curve

Most products go through three distinct phases from inception to wear-out. Figure 26-1 depicts a typical product life cycle curve. The curve,

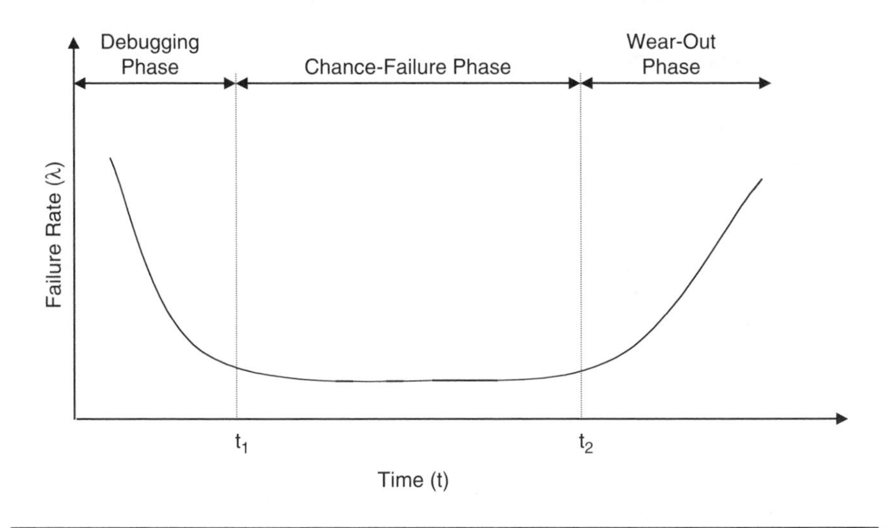

Figure 26-1 Product Life Cycle Curve

sometimes referred to as "bathtub" curve, shows a plot of the failure rate λ as a function of time. It consists of the debugging phase, the chance-failure phase, and the wear-out phase.

The *debugging phase*, which is also called infant-mortality phase, is characterized by a time in the life of the product when there is a drop in the failure rate as early failures are identified and worked out, particularly during prototype testing or pre-shipment testing. During this phase, the curve is exponential. Some of the early failures are due to any number of factors including inadequate materials, incorrect installation, or errors in the manufacturing process.

The *chance-failure* phase occurs between time t_1 and t_2, at a constant rate. Failure during this phase occurs randomly, and may be due to misapplication or misuse. The assumption of a constant failure rate is valid for most products; however, some products may have a failure rate that increases with time. The final phase of the cycle is the *wear-out phase*, and is characterized by a rise in the failure rate that occurs as the product ages and wears out. Normal wear could result in misalignment, loose fittings, and component interference, and could lead to an increase in the failure rate.

Before we introduce the components of reliability, let us first understand the causes of unreliability.

Causes of Product Unreliability

Several factors can act in concert or individually to affect the reliability of a product. The following are among the most common factors:

At the Manufacturer's Site

- Improper design of the product
- Inferior construction materials
- Faulty assembly
- Faulty manufacturing
- Inappropriate testing, leading to false results and wrong conclusions
- Damage during shipment
- Inadequate packaging, leading to damage during shipment

At the Hands of the User

- Improper startups
- Product abuse
- Lack of maintenance
- Misapplication of product (e.g., using a pen to pry open a box)

MEASURES OF RELIABILITY

One of the purposes of reliability tests is to determine if there are recognizable patterns of failure during the life cycle of a product. Reliability tests seek to answer three important questions: What failed? How did it fail? And how many hours, cycles, actuations, or stresses was it able to bear before failure?[2] There are a number of tests for determining the reliability of a product, including failure-terminated, time-terminated, and sequential tests. *Failure-terminated* tests are ended when a predetermined number of failures occur within the sample being tested. The decision to accept or reject the product hinges on the number of products that have failed during the test. A *time-terminated test* ends when a pre-established number of hours is reached. The decision to accept or reject the product is based on the number of products that failed before reaching the time limit. A *sequential test* is based on the accumulated results of the tests performed.

Failure Rate, Mean Life, and Availability

When the performance of a system is a function of time, such as the number of hours a light bulb is expected to burn, then reliability is measured in terms of mean life, failure rates, availability, mean time between failures, and specific mission reliability.[3] The data on failures are accumulated as the system is put to use. This accumulated data are used to estimate failure rate and the mean life of the system. The failure rate λ is the probability of a failure occurring during a specified period of time or cycle. The failure rate can be estimated from test data as follows:

$$\lambda_{est} = \frac{r}{\sum t + (n-r)T}$$

Where

λ = failure rate, which is the probability that a unit will fail in a stated unit of time or cycles

r = number of test failures

t = test time for a failed item

n = number of items tested

T = termination time

From the value of λ above, the average life θ can be estimated.

$$\theta_{est} = \frac{1}{\lambda}$$

Failure rate may be expressed in terms of failures per hour and is usually represented by the symbol lambda, λ. For many products and components, the value of lambda can be quite small, therefore, in some cases, it is expressed in terms of failure per one million hours or in scientific notation: 10^6 hours. Other forms of expressing failure rate apply, such as number of transactions — as in number of failures per million ATM transactions. It is important to note that the value of lambda represents an average value.

Example 26-1

A manufacturer of scooter motors is interested in improving the reliability of its motors. Twelve scooter motors are being tested using a time-terminated test. The test is concluded when each motor completes a total of 220 hours of continuous operation. During this time, it was observed that four motors failed prior to reaching the 220 hours. The four motors failed at times 190 hours, 140 hours, 156 hours, and 205 hours respectively. What are the failure rate λ and the average life θ?

$$\lambda_{est} \quad \frac{4}{(190 + 140 + 156 + 205) + (12 - 4)\text{:}}$$

$$= \frac{4}{2451} = 0.0016$$

From this, the average life θ can be estimated.

$$\theta_{est} = \frac{1}{\lambda} = \frac{1}{0.0016} = 625 \text{ hou}$$

$$\theta = \text{MTTF} = 625 \text{ hours}$$

Mean times between failures (MTBF) and *mean time to failure* (MTTF) are used to express reliability as a function of time. For repairable equipment, θ is also equal to the mean time between failure (MTBF). There will be a difference between MTBF and MTTF only if there is a significant repair or replacement time upon failure of the product.[4] As is the case with many critical medical devices and equipment, the amount of time a device or equipment is available for use is very important. In such situations, reliability can be judged in terms of the amount of time an equipment or a device is available. Hence,

$$\text{Availability} = \frac{MTTF}{MTTF + \text{mean time to repair}}$$

MTBF values can be used in place of MTTF.

Example 26-2

The motor in a medical device has a mean time between failure (or average life) θ of 1420 hours. When the motor fails, the device malfunctions, and requires 48 hours to repair it. Determine the availability of the device.

$$\text{Availability} = \frac{MTTF}{MTTF + \text{mean time to repair}}$$

$$= \frac{1420}{1420 + 48} = 0.9$$

The medical device is available 97% of the time.

Example 26-3

Find the failure rate for five items that are tested-to-failure, based on the following test cycles: 1750, 2035, 3780, 5892, and 8019.

Since this is a failure-terminated test, the value of lambda can be determined as follows:

$$\lambda = \frac{r}{\sum t}$$

$$= \frac{5}{1750 + 2035 + 3780 + 5892 + 8019}$$

$$= 0.000233$$

SYSTEM RELIABILITY

There are some who believe that the more features a product has the better its quality. While that may be true in some cases, it also depends on how one defines quality. Your perspective might depend on whether you define quality in terms of the frequency with which a product fails, or how long it performs without failure, or all the things the product is able to do, etc. One thing is certain: as products become more complex (fitted with more components and features), the chance that they will develop problems increases. There was a time when a camera was a simple device for taking pictures. You were required to manually load the film, manually advance the film, manually attach the flash, point, and shoot with the click of a button. Today, the camera comes with a 100-page document that tells the reader about dozens of components and features. Modern features include an autoloading device, built-in-flash, auto film advance, autofocus, photo date, distance adjustment, light adjustment, etc. The basic cell phone was a device for making and receiving phone calls. Today, some cell phones are capable of text messaging, producing and sending digital pictures, and providing access to the Internet. The reliability of these systems is ultimately affected by the number of components in the systems. The reliability of the entire system will also be affected by the manner in which the components are arranged. Components are arranged in series, parallel, or a combination. Figure 27-2 illustrates the various arrangements.

Note that the "R" values are the probability that the components will work. When components are arranged in series, the reliability of the system is the product of the reliability of the individual components. For the arrangement shown in Figure 26-2(i), the reliability of the system, R_s, is computed as follows:

$$\begin{aligned} R_s &= (R_A)(R_B)(R_C)(R_D) \\ &= (0.975)(0.997)(0.905)(0.990) \\ &= 0.871 \end{aligned}$$

Note that the reliability of the system, R_s, is lower than the individual reliabilities of the components. When a system is arranged in series, if a component does not function, the entire system does not work.

For the arrangement shown in Figure 26-2(ii), the components are arranged in parallel. In this case, if a component does not function, the component continues to function using another component until all parallel components fail. Thus, for parallel components, the system reliability is determined as follows:

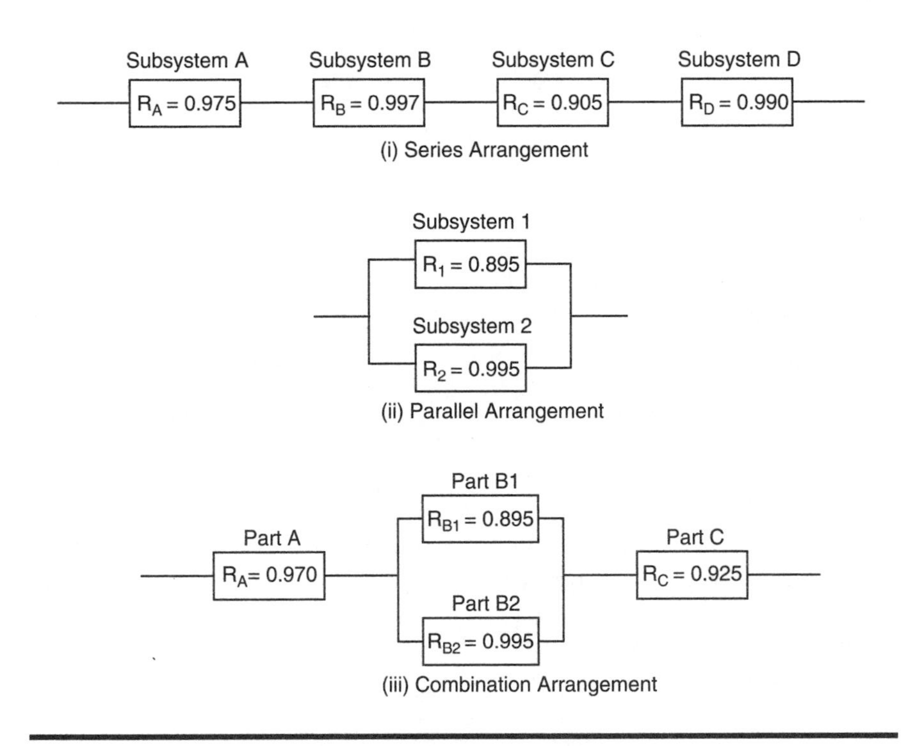

Figure 26-2 Components Arrangement

$$
\begin{aligned}
R_s &= 1 - \text{Probability that components 1 and 2 will not function} \\
&= 1 - (1 - R_1)(1 - R_2) \\
&= 1 - (1 - 0.895)(1 - 0.995) \\
&= 0.9995
\end{aligned}
$$

As the number of components in parallel increases, the reliability of the system increases. This is due to the fact it would take a failure of all of the additional parallel components in order for the system to fail. Notice also that the reliability of the system is greater than the reliability of the individual components.

It would be nice to make every system adopt the parallel arrangement; however, this is not realistic. Most complex products tend to be a combination of series and parallel arrangements of components. The arrangement presented in Figure 26-2(iii) shows a combination of series and parallel arrangements of components. The reliability of the system is determined as follows:

$$R_s = (R_A)(R_{B1,B2})(R_C)$$

$$= (R_A)[1 - (1 - R_{B1})(1 - R_{B2})](R_C)$$
$$= (0.970)(0.9995)(0.925)$$
$$= 0.897$$

Implications for Design

- The simpler the design, the more reliable it is.
- The fewer the number of components, the higher the reliability of the system.
- The reliability of a system can be improved by having a backup or redundant component. When the primary component fails, the backup component is activated. This idea is reflected in the parallel arrangement of components.
- When the cost of failure is very high, we can achieve higher reliability with a more robust design that would withstand any misapplication.
- Reliability can be improved when a product is designed for easy maintainability.
- Better packaging can preserve the reliability of a product, since products get damaged during shipment. The mode of transporting the product to the customer can also be a factor in enhancing the reliability of a product.

EXERCISES

Questions

26-1 How would you define reliability with respect to the following?
  a. Mail delivery service
  b. Courier service package delivery
  c. Blood pressure monitoring device
  d. A television set

26-2 What are the factors that would affect the reliability of a ceiling fan for home use?

26-3 How would you analyze the reliability issues affecting your local fire department?

26-4 List the features of the following modern devices and discuss how these features might affect their reliability.
  a. Photocopy machine
  b. Washing machine
  c. Refrigerator
  d. Camera

26-5 Give three examples of products that are often used in ways that are different from what the manufacturers intended or specified. How might the misapplication affect the reliability of the products?

Problems

26-1 A manufacturer of a pool pump wants to improve the reliability of its pool pump motors. Fifteen pool pumps are being tested using a time-terminated test protocol. The test is concluded when each motor completes a total of 325 hours of continuous operation. It is observed that six of the motors stopped working prior to reaching the 325 hours. The six motors stopped working at times 302 hours, 270 hours, 291 hours, 205 hours, 311 hours, and 261 hours respectively. Determine the failure rate λ and the average life θ.

26-2 The motor inside a slide projector has been targeted for a reliability study. Twenty projector motors are being tested using a time-terminated test. Each motor is required to complete the test by running for a total of 150 hours continuously. Three motors failed during the test, at times 95 hours, 132 hours, and 140 hours respectively. What are the failure rate λ and the MTTF?

26-3 The motor in an O_2 analyzer device has a time between failure θ of 1720 hours. When the motor fails, the device is inoperable, and requires 72 hours to repair it. What is the availability of the device?

26-4 Determine the failure rate for 7 items that are tested-to-failure, based on the following test cycles: 1975, 2460, 4550, 5280, 6148, 7460, and 8210.

26-5 What is the failure rate for 4 items that are tested-to-failure, based on the following test cycles: 2804, 3042, 5201, and 5840.

26-6 A system has five components, A, B, C, D, and E, with reliability values of 0.95, 0.92, 0.88, 0.97, and 0.96, respectively. If the components are arranged in series, determine the reliability of the system.

26-7 A product has 20 components arranged in series. Each of the components has a reliability of 0.98. What is the reliability of the system? Suppose the new model of this product scheduled for release next year, has only 15 components, each with the same reliability of 0.98. How would the reliability of the system change? Comment on the difference.

26-8 What is the reliability of the three systems shown on the following page?

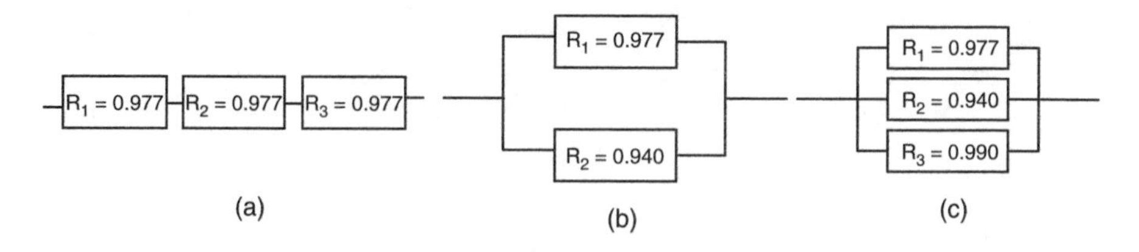

26-9 Determine the reliability of the system shown below, given that the reliabilities of the components A, B, C, D, E, and F are 0.96, 0.94, 0.88, 0.87, 0.93, and 0.99 respectively.

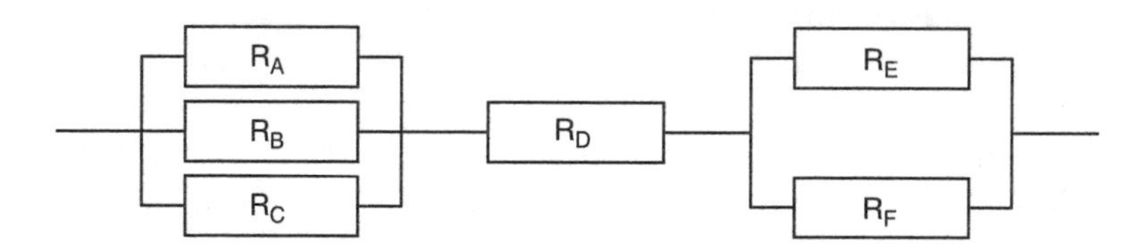

26-10 What is the reliability of the system shown below, given that the reliabilities of the components A, B, C, and D are 0.98, 0.95, 0.91, and 0.90 respectively?

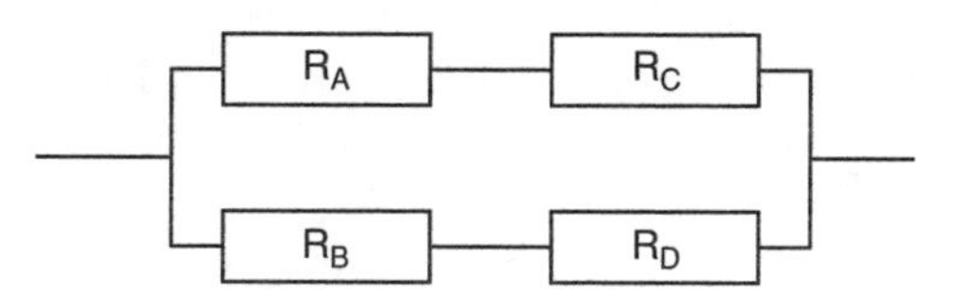

ENDNOTES

1. Besterfield, D. H. *Quality Control,* 5th ed., Upper Saddle River, NJ: Prentice Hall, 1998, pp. 391–394.
2. Summers, D. C. S. *Quality,* 2nd ed., Upper Saddle River, NJ: Prentice Hall, 2000.
3. Summers.
4. Mitra, A.. *Fundamentals of Quality Control and Improvement,* 2nd ed., Upper Saddle River, NJ: Prentice Hall, 1998.

REFERENCES

Neubeck, K. *Practical Reliability Analysis,* Upper Saddle River, NJ: Pearson Education, Inc. Prentice Hall, 2004.

Omachonu, V. K. *Health Care Performance Improvement,* Norcross, GA: Engineering and Management Press, 1999.

Sower, V. E., Savoie, M. J. and Renick, S. *An Introduction to Quality Management and Engineering,* Upper Saddle River, NJ: Prentice Hall, 1999.

27

INTRODUCTION TO SIX SIGMA

The six-sigma approach is a systematic application of business and statistical concepts and techniques for the purpose of reducing process variations and preventing deficiencies in a product. Six Sigma is both a technique and a philosophy based on the desire to eliminate waste and improve performance as much as is technically possible. Motorola introduced this innovative development to quality in the mid-1980s, in its quest to reduce defects of manufactured electronic products. The main goal of Six Sigma is to improve the performance of processes to the point where the rate of defect is 3.4 per million or less. The concept was designed for use in a high-volume manufacturing or service environment. A defect could be any of the following:

- A faulty part
- Incorrect customer bill
- Turnaround time for x-ray

The name Six Sigma is derived from the statistical concept of standard deviation, usually denoted by the Greek letter sigma (σ). The variation in a process or in the output of that process is typically measured in terms of the number of standard deviations from the mean (Figure 27-1). The following are examples of Sigma values and the corresponding Defects Per Million Opportunities:[1]

Sigma values	*DPMO (statistical)*	*% Acceptable (statistical)*	*DPMO (Motorola)*	*% Acceptable (Motorola)*
1	317,400	68.26	697,700	30.23
2	45,400	95.46	308,733	69.1267
3	2,700	99.73	66,803	93.3197
4	63	99.9937	6,200	99.38
5	0.57	99.999943	233	99.9767
6	0.002	99.9999998	3.4	99.99966

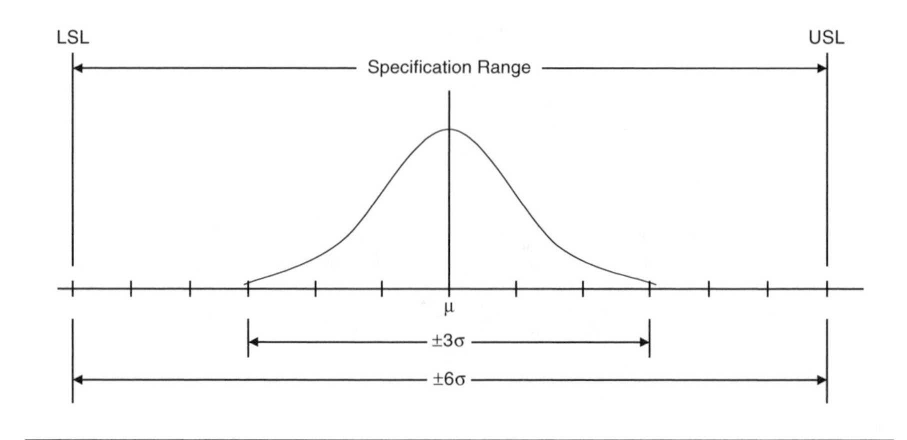

Figure 27-1 Specification Achieved at the Six Sigma Level

It is clear that the numbers used by Motorola in its Six Sigma program are significantly different from those derived from statistical knowledge. For example, at Three Sigma, there will be 2,700 failures out of one million opportunities. This represents a success rate of 99.73%. At Six Sigma, the prediction is that 0.002 failures will occur out of one million opportunities; that is a success rate of 99.9999998%. However, the most often cited number of failures per million opportunities at Six Sigma is 3.4. This is a significant difference. Statistics predicts one failure in 500 million opportunities at Six Sigma. Motorola uses 3.4 defects out of one million, or 1,700 times more failures that what is predicted by statistics. The question is, which is correct?

Most organizations operate at between 3 and 4 sigma, that is, between 66,800 and 6,210 DPMO (Figure 27-2). All processes are not equally

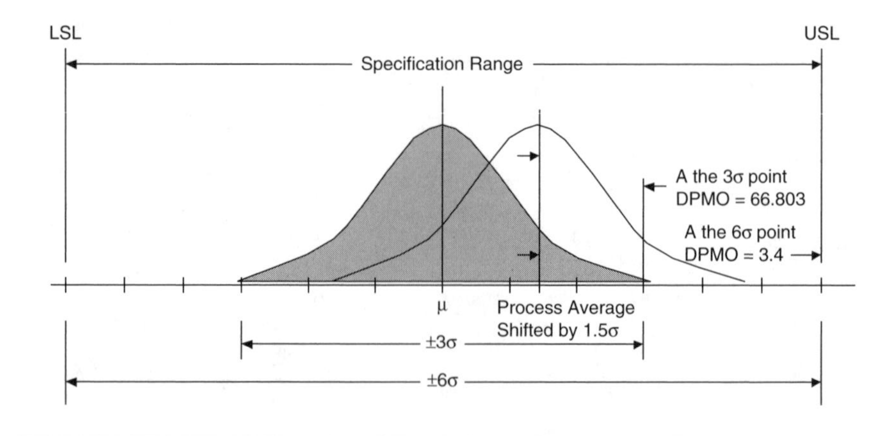

Figure 27-2 Specification Achieved at a Shift of 1.5 Six Sigma Level

important; therefore, it would be foolish to try to achieve Six Sigma levels of performance for every process in the organization. The focus of Six Sigma should be directed at the mission-critical areas. The criticality of a process should be driven by the requirements and needs of the customers. The critical aspects of a product, service, or process are referred to as the "critical-to-quality" requirements or CTQs for short. The Six Sigma process translates customers' needs into separate tasks and defines the optimal specification for each task depending on how each task interacts with others. Once the critical processes and tasks are defined, depending on the analysis and improvement interventions, the process of Six Sigma can be used to drive the performance of products, services, and process to breakthrough levels.

Some Examples of "Defects per Million"

- At Six Sigma, there are 5.4 deaths per million caused by anesthesia during surgery.
- At Five Sigma, there are 230 airline fatalities per million.
- At Two Sigma, there are 580,000 patients with depression who are not detected or treated.

THE HISTORY OF SIX SIGMA

In 1981, Bob Galvin, then chairman of Motorola, challenged his company to achieve a tenfold improvement in performance over a five-year period. Motorola became profoundly successful in its use of Six Sigma. As other organizations studied its success, Motorola realized the need to further extend its strategy. At Motorola, Six Sigma has been and still is defined as a quality improvement program with a goal of reducing the number of defects to as low as 3.4 parts per million opportunities. In the ten years following Garvin's challenge to his company, and as a result of the implementation of Six Sigma, Motorola claims to have saved over $400 billion.[1] Motorola won the Malcolm Baldrige National Quality Award in 1988 for its pioneering effort in the development and application of the Six Sigma concept. As the success of Six Sigma became widely known, other organizations embraced the challenge. In 1995, Jack Welch, CEO of General Electric, committed his company to the Six Sigma process. Shortly thereafter, Allied Signal became the other large company to take on Six Sigma. Despite the fact that Six Sigma was still linked to total quality management, it was beginning to evolve as a legitimate business strategy.

THE SIX SIGMA CONCEPT

When the concept of Six Sigma was first launched at GE Aircraft Engines, it was presented as a four-step methodology — Measure, Analyze, Improve, and Control (MAIC). Recently, the define phase has been added to recognize the importance of having a properly scoped project. The practice of Six Sigma takes the form of projects carried out in phases generally recognized as Define-Measure-Analyze-Improve-Control (DMAIC).

Define Phase

The key questions to ask during the Define phase are: Who are the customers and what are their priorities? This phase is characterized by the identification, evaluation, and selection of projects, preparation of the mission, selection and launching of the team. The Define phase includes the identification of the CTQs (critical to quality characteristics) that the customer considers to have the most impact on quality.

Measure Phase

In this phase, select the most appropriate output quality characteristics to be improved; then measure the scope of the problem or establish what is unacceptable performance or a defect for such characteristics. Document the process, document potential failure modes and effects, and gather preliminary data to evaluate current process performance and capability.

Analyze Phase

Plan for data collection, perform an analysis of the data, analyze the root causes of defects or errors, and establish and confirm the "vital few" determinants of performance.

Improve Phase

Design and conduct experiments to determine the mathematical cause-and-effect relationships, and optimize the process. The goal in the Improve phase is to reduce the defect rate or number of defects using simple but powerful statistical tools and techniques. For some processes, several rounds of improvements may be necessary to achieve a desired process performance or capability.

Control Phase

In the Control phase, sustain the gains that have been achieved from the improvement phase. Design controls necessary to maintain the gains, and continue to monitor.

SIX SIGMA FEATURES

- The DMAIC framework allows for the logical integration of techniques such as Quality Function Deployment (QFD), Failure Mode and Effect Analysis (FMEA), Design of Experiment (DOE), and Statistical Process Control (SPC).
- Like total quality management, the approach for Six Sigma is "top-down."
- Six Sigma is gaining wide acceptance in the service sector, especially when applied to transactional situations.
- Six Sigma is driven by the requirements of the customers. This can be seen through its Critical to Quality (CTQ) focus.
- It emphasizes its project-by-project feature in its implementation.
- The outcomes of Six Sigma projects are usually expressed in financial terms.
- Six Sigma demands an elaborate training and certification regimen leading to a certification hierarchy, such as Green Belts, Black Belts, Master Black Belts, etc.
- The Six Sigma process is driven by performance metric defined in terms of sigma levels and defects per million opportunities.

Customer-Focused

Six Sigma translates a customer's needs into individual tasks and defines the optimal specification for each task, depending on how the tasks interact. The focus quickly turns to how to achieve breakthroughs using the project-by-project approach. Breakthrough improvement is needed to drastically reduce chronic waste, i.e., waste that is systemic and built into the operations of the organization. The cost attributed to this chronic waste is what Dr. Juran refers to as the *Cost of Poor Quality* (CPQ).

BENEFITS OF SIX SIGMA

The following are the benefits of implementing Six Sigma, as reported by some organizations:

Motorola (1987–1994)[2]

- Reduced in-process defect levels by a factor of 200.
- Reduced manufacturing costs by $1.4 billion.
- Increased stockholders' share value fourfold.

Allied Signal (1992–1996)[3]

- Reduced new product introduction time by 16%.
- Reduced manufacturing costs by more than $1 billion.

General Electric (1995–1998)[4]

- Company-wide savings of over $1 billion.

Samsung Electronics (1999–2001)

- Significant savings and financial benefits in all 16 of its business units in South Korea and internationally.[5]

Honeywell (after its merger with Allied Signal) (1999–2001)

- Reportable cases with respect to safety were reduced 43% and lost workday cases by 50% in 1999 compared to the previous year.[6]

Service and transactional processes

- Reduced medication and laboratory errors and thereby improved patient safety.[7]
- Significant savings in process timeliness, improvements in cash management and increased customer loyalty and satisfaction.[8]

EXERCISES

27-1 How does the Six Sigma process of DMAIC compare to the PDSA concept described earlier in this book?

27-2 Give five examples of transactional process applications for Six Sigma in the service sector.

27-3 In what ways is the Six Sigma philosophy different from Total Quality Management?

27-4 Provide some examples of "defects per million" within the following industries:

- Banking
- Healthcare
- Transportation
- Education
- Law Enforcement

ENDNOTES

1. Harry, M. and Schroeder, R. *"Six Sigma": The Breakthrough Management Strategy Revolutionising the World's Top Corporations,* New York: Doubleday Currency, 2000.

2. Air Academy Associates, LLC. "Six Sigma For Manufacturing And Non-Manufacturing Processes," available at http://www.airacad.com/papers/Six-sigma.html. Accessed February 1, 2004.
3. Air Academy Associates.
4. Air Academy Associates.
5. Chua, R. C. "What You Need to Know About Six Sigma," *Productivity Digest* (2001), pp. 37–44.
6. Chua, pp. 37–44.
7. Buck, C. "Applications of Six Sigma to Reduce Medical Errors," Milwaukee: ASQ Congress Proceedings, 2001, pp. 239–242.
8. Rucker, R. "Citibank Increases Customer Loyalty With Defect-Free Processes," *The Journal for Quality and Participation,* Vol. 23, No. 4 (2000), pp. 32–36.

REFERENCES

Anthony, J. and Banuelas, R. "Key Ingredients For the Effective Implementation of Six Sigma Program," *Measuring Business Excellence,* Vol. 6, No. 4 (2002), pp. 20–27.

Chassin, M. R. "Is Health Care Ready for Six Sigma?" *The Milbank Quarterly,* Vol. 76, No.4 (1998), pp. 565–591.

Goh, T. N. "A Strategic Assessment of Six Sigma," *Quality and Reliability Engineering International,* No. 18 (2000), pp. 403–410.

Henderson, K. M. and Evans, J. R. "Successful Implementation of Six Sigma: Benchmarking General Electric Company," *Benchmarking: An International Journal,* Vol. 7, No. 4 (2000).

28

HEALTHCARE SERVICE EXCELLENCE

Healthcare organizations are spending hundreds of billions of dollars on marketing, advertising, and promotional efforts aimed at attracting new customers, while many of the existing customers routinely slip away through the back door. This is like a metaphorical leaky bucket, with much of the emphasis on the volume of water being poured into the bucket, while continuing to ignore the substantial leak. Little or no effort is made to seal the leak. In managed care organizations, it is not uncommon to speak of disenrollment figures of anywhere from 2 to 25% per month. Hospital patients are asking their physicians not to send them back to certain hospitals, some managed care members are asking not be sent to certain hospitals, and some homebound patients are urging their physicians and legal guardians not to do business with certain home healthcare agencies.

Posterity will remember the decades of the 1980s and 1990s as the era of incessant debates over the value of healthcare services. This era gave rise to healthcare customers known for their fierce intolerance toward service mediocrity. Customer service has become the new yardstick for measuring performance in healthcare organizations.

SERVICE IS THE COMPETITIVE EDGE

In the era of managed care, it is quite common to find several physicians who provide medical services for a number of managed care organizations. The price disparities among the managed care organizations are increasingly less of a distinguishing factor in the choice of a managed care company. If price and medical care can be considered constants, then the only distinction lies in the superiority of the service provided. Similarly, most hospitals can treat headaches, asthma, colds, diabetes, and heart diseases. The insurance co-payments and price are only slightly different or constant. Service excellence has become the new variable.

Research in service organizations has concluded that it is five to six times more expensive to get a new customer than to keep an old one. In managed care organizations, most disenrollment occurs within the first three to six months of membership. The cost of getting a new member includes the following:

- Advertising cost
- Sales commission for brokers
- Brokers salaries
- Marketing costs
- Underwriting costs
- Costs of investigating pre-existing conditions

Customers are complaining loudly and litigiously about poor customer service. In today's fiercely competitive market, a healthcare organization that ignores its customers is embracing a prescription for failure. According to a 1985 study conducted for the U.S. Office of Consumer Affairs by TARP, a Washington, D.C., consultant:[1]

- The average business never hears from 96% of its unhappy customers. For every complaint received, the average company in fact has 26 customers with problems, 6 of which can be described as "serious" problems.
- Those who complain are more likely than those who don't to do business with the organization that upset them, even if the problem isn't satisfactorily resolved.
- Of the customers who register a complaint, between 54% and 70% will do business with the organization again if their complaint is resolved. That figure jumps to a staggering 95%, if their problem is resolved on the spot.
- Each unhappy customer will tell his or her story to at least nine people.
- 13% of those unhappy former customers will tell their stories to more than 20 people.
- Customers who have complained to an organization and had their complaints satisfactorily resolved tell an average of five people about the treatment they received.

Healthcare is a service business. The labor involved in providing a service is principally emotional labor as distinct from physical labor. There is some physical labor in the business of providing healthcare services; however, the expenditure of physical energy must often be balanced against the need for emotional energy. Consider the following:

Physical Labor Activities and Their Corresponding Emotional Labor Component

Physical Labor	*Emotional Labor*
Transporting a patient from triage to x-ray	Talking to the patient on the way
	Explaining the turns ahead of time
	Showing some courtesy and respect
	Showing empathy
	Showing compassion

Emotional labor is far more challenging than physical labor, primarily because of the human element. There are only a handful of naturally gifted people who possess the emotional capacity to do both the physical and the emotional very well every time. For instance, a service provider who has an unstable temperament should never be allowed to interact with patients or their relatives. This person is not necessarily bad; but this person should be put in the rear of a dimly lit room in front of a computer terminal, with absolutely no opportunity to deal with patients and their relatives. No offense to computer people. In fact, some of my best friends are computer *nerds*.

Have you noticed how much noisier the waiting room areas are in hospitals, clinics, and healthcare facilities in general? There seems to be an increase in the incidences of confrontation, arguments, and an overall intolerance for mediocrity by the customer. More and more patients are demanding to speak with the supervisor and or manager about a customer service type problem.

One thing is abundantly clear. Today's customers are far more willing to exert their influence to determine which organizations stay in business and which organizations die.

Albrecht and Zemke, in their book *Service America,* credit the TARP research in presenting the following about the value of a customer to an organization: A brand-loyal automobile customer represents an average lifetime revenue of at least $140,000. In banking, the average customer represents an annual profit of $80. Appliance manufacturers figure brand loyalty is worth $2,800 over a 20-year period. Your local supermarket will count on you for $4,400 each year, and $22,000 for the five years you live in the same neighborhood.[2]

Implications for a Hospital

The average hospital takes in approximately $4,000 in revenues for normal births. A family of three children would bring that figure to close to $14,000 over a five-year period. A pleasant experience in and out of the delivery room may produce additional income for a hospital in the form of referrals, word of mouth, etc. There is no doubt that a happy healthcare customer is worth more than the value of his/her purchase. The real value of a patient lies in the long-term value of both the revenue and profit stream from all his service purchases. This becomes especially important if the patient could potentially purchase a range of different services (such as other outpatient services, consultations, etc.) from the hospital.

In the healthcare industry, it is suspected that less than 4% of the unhappy customers will dare to complain (Technical Assistance Research Program, TARP, Washington, D.C.). Patients are generally less willing to complain about the person responsible for their medical well being for fear of retribution. Some managed care organizations estimate that a satisfied member will bring in as much as $100,000 in revenues during his or her lifetime. If so, why have a fight with a member over the late payment of a $25 premium or quibble about some little something that the member thinks is not right? Consider the following key questions for a hospital:

- What is the cost of an unoccupied hospital bed?
- What is the revenue (over a patient's lifetime) from a satisfied patient?

WHY CUSTOMER SERVICE IN HEALTHCARE?

Intuitively, it is easy to see how good customer service benefits the organization; too often, healthcare employees are asked to attend workshops and seminars on *customer service* without an attempt to explain how it benefits them. Any serious attempt at customer service training should first address its implications for those charged with the responsibilities of carrying out its mandates — the employees.

What's In It for the Employee?

- ***More energy*** — Service excellence requires less emotional energy than service mediocrity. No one is happy to have to deal with angry and abusive customers routinely. It creates an atmosphere of hostility and confrontation that slowly erodes the joy in work.

- ***Better relationships*** — Excellence in service builds harmony and friendship between you and your customer. Some meaningful friendships can develop during the process of providing a service to a customer. Sometimes, those friendships can become the most important assets we have.
- ***Less stress*** — Stress is a known killer. It slowly weakens the body's immune system and can ultimately lead to other more serious problems.
- ***Increased job satisfaction*** — One critical component of job satisfaction is being happy with your work. Better customer service often means greater satisfaction with your work.
- ***Job security*** — It used to be that if you belonged to the union or were liked by your boss, you had job security. The only realistic source of job security today is the customer. He and he alone determines who has a job and who doesn't.

What's In It for the Organization?

Healthcare customer service initiatives generate significant benefits for organizations including the following:

- Fewer complaints, rework, and litigation
- A boost in customer retention and therefore profitability
- Organization stays in business
- Reduced employee turnover rate
- Reduced absenteeism
- Reduced "mental health" days
- Increased productivity
- Customer delight and loyalty
- Attractiveness to prospective employees

According to Bell and Zemke, studies done by business economists show that companies rated high on the quality of their customer service enjoy the following benefits:[3]

- Keep customers longer — 50% longer or more.
- Lower sales and marketing costs — 20 to 40% lower.
- Higher return on sales — 7 to 12% higher.
- Better net profits — 7 to 17% better.

What's In It for the Community?

- More jobs for the community
- Better service for residents of the community

THE HEALTHCARE CUSTOMER

There are some who are offended by the use of the term "customers" to describe patients. *Webster's Dictionary* defines a customer as "a person who purchases goods or services from another; buyer; patron. Informal — a person one has dealings with." The following attributes must be present in order for a person to merit the label of "customer."

Who Is a Customer?

- A person (or organization) paying to receive your services or goods, who has the right to pass judgment on the quality of your goods or services.
- A person (or organization) entitled to your service or goods as a result of a legitimate arrangement, who has the right to pass judgment on the quality of your services or goods.
- A person (or organization) who can be influential in bringing or denying you new customers as a result of his or her level of satisfaction with your goods or services.
- A person (or organization) with whom you have a partnership or agreement to reach a common objective, who has the ability to terminate the agreement based on his or her level of satisfaction with your performance of the terms of the agreement.
- A person (or organization) who, by continuing to buy or use your product or services, will directly influence your organization's bottom line and overall success.

There are two types of customers, as follows:

- ***External Customers*** — Those who are outside the organization and meet some of the preceding attributes.
- ***Internal Customers*** — Those who are inside the organization (staff, co-workers, etc) and meet some of the preceding attributes.

Hospital External Customers

- Patients
- Physicians
- Relatives and visitors of patients
- Managed care organizations
- The government (state, local, federal)
- Accreditation bodies

- The community
- Nursing homes
- Police agencies
- Fire fighting agencies
- Other allied healthcare organizations

Hospital Internal Customers

- Staff/co-workers (clinical and non-clinical)
- Physicians

Managed Care External Customers

- Enrolled members
- Hospitals
- Providers (physicians)
- Relatives of the members
- Accreditation/regulatory bodies
- Other allied healthcare oganizations

Managed Care Internal Customers

- Staff/co-workers (clinical and non-clinical)
- Physicians

Home Healthcare External Customers

- Homebound patients
- Patients' relatives or significant others
- Hospitals
- Managed care organizations
- Nurses
- Physicians
- Accreditation/regulatory agencies
- Other allied healthcare organizations

Home Healthcare Internal Customers

- Staff/co-worker (clinical and non-clinical)
- Physicians

Understanding the Healthcare Customer

Hospital Patients

Any of the following may be true about hospital patients:

- Sick
- Afraid of death
- Elderly
- Anxious
- Varying degree of educational experience
- Angry over what is happening to him/her
- A person with little or no family support
- A person in denial
- Blaming self
- Feeling self-pity
- Chemically dependent
- A person who has talked to two or more people about his/her condition before coming to the hospital
- Wants to know what is happening to him/her and why
- Wants to know what will happen to him/her now
- Worried about the financial impact of the illness on him/her, and on the family
- A person afraid to ask questions about the condition
- A person seeking reassurance
- A person worried about what the onset of illness would mean
- Concerned about how illness might affect income-earning ability
- A person seeking information
- A person too embarrassed to admit he/she doesn't know
- A person who used public transportation to get to the hospital
- A person with a disease

HOW TO USE CUSTOMER PROFILES

Professor Noriaki Kano, a notable quality management scholar in Japan, once told a story about a famous camera manufacturer in Japan who successfully used its customers' profiles in the redesign of its cameras. When one asks the question, "Who uses a camera?" ordinarily, the answer will be, "People." That answer however, does not immediately translate into customers' needs, wants, and expectations. Let's examine the following profiles for average camera users:

- A person who might be technically challenged — may not be adept at loading a roll of film in a camera
- A person who does not own a flash
- A person who owns a flash, but has a tendency of forgetting to take it along with the camera
- A person who dislikes the bulkiness of a flash, and so tends not to carry one
- A person with an imperfect vision, who may have difficulty judging sharpness and clarity as they pertain to objects viewed through the lens of the camera

To arrive at these customer profiles, the camera manufacturer conducted a survey by inspecting several pictures that were submitted for processing by the average camera user. Three types of problems were detected from the survey, as follows:

1. After the submitted pictures were developed, some of the prints came out blank. Usually, this would occur if the film was not loaded correctly. The film sat inside the camera without advancing after the previous picture was snapped.
2. Some of the prints had images that were out of focus. In a manually focused camera, this might be due to an imperfect vision or the inability accurately gauge distance.
3. Some of the prints had insufficient light. One explanation for this could be that the camera operator did not have or use a flash when one was required.

In response to the first problem, the camera manufacturer redesigned its camera to have an automatic loading device. In response to the second problem, it redesigned its camera to have an auto-focus. The third feature added to the camera was a built-in-flash. So, what was wrong with this company's original camera? Nothing! In fact, some "Inspector Thirteen" determined that the original camera was "acceptable." It met the requirements for quality assurance, but did not fulfill the needs of the customers. Passing an inspection conducted by a regulatory agency does not imply or create customer satisfaction. Conformance to standards does not create customer satisfaction!

Patient Profiling

A Medicare (65 years or older) patient *may* exhibit any of the following tendencies:

- Hard of hearing
- Forgetful
- Vision problems
- Frail
- Afraid of death
- Slow in performing certain activities

Service Design

A service provider (once he/she determines the presence of any of the preceding tendencies) would make some adjustments in the service process to reflect the profile. The following service design features may be necessary:

- Speaking a bit louder (not screaming or yelling)
- Repeating information already provided (allowing some redundancy)
- Writing information down on paper in addition to verbalizing it
- Asking the patient to repeat the information back to the healthcare provider
- Exercising patience in dealing with the patient
- Relying more on demonstration than on verbal instructions
- Getting relatives involved in the treatment process
- Providing calming and reassuring words

The "Find It, Fix It" Approach to Medicine

The mentality of clinicians toward diseases has always been one of "find it, fix it." Lost within this mindset is the equally important notion of the "person." The problem with the "find it, fix it" approach to medicine is that it ignores the human while focusing only on the disease. As a consequence, the disease may be cured, but the patient remains scared by the loss of dignity, lack of respect, and an overall unpleasant experience. The patient's overall experience can be enhanced only when the service provider knows something about the *disease* and something about the *person*. History taking provides information about the disease, with very little information about the person. It is dangerous to treat the person and the disease as one. In the camera manufacturer's example, one could conclude that when the camera manufacturer produced a camera that passed inspection, the manufacturer focused only on the *disease*, but not the *person*. However, when it redesigned its original camera, it focused on both the *disease* and the *person*. Some clinicians may describe a patient as "the hernia in Room 312." It often seems logical to put the *disease* before the *person*. However, success in treating the disease depends to a great extent on how much you know about the person. Patients do self-diagnose. They generally talk to a friend, relative, neighbor, or significant other prior to coming to your healthcare facility. Patients tend to have some theory as to what is wrong with them, what causes it, and how it is treated. It may be impossible to find out what the patient's theory is (no matter how erroneous) unless the service provider tries to learn something about the "person." Here are some reasons to separate the person from the disease, and for addressing both the person and the disease:

- Calling a patient by name shows that you recognize the person and have respect for his or her individuality.
- If this is your first time meeting this patient, the patient wants you to know that he or she did not always look, talk, and act like this.
- Patients tend to trust you more if they feel you know, hear, and respect them. They are more likely to believe in your treatment plan.

Some Suggestions on How to Handle the Disease and the Person

1. The first time you see a patient, see the patient with his or her clothes on. That might mean first welcoming the patient, then giving the patient a minute or so to undress.
2. Learn the patient's name and address the patient by name.
3. Knock before entering the patient's room or an examination room where the patient is waiting for you.
4. Make an effort to touch the patient with your hand, as in placing your hand on the patient's shoulder. Touching represents one way to validate a person's humanity. It is indeed the silent language.
5. Give the patient an opportunity to tell you about him or her self. Listen attentively, without being preoccupied with reading through the patient's charts at the same time. Be sure to remove all physical barriers between you and the patient such as a table or desk. Come out from behind the table or desk and eliminate any operational distance that might separate you from the patient.
6. Endeavor to find out the level at which the patient operates and make the effort to speak to the patient at that level. Avoid technical jargon, but if you must use it, be sure to provide multilevel explanations on its meaning.
7. Avoid chastising or judging the patient. Focus only on the behavior or act, not the person.

Today's healthcare consumer has a distinct list of expectations. Hospitals that make a conscious effort to respond to these expectations are certain to emerge as the leaders among health-care organizations in the 1990s. Perhaps the critical test of how well a hospital meets or exceeds the expectations of its consumers may be found in the answers to the following four questions:

Patients

- Would you enthusiastically recommend this healthcare organization to someone you
- If you have a say in the matter, would you come back to this facility for treatment?

Healthcare Professionals

- Should you need medical treatment, would you (as a physician, nurse, or other clinical professional) make this healthcare facility your first choice?
- Would you (as a physician, nurse, or other clinical professional), select this hospital as your place of practice?

If an organization cannot obtain "yes" answers to these questions, then it should reexamine its commitment to service quality. Responses such as "maybe," "perhaps," and "possibly" are simply not good enough. Such responses only falsely reinforce an already fragile commitment to quality. One of the goals must be to obtain a "yes" response to each question. While a "yes" response may not be the absolute test for good quality, it could mean an initial indicator of some level of commitment.

Quality of Conformance

Two major organizations responsible for providing guidelines for healthcare quality standards are The American Nurses' Association (ANA) and the Joint Commission on the Accreditation of Health Care Organizations (JCAHO). These two organizations, together with state licensing agencies, also set standards for nursing care.

The quality scoring procedure involves an independent group of staff members assigned to coordinate and implement the quality assurance program. The following six specific objectives of the nursing process are taken into account.

1. Plan of care is formulated
2. Physical needs of the patients are met
3. Non-physical (psychosocial, emotional, and social) needs of the patients are met
4. Achievement of nursing care objectives is evaluated
5. Unit procedures are followed for the protection of all patients
6. Delivery of nursing care is facilitated by administrative and managerial services

There are major differences between the Quality Assurance (QA) approach to quality and the Total Quality Management (TQM) approach to quality. Some of the differences are outlined here:

- QA is driven by standards prescribed by the JCAHO; TQM puts every employee in the organization in charge of quality.
- QA stresses conformance to standards (quality in fact); TQM stresses quality in fact and quality in perception.
- QA assumes that JCAHO knows best and should dictate what good quality means; TQM is driven by a definition of quality based on the needs, wants, and expectations of the customers.
- QA in some organizations means preparation for an announced visit from JCAHO, by staging a show to reflect good quality assurance; TQM is a never-ending activity of continuous improvement.
- An on-going QA activity involves a group of in-house professionals who make sporadic visits to the units to check for compliance and identify problem areas; TQM involves everyone in the organization (telephone operators, receptionists, secretaries, nurses, unit support staff, medical records clerks, etc.).
- Most QA activities are driven by episodic measurements. Data gathering is usually carried out in response to an episode or a crises; TQM emphasizes the need to focus on the process generating the episodes and to put mechanisms in place to better control the process.
- QA relies on judgmental and subjective evaluations; TQM emphasizes management by facts (numbers).

Effective monitoring of patient care quality must begin with a complete understanding of who the customers are. This question is addressed next.

Two Components of Quality

Quality consists of two interdependent parts: quality in fact and quality in perception. The first involves meeting your own specifications (conformance to standards), and the second part is meeting the expectations of your customer. Neither of these in itself will carry a hospital far. To deliver healthcare exactly as JCAHO intends will be to no avail if your patients believe you are providing inferior service. Also, the quality of the services provided by the support functions, such as accounting or billing, does not necessarily improve because of your adherence to the requirements of JCAHO.

Theories of Service Quality

According to Carol King, several theories attempt to explain how customers evaluate the quality of service:[4]

1. The first theory states that there are two sides to the customer's perception of service quality: First, whether the primary or "core" service is performed; second, whether the surrounding or secondary services are performed satisfactorily. This theory holds that if the primary function is not performed satisfactorily, customer satisfaction cannot be recovered by high performance levels in the secondary functions.
2. The second theory separates the hard functions — the technology of the service — from the soft functions, the manner in which the service is performed. The customer's attitude toward a breakdown in the technology can be influenced by the manner in which such a situation is handled, but the effects of poor interactions with the service provider cannot be overcome by technically competent performance.
3. The third theory states that the service transaction is not a fixed entity, but rather a process, and the customer's evaluation of satisfaction can change over the course of the encounter.
4. A fourth theory pertains to the degree of perceived risk and the related cost of the service. The intangible nature of service presents a risk to the customer, in that he cannot see and evaluate in advance what he is going to get. In addition, services performed generally involve either the customer personally or his property. When little risk or low cost is involved, customers apply less stringent standards of evaluation than when high risk or high cost is involved.

Each of these theories will now be examined in the context of quality in healthcare. For example, from the first theory, the two sides of a patient's perception of quality may be stated as follows:

- ***Primary service*** — Was the surgery performed correctly or successfully?
- ***Secondary service*** — Were the nurses friendly? Was the billing accurate? Were the bed linens changed regularly?

Very often, hospital, clinical, and administrative staff will define quality in the context of the primary service alone. However, the fact remains that patients generally expect a hospital to have competent and professionally trained clinicians who will follow the proper procedures in their work. Except for patients who are familiar with the workings of a hospital, most patients do not know what constitutes a "proper procedure." What patients do know and can respond to is the manner in which a service is performed (for example, the behavior of the individuals providing the service). The quality of the interaction between the care provider and recipient is critical in health care. King calls this factor the "quality of behavior." According to J. A. Johnson,

> Despite all the new technology in hospitals, health care remains a humanistic activity. People's experiences at the hands of a healthcare professional in times of vulnerability are intensely personal. A patient needs that human touch, caring, and compassion. Consumers have come to expect not only the latest technology and highly competent professionals, but a caregiving culture as well.[5]

In many cases, patients change physicians, nurses, clinics, and even hospitals, not because of the poor clinical quality but because of poor "quality of behavior." Health care is a humanistic profession and the labor involved is a labor of love. A. R. Hochschild refers to this type of labor as "emotional labor," to distinguish it from physical labor and mental labor.[6] Healthcare dollars buy not only the medical expertise and ability of the practitioners, but also their attitude.

The second theory concludes that technical or clinical expertise cannot compensate for a caregiver's poor interaction with the patient or other non-care activity. However, good quality interactions with the recipient of service may positively influence a patient if there is inadequacy in the technical aspects. A quality program must give sufficient attention to the hard (technical) functions as well as the soft (non-technical) functions of the care delivery process. Quality should not be judged only in terms of the visible aspects of patient services. Many of the services provided to the patient are not seen; examples of such "invisible" services are medical records and information management. Still, there are other services that the patient cannot evaluate while care is being provided, but will have the opportunity of evaluating later, e.g., billing and follow-up care. Quality must be viewed in the context of a patient's total experience.

The premise of the third theory is somewhat similar to what Albrecht and Zemke describe as "moment of truth." The term "moment of truth," originally coined by Jan Carlzon, president of Scandinavian Airlines, is defined as "any episode in which the customer comes into contact with any aspect of the organization and gets an impression of the quality of its service."[7]

According to this definition, every "moment of truth" provides an opportunity for the care provider to make a good impression on the care recipient, and thus influence the evaluation of care received by the patient. The implication is that if you, as a provider of service, get it wrong at your point in a patient's chain of experiences, you will likely erase from the patient's mind any other good experiences he or she may have had before the encounter with you. But if you get it right, you have a chance to undo all the wrong things that may have occurred during the patient's experience prior to you. You really are the "moment of truth" in the Provider-Receiver (PR) encounter. When PR encounters go unmanaged, the quality of service deteriorates to unacceptable levels.

This point can be further illustrated by examining a hospital whose emergency room is divided into six departments, as follows:

1. The initial examination station to treat minor problems or make diagnosis (INITIAL EXAM)
2. An x-ray department (X-RAY)
3. An operating room
4. A cast fitting room
5. An observation room for recovery and general observation before the final diagnosis or release
6. An out-processing department where clerks check patients out and arrange for payment or insurance forms

A patient in this example can be expected to have anywhere from two to four PR encounters during a visit to the emergency room. The patient's perception can be altered or influenced by any contact made in any of the six departments of the emergency room.

Failure to adequately manage PR encounters may lead to patient abuse and neglect, particularly in nursing homes. Rosander (1988) notes that patient abuse and neglect (in nursing home care) include a wide variety of practices: removing trays before meals are eaten, threatening slow eaters, delays in answering the call light, refusing to give patients something they like to eat such as a slice of bread, isolating patients in a corridor and leaving them for hours, refusing to talk to patients, or employees taking long coffee breaks regardless of the patients' needs.[8] An understanding of the expectations of the customers is a necessary starting point in defining the needs of customers.

The fouth theory addresses the relationship between perceived risk and quality. Simply stated, the greater the perceived risk in a transaction, the greater the customer's insistence on quality.

How a Patient's Expectations Are Formed

- ***Needs*** — The need to alleviate a health-related problem, be cured, or slow the growth process of a certain illness is critical in the definition of patient expectations.
- ***Experience*** — What the customer has encountered (in the past) in the delivery of health care. Past experience is perhaps the strongest determinant of present expectation.
- ***Knowledge of technology and environment*** — How much the patient knows about what types of services are available, the scope of such services, including risks, and the technological level at which the services are available. Also important is knowledge of the experiences of others, such as relatives, friends, and visitors.
- ***Competitors' offerings*** — The knowledge of what other healthcare organizations are offering, or promising to offer, can have a tremendous impact on the expectations of customers.
- ***The reputation of the healthcare organization*** — A healthcare organization that enjoys a good reputation will also have to deal with patients who come to it with high expectations. The same argument holds for organizations with a poor reputation.

Information Received from Relatives and Friends

A patient's expectation is strongly influenced by what he or she is told by trusted friends or a significant other. Word of mouth is still a powerful form of advertising. Expectations are formed based on this type of information. Several studies have suggested that an increasing number of healthcare consumers are obtaining their information from friends and relatives Although the influence of physicians over patients remains strong in the selection of hospitals, more patients are insisting on having a say in matters affecting hospital selection.

Regulatory and Legal Requirements

All consumers tend to expect protection by the law or through the governance of regulatory agencies. For example, patients expect a hospital to be a clean and safe place. The law and regulatory agencies also call for it.

In order to understand a patient's expectations, it is helpful to categorize them in three phases: before going to the hospital, during the hospital visit, and after leaving the hospital. The following are examples of the expectations of a person going into the hospital for a procedure:

Before Entering Hospital

- The condition/disease will be accurately diagnosed.
- The condition/disease will be adequately explained.
- The risks inherent in the procedure(s) will be explained.
- Charges will be satisfactorily explained.
- The information given will be reliable and complete.

During Hospital Stay

- The procedure will be performed by a team of competent medical and nursing professionals.
- The procedure will take place as promised, predicted, or explained.
- The medical and nursing staff will show compassion, warmth, and care during hospitalization.
- The recovery will progress as predicted, promised, or explained.
- The staff will do everything possible to bring about complete recovery.
- An acceptable level of hygiene and cleanliness will be observed by the hospital staff.
- Meals will be served at an acceptable temperature, in adequate quantity, with proper nutritional balance, and at the right time.
- Call lights will be answered promptly.
- He/she will get better or be cured.
- Instructions for rehabilitation and medication will be clear and complete.

After Leaving the Hospital

- There will not be a repeat visit to the hospital for the same problem.
- The bill from the hospital and professional staff will be accurate and adequately explained.
- Recovery will take place as explained.
- The hospital and its professional staff will respond promptly if problems develop again.

The Art of Caring

Parkland Associates (Park Ridge, Illinois) conducted a study of patient data from nearly 200 hospitals in 43 states which revealed that nursing care is the most important factor patients consider when recommending a hospital.[9]

Leebov (1988) identifies four primary reasons why healthcare organizations should focus on a patient's perception and satisfaction with the services provided:[10]

- ***The humanistic reason*** — Patients deserve excellent quality of care and service because they are often quite vulnerable. They come with anxiety about their physical, emotional, and economic well-being. Excellent service not only enhances quality of care, but also helps allay the anxiety that comes with being hospitalized.
- ***The economic reason*** — Patients are customers. They now have more options and are expecting value for their money.
- ***The marketing reason*** — Patients can be good or bad for public relations, depending upon the experiences they have while receiving services.
- ***The efficiency reason*** — Satisfied patients are easier to serve, whereas dissatisfied patients consume more valuable staff time that could be used serving others.

THE QUALITY OF BEHAVIOR

In a 1985 Gallup survey for the American Society for Quality Control, it was revealed that employee behavior and attitudes are the major determinant of the quality of services. The survey was based on interviews with 1,005 persons. A. C. Rosander identified three classes of human traits that can affect quality of service: behavior, attitudes, and appearance.[11] Using the examples suggested by Rosander, the following cases are developed for healthcare on two of these three classes.

Behavior

- ***Acting promptly*** — A patient (or patient's family member) should be able to see promptness in response to pain, to a request, to a patient's call for assistance, or to a need. This attribute is critical for nurses, physicians, laboratory testing services, the pharmacy, etc.

- ***Listening carefully*** — Sometimes physicians are perceived as not listening enough to their patients. A patient or family member should be able to perceive the doctor as one who listens attentively to health-related complaints and questions. It is equally important for nurses to listen carefully to the patients and also for nurses and physicians to listen carefully to each other.
- ***Being attentive*** — For a patient whose life is on the line, being attentive is not just an indication of good quality, it is a critical attribute. The need to be attentive is especially critical during the performance of surgical operations, administering medication (oral or intravenously), and monitoring cardiac activity.
- ***Acting with understanding*** — Nursing is a humanistic profession — one that calls for understanding and compassion on the part of the provider of care. The provider of care must act in a manner that shows empathy with the pain and suffering of the afflicted. When a nurse says to a screaming patient, "You don't have to scream at me; I didn't cause your injury," it shows a lack of understanding and compassion.
- ***Making "to-the-point" explanations*** — Nurses are expected to demonstrate sufficient knowledge of their jobs to provide precise and adequate explanations to patients, relatives, and physicians. It is not always easy to provide a to-the-point explanation to the family of a patient whose chance of survival after an operation is less than desirable.
- ***Avoiding unusual ways of talking*** — Appropriate linguistic skills on the part of the service provider is often perceived as good quality by the customer. This is no less true in health care. Nurses, physicians, nursing unit support services staff, and x-ray and laboratory technicians are all expected to avoid unusual ways of talking around or away from the patients and their relatives.
- ***Showing ability to do the job*** — When a service provider demonstrates a lack of confidence and knowledge, he or she is usually perceived as not possessing the ability to do the job. Sometimes, after consecutive 12-hour shifts, fatigue sets in and it becomes harder to demonstrate "ability."
- ***Getting along with people*** — In general, customers perceive the ability to get along with people (on the part of the provider of service) as an indication of the presence of quality in the service being delivered.

Attitudes

Attitudes include being courteous, friendly, mannerly, kind, conversational, alert, accurate, responsible, and compassionate.

Reports from Healthcare Surveys

An article by Mary Koska in *Hospitals*, reported the results of a survey of 663 hospital chief executive officers (CEOs). The survey was conducted by the consulting firm of Hamilton/KSA from Atlanta, Georgia. The survey asked CEOs to rank, in order of significance, ten factors that contribute to a hospital's quality of care.

Koska reported Barry Moore, president of Hamilton/KSA, as saying, "Truly effective nursing care has two parts — the clinical component and the caring component — neither of which can stand alone and still be considered high-quality care." According to Moore, "It's almost impossible for patients to measure technical quality. They generally assume it will be there…on the other hand, what patients can easily measure, they can criticize." For example, food service, which was ranked as a significant contributor to high-quality care by 61% of CEOs surveyed, may be the quickest way to impress patients. According to Moore, poor food service, noise, inadequate explanations, and rudeness of the staff are the most common complaints that patients have about hospitals.[12]

S. R. Steiber of SRI Gallup Poll reported the results of a national poll conducted for *Hospitals* magazine, in which consumer satisfaction was found to be influenced more by concern shown for the patient than by clinical care. The 414 respondents had either been hospitalized themselves within the past two years or had immediate family members who had been hospitalized. According to Steiber, traditional analyses of patient satisfaction generally looked only at how patients rate different services, such as food, cleanliness, and parking. But these scores told hospital executives only how well the hospital performed those individual tasks. Such surveys did not indicate the degree to which consumers associated these services with quality of care. Steiber noted that, "The single most important action hospital executives can take to maintain quality from the patient's perspective is to deliver a satisfactory experience."[13]

MOVING TOWARD A TOTAL SERVICE QUALITY MANAGEMENT PROCESS

According to Albrecht and Zemke, service management is a total organizational approach that makes quality of service, as perceived by the customer, the number one driving force for the operation of the business.[14]

If hospitals are to become competitive and guarantee themselves long-term survival, they should look more closely at the concept of Total Service Quality Management (TSQM). I offer the following definition of TSQM:

> TSQM is a concept that defines quality in the context of a customer's experience. The customer's experience and subsequent perception of quality is affected by both the tangible and the intangible components of the services provided as well as what happens after the customer departs physically from the system providing the service. TSQM begins with top management commitment and must be instituted at all levels of the organization.

Critical Factors in TSQM

Total Service Quality Management is not a departmental activity. It is a hospital-wide concept embedded in every aspect of patient care. It is not an exclusive function of the "customer service department" nor the "patient complaints department." Everyone has a responsibility to ensure that things turn out right for the patient or the customer. Unfortunately, as most QA Departments in hospitals learn, the implication of departmentalizing QA is that some employees feel that they can afford to mess up, since there is a central department that can fix their mistakes. Notwithstanding, the whole organization should act like one huge customer service or quality assurance department.

A hospital is typically managed on the basis of individual functions, such as nursing, pharmacy, and housekeeping. Consequently, no single individual or group is really accountable for building quality into the patient's experience. This is perhaps the most compelling reason why the TSQM concept should encompass every individual and every level in the organization.

Service is not only a potent weapon in a hospital's competitive arsenal; it is the driving force behind profitability. Productivity must be seen as a subset of quality. What is needed now is a new paradigm — one that reverses the old emphasis on cost reduction and productivity. Hospitals can only become or remain profitable if they first identify the right things and then do them right. Johnson notes that, "Hospitals are much more than buildings and machines, they are human organizations meeting human needs; and to remain successful in a competitive healthcare market, they must outperform their competitors on the human dimension."[15]

Marketing Research

Once a healthcare organization has answered the questions from the preceding section — who are its customers and what are their wants and needs — it must determine how it is rated against its competitors, and, if necessary, what it must do to win over the customer. There is only one way to convince the customer that your organization is better: by offering a service that meets or surpasses the expectation of the customer and providing a service that is decisively superior to what the competitors are offering. It is important to determine for example, why some physicians do not recommend your hospital to their patients, or why certain patient groups perceive your nurses as being unfriendly (when compared to the other hospitals), or how your emergency room services compare against that of your competitors (in the eyes of the customers).

Marketing research involves the systematic gathering, recording, and analyzing of information about specific issues affecting the product or services being provided. It is the information link between the customer and the provider of the service. Marketing research should be based on objectivity, accuracy, and completeness. Objectivity means that the research is conducted in an unbiased and open-minded manner. Accuracy refers to the use of valid research tools and/or instruments that are carefully constructed. Each aspect of research, such as the sample chosen, questionnaire format, and tabulation of responses, must be carefully planned. Completeness refers to the comprehensive nature of the research. Erroneous conclusions may be reached if the research does not probe deeply or widely enough.

When, for example, a hospital asks its patients to rate its (the hospital's) parking facility on a scale of 1 to 5 (1 being poor, and 5 being excellent), what exactly is it asking? It is merely asking its patients to rate how well it is doing with respect to parking, but not necessarily asking whether or not parking is important to the patient, or for that matter how important parking is compared to, for example, warm meals. It is advisable to hire the services of an outside research firm if a healthcare organization finds that it does not have the in-house capability to carry out good market research.

The concept of TSQM in healthcare can be grasped more easily if adequate attempts are first made to understand some of its theoretical foundations. The theoretical foundations are presented in the following section.

The Dimensions of Quality in Healthcare

1. ***Caring*** — The patient and the disease are not one. Separate the person from the ailment. Show compassion in providing care.
2. ***Dignity*** — Preserve the patient's dignity in every interaction. Call the patient by name. Knock before entering the patient's room. Do not talk down to the patient.

3. ***Empathy*** — Put yourself in the patient's shoes.
4. ***Information*** — Assume the following unasked questions and provide the answers: What is happening to me? What has happened to me? Why has it happened to me? What will happen to me now? Will it hurt, and for how long? How much will this affect my life?
5. ***Knowledgeability*** — Be able to answer all relevant questions about the disease, the diagnosis, the symptoms, treatment, side effects, and cure.
6. ***Responsiveness*** — Address the issues of waiting time, promptly acknowledging a person's presence once the person arrives for an appointment.
7. ***Professionalism*** — Maintain appropriate linguistic skills, appearance, listening skills, and manners.
8. ***Accessibility*** — Enable patients to easily reach the service providers — in person or via telephone.
9. ***Communication*** — Speak and write clearly, and address the patient at his or her level. Avoid medical jargon that only confounds the patient. Also use posted signs to facilitate patients' flow and access.
10. ***Convenience*** — Strive for the patient's convenience in all things — parking, telephones in waiting areas, appointment times, etc.

SUMMARY

Quality management is not a departmental activity. The responsibility for quality lies with everyone — from the secretary who avoids typing errors, to the hospital's telephone operator who must effectively handle the inquiries of potential patients and their relatives and friends, to the nurse who must enter correct and accurate information into patient charts and avoid medication errors, and so on. No hospital can evade this quality challenge: profit-making or non-profit hospitals, general or specialist hospitals, private or public hospitals, government or non-government. All must face the task of responding effectively and efficiently to patients who expect quality service in the delivery of care as part of the hospital's product. Some hospitals are well aware of this need and are generating measured responses. For many others, the need to be patient-centered and service-quality driven comes as a major surprise. Nevertheless, this need cannot be ignored; it is not a whimsical trend that will suddenly disappear. It is the new standard by which today's patients measure the performance of healthcare organizations.

It may be useful to envision the patient as carrying around a mental "report card," in which he or she records a perceived score for the services received from the system. This score helps the patient decide whether to return or go to another healthcare organization. It also helps the patient decide whether or not to recommend a particular facility to someone else. As health care moves into a new era of competitiveness, it is critical that providers learn as much as possible about the all-important, but invisible, report card. In the final analysis, a healthcare organization's ability to consistently score impressive marks on the patient's report card ultimately depends on how much it knows about the patient's evaluation criteria. It is not those who provide the service, but those whom it serves, who have the final word on how well the service fulfills needs and expectations.

EXERCISES

28-1 Give a list of five customer quality requirements that can be categorized as soft and hard factors.

28-2 What mechanisms are in place at your facility for obtaining feedback from your healthcare customers?

28-3 How does your organization determine the requirements of its internal customers?

28-4 When was the last time your organization conducted a survey to determine how it performs against its competition?

ENDNOTES

1. U.S. Office of Consumer Affairs, *Complaint Handling in America,* Washington, D.C.: Government Printing Office, 1986.
2. K. Albrecht and R. Zemke, *Service America,* Homewood, IL: Dow Jones Irwin Publishers, 1985.
3. Bell and Zemke, 1992.
4. Carol A. King, "Service Quality Assurance Is Different," *Quality Progress,* June 1985, 18(6), p. 14.
5. J. A. Johnson, "Service Management, a Strategy for Excellence in Health Care," *Business Perspectives,* July 1988, p. 13.
6. A. R. Hochschild, *The Managed Heart: The Commercialization of Human Feeling,* Berkeley: University of California Press, 1983.
7. Albrecht and Zemke.
8. Rosander, 1985.
9. Parkland Associates, *AHA News,* December 17, 1990.
10. W. Leebov, *Service Excellence: The Customer Relations Strategy for Health Care,* Chicago: American Hospital Publishing, Inc., 1988 p. 109.
11. A. C. Rosander, *The Quest for Quality in Services,* Milwaukee: Quality Press and White Plains, NY: Quality Resources, 1989, p. 55.
12. Mary T. Koska, "Quality — Thy Name is Nursing Care, CEOs Say," *Hospitals,* February 5, 1989, p. 32.
13. S. R. Steiber, "How Consumers Perceive Health Care Quality," *Hospitals,* April 5, 1988, p. 84.

14. Albrecht and Zemke.
15. J. A. Johnson, "Service Management, a Strategy for Excellence in Health Care," *Business Perspectives,* July 1988, p. 13.

REFERENCES

Accreditation Manual for Hospitals, Chicago: Joint Commission on Accreditation of Hospital Organizations, 1990, p. 310.

American Nurses' Association, *Standards for Nursing Services,* Kansas City, MO: American Nurses' Association, 1973.

Aydelotte, M. K., "Staffing for high quality care," *Hospitals,* January 16, 1973, pp. 58–60.

Batchelor, G. J. and R. Graham, "Quality management in nursing," *Journal of the Society for Health Systems,* Vol. 1, No. 1 (1989), p. 63.

Bennett, A. C., *Productivity and the Quality of Work Life in Hospitals,* Chicago: American Hospital Association, 1983.

Crosby, P. B., *Quality is Free,* New York: McGraw Hill Book Co., 1979.

Day, R. G., "Quality cost and productivity," *Quality Progress,* July 1988, p. 59.

Donabedian, A., "Some Basic Issues in Evaluating the Quality of Care," in *Issues in Evaluation Research,* Pub. No. G1224M, New York: American Nurses' Association, 1976, p. 7.

Donabedian, A., *The Definition of Quality and Approaches to Its Assessment, Vol. 1: Explorations in Quality Assurance and Monitoring,* Ann Arbor, MI: Health Administration Press, 1980.

Egdahl, R. and P. Gertman, *Quality Assurance in Health Care,* Rockville, MD: Aspen Publishers, 1976.

Franz, J., "Challenge for Nursing: Hiking Productivity without Lowering Quality of Care", *Modern Healthcare,* September 1984, pp. 60–68.

Garvin, D. A., *Managing Quality,* New York: The Free Press, 1988.

Graham, N., *Quality Assurance in Hospitals: Strategies for Assessment and Implementation,* Rockville, MD: Aspen Publishers, 1982.

Hostage, G. M., "Quality control in a service business," *Harvard Business Review,* July-August 1975.

Jelinek, R. C., R. Dieter Haussman, S. Hegyvary, and J. Newman, Jr., *A Methodology for Monitoring Quality of Nursing Care,* DHEW Pub. No. (HRA) 76-25, Washington, DC: Government Printing Office, 1974.

Luke, R. D., J. Krueger and R. Modrow, *Organization and Change in Health Care Quality Assurance,* Rockville, MD: Aspen Publications, 1983.

Meisenheimer, C. G. (ed.), *Quality Assurance : A Complete Guide to Effective Programs,* Rockville, MD: Aspen Publications, 1985.

Millenson, M. L., "A Prescription for Change," *Quality Progress,* May 1987, p. 18.

Miller, M.C. and R. Knapp, *Evaluating Quality of Care,* Gaithersburg, MD: Aspen Systems Corp., 1979.

Miller, M.C. and R. Knapp, *Evaluating Quality of Care: Analytical Procedures, Monitoring Techniques,* Germantown, MD: Aspen Publishers, 1979.

Mowry, M. M. and R. Korpman, *Managing Health Care Costs, Quality, and Technology: Product Line Strategies for Nursing,* Rockville, MD: Aspen Publishers, 1986.

Omachonu, V. K., *Total Quality and Productivity Management in Health Care Organizations,* Norcross, GA: Industrial Engineering and Management Press, and Milwaukee, WI: ASQC Quality Press, 1991.

Omachonu, V. K., "Quality of care and the patient: New criteria for evaluation," *Health Care Management Review,* Fall 1990.

Omachonu, V. K. and R. Nanda, "IEs in Health Care Management Must Emphasize Client-Centered Projects Which Can Contain Cost While Supporting Quality," *Industrial Engineering,* Vol. 20, No. 10. (1988).

Omachonu, V. K. and M. Beruvides, "Improving Hospital Productivity: Patient-, unit-, and hospital-based measures, *International Industrial Engineering Conference Proceedings,* Norcross, GA: Institute of Industrial Engineers, 1989.

Peters, T. J. and R. H. Waterman, Jr., *In Search of Excellence,* New York: Warner Books, 1984.

Rosander, A. C., *Applications of Quality Control in the Service Industries,* Milwaukee: Quality Press and New York: Marcel Dekker, Inc., 1988.

Rush-Presbyterian-St. Luke's Medical Center and Medicus Systems Corporation, *A Methodology for Evaluation of Quality of Life and Care in Long-Term Care Facilities,* Chicago: Medicus, 1974.

Sahney, V. K., J. Dutkewych and W. Schramm, "Quality Improvement Process: The Foundation for Excellence in Health Care," *Journal of the Society for Health Systems,* Vol. 1, No. 1 (1989), 17.

Sasser, W. E., R. Olsen and D. Wyckoff, *Management of Service Operations, Text and Cases,* Allyn and Bacon, Inc., 1978, p. 180.

Schroeder, P. and M. Regina, *Nursing Quality Assurance: A Unit-Based Approach,* Gaithersburg, MD: Aspen Systems Corp., 1984.

Thompson, P., G. DeSouza, and B. Gale, "Strategic management of service quality," *Quality Progress,* June 1985, p. 20.

Townsend, P. L., "Insurance firm shows that quality has value," *Quality Progress,* June 1985, p. 42.

Ullmann, S. G., "The impact of quality on cost in the provision of long-term care," *Inquiry,* Fall 1985, pp. 293–302.

Werner, J. P., "Measuring Quality in the Health Care Delivery Business: Can It Be Done?." *Industrial Management,* Sep./Oct. 1988, p. 22.

INDEX

A

Absolute weight, QFD, 339
Acceptance of TQM, 12–13
Acceptance sampling, 245
Accounting systems, see also Activity-based Costing (ABC)
  activity-based accounting, 34, 40n28
  control, 74
  cost of quality, 6, 214
  information needs, 45–47, 198
  strategic quality planning, 78n25, 221n29
Accredited registration services, 358
Activity analysis, productivity, 198–200
Activity-based Costing (ABC), 214–218, 220n25
Albrecht, K., 453, 466, 471
Alcoa company, 157n10
Allied Signal company, 31, 445, 448
Alphabet management, 49
Alternative technology, 415, 422
American Airlines, 31, 67
American Express company, 47
American Leadership Forum, 23
American Management Association (AMA), 16, 29, 60
American National Standards Institute (ANSI), 348
American Nurses Association (ANA), 462
American Productivity and Quality Center
  benchmarking, 141, 151
  information, 49
  quality and productivity, 187
American Society for Quality Control (ASQC), 4, 347, 350
American Society for Testing Materials (ASTM), 277
American Supplier Institute, 331
Ames Rubber Corporation, 83
Amtrak, 122–123
Andrews, Professor (Harvard Business School), 61
Apple Computer company, 41
Appley approach, 29, 39n13
Applications System/400, 368
Appraisal costs, 208
Armstrong World Industries, 115
AT&T
  benchmarking, 146, 152, 157n13
  process control, 106
  TQM adoption, 370n8
  variables control charts, 288
Attitudes, healthcare customer service, 471
Attributes control charts
  basics, 295–296, 309
  fraction non-conforming, 296–301
  non-conformities, 304–309
  number non-conforming, 301–303
  process capability, 394
Australia, ISO 9000, 349
Automation investment, 56n18
Availability, reliability, 435–437
Avon company, 72

B

Baldrige, M., 187
Baldrige Award, see Malcolm Baldrige National Quality Award
Banking industry
  customers, 122, 129–130, 132
  information systems, 51, 105
  organizational structure, 166
  process control, 105
Bar graphs, 257–258

Baxter Healthcare Corporation, 149
Behavior, healthcare services, 469–471
Beliefs, see Culture
Bell, D., 50
Benchmarking
basics, 52, 141, 144
benefits, 145
best-in-class, 151–153
bottom line, 144–145
case study, 156
competitive evaluation, 109, 191
cultural benefits, 145
firms, types of, 157n4
Ford, 143
gap analysis, 154
historical developments, 141–143
human resources, 145
ISO 9000 implementation, 386
Motorola, 143
operational analysis, 148
performance improvement, 145
pitfalls, 154–155
processes, 149–151, 415, 421–422
self-analysis, 154
strategic analysis, 146–147
Xerox, 142–143
Benetton group, 143
Best-in-class, 151–153
Best practice companies, 153
Betti, J., 59
Bias errors, 246–247
Bic Pen company, 72
Binomial distribution, 296, 301
Black & Decker company, 67, 108
Bloomingdale's company, 65
Boeing Aircraft company, 152, 165
BOHICA syndrome, 26
Bottom line, benchmarking, 144–145
Bowen, D., 91
Break points, 126–127
British Science and Engineering Research Council study, 219n15
British Standards Institute, 356, 375
BSQ Group, 165
BUBBA syndrome, 78n14
Budd Company, 108, 331
Building Products Operations Division, 115
Business performance relationship, 13–15
Business processes, 56n18, 151
Buyer-supplier relationships, 133–135, 139n26
Bytex Corporation, 101

C

Cadillac Division of General Motors, 368
Caine, R., 347
Camp, R.C., 144–145
Campbell USA company, 170
Canada, ISO 9000, 349
Canon company, 142
Capability indexes, 399–403
Capital equipment, 197–198
Capital formation, 188
Carlzon, J., 466
Case studies
benchmarking, 156
customers, 136–137
human resource management, 93–94
information needs, 54–55
leadership, 37–38
organizational structure, 177–178
process quality management, 115–116
productivity, 200–201
strategic quality planning, 76–77
Caterpillar company, 31, 67
Cause-and-effect diagrams
basics, 252, 262–263, 265–269
usage in Japan, 117n25
Cause enumeration cause-and-effect diagrams, 269
c charts, see Non-conformities control charts
Certification, ISO 9000, 355–357, 359
Chance-failure phase, 434
Chase Manhattan Bank, 132
Check sheets, 117n25, 252–255
Chemical process, 412–413
Chief competitor, QFD, 338, 341
Christensen, Professor (Harvard Business School), 61
Chrylser/DaimlerChrysler, 371
Circle graphs, 259
Clausing, D., 331
Cleveland Twist Drill company, 179n14
Clustered sampling, 244–245
Coca-Cola company, 35, 165
Code of Hammurabi, 98–99
Colleges, 13, 152
Colony Fasteners, Inc. (CFI), see also Process capability
alternative technology, 415, 422
basics, 403–404
benchmarking, 415, 421–422
company requirements, 424–427

customers, 405–408, 415–416, 422–423
delivery, 410–416
design, 405–410
evaluation, 409–410
improvement, 409–410, 414–416, 421–423
inspection costs, 427
introduction, 405–410
measurement plan, 413–414, 421
performance, 419–421, 425–427
procurement, 426
production, 410–416
review of processes, 408–409
suppliers, 423–427
support services, 416–423
Communication, 27–30
Compaq company, 146, 170
Compensation systems, 89–91
Competition
customer defection, 105, 131
evaluation, 109, 191
quality function deployment, 338, 340–341
Competitive Advantage, 161
Competitive Strategy, 67
Computer-Aided Manufacturing-International (CAM-I), 46
Computer industry, 351, 362n11
Computer-integrated manufacturing (CIM), 49, 50
Conflicts, customer needs, 124–125
Conformance quality, 15, 462–463
Consumer Protection Act of 1972, 433
Control, 35–36, 73–74
Control charts, 269–271, 313–317, see also Attributes control charts; Variables control charts
Control limits, 278–279
Cooperative sharing agreements, 152
Coopers & Lybrand company, 202n15
Corning, Inc., 93, 129
Corrective action, 322–323, 354
Corrective Action Plan, 425
Cost avoidance, 195
Cost of Doing Business factor, 425
Costs
accounting systems, 214
activity-based costing, 214–218
basics, 205–206, 218
categories, 208–211, 219n15
conformance quality, 15
decision factor, 69, 78n21
inflation, 188, 202n14
information, 213–214
ISO 9000 certification, 359, 363n23
measurement, 211–213, 220n19
multiproduct lines, 216–217
poor and non-quality, 5, 10, 14–15, 21n30, 101, 184, 206–207, 219n18
process capability, 427
productivity, 150
reduction, 14–15, 195
strategic quality planning, 217
viewpoints, 207–208
Crandall, R., 31
Criteria for quality programs, see Program criteria
Crosby, P.
approach, 10–12
conformance quality, 15
cost of quality, 205, 207–208
culture beliefs, 31
customers, 123
human resources, 91
organizational structure, 159, 172
Cross-functional process, 228
Cross-functional teams, 85–86, 175–176
Culture
basics, 5
benchmarking, 145
leadership, 30–33
organizational structure, 160
process quality management, 97
Cumulative frequency distribution curves, 260–261
Customers
Baldrige Award, 370n2
basics, 121–123
break points, 126–127
buyer-supplier relationships, 133–135
case studies, 136–137
conflicts, 124–125
decision factors, 69, 78n21
defection, 105, 131
defining criteria, 14, 21n29, 51–52, 66, 125
design process, 118n29
dimensions, 137n5
employee input, 128–129
focus, 125–127

healthcare customer service, 122, 124, 452, 456–458
indicators, 16–17, 127–129
information, 51–52
marketing and sales, 130
measurement, 129–130, 167
organizational structure, 167
process capability, 405–408, 415–416, 422–423
processes, 105–106, 123–124
profiles, 52, 458–469
profitability, 132–133
quality function deployment, 333, 335–336
retention, 131–133
sales process, 131
service quality, 74, 131–132
themes, 127
value, 166, 178n10

D

DaimlerChrysler, 371–373
Dashboard, 232
Data
acceptance sampling, 245
basics, 231, 247
best plants, 56n17
bias errors, 246–247
clustered sampling, 244–245
collection methods, 233
dashboard, 232
dispersion errors, 246–247
displays, 237–241
errors, 246–247
graphical displays, 237–238
information, 231–237
numerical description, 238–241
objective, 234
presentation, 237–241
process quality management, 104
QI stories, 328–329
random sampling, 242
reliability, 235
sampling, 241–247
selected sampling, 245
significance, 232–233
simple random sampling, 242–244
stratification, 235–237
stratified random sampling, 244
subjective, 234
systematic sampling, 245
tabular displays, 237
types, 234
Databases, 151–152, see also Profit Impact of Market Strategy (PIMS)
Data linkages, 47–49
Dazzle factor, 31
Debugging phase, 434
Decision Data Computer Corporation, 85, 173
Decision making, 43
Defect-cause check sheets, 254–255
Defection of customers, 105, 131, see also Customers; Retention of customers
Defect-location check sheets, 253
Defects, 295–296, see also Six Sigma
Delivery, process capability, 410–416
Delta Dental Plans, 126
Demanded weight, QFD, 339–340
Deming, W. Edwards
approach, 7–10, 12, 82
benchmarking, 141
certification, 356
consistency, 64
cost of quality, 207
culture beliefs, 31, 59
customers, 123, 133
ISO 9000, 356
organizational structure, 159
process quality management, 100, 102
statistical quality control, 103
Yale University honorary degree, 12
Deming chain reaction, 7
Deming cycle
process capability, 414
processes and quality tools, 223
statistical quality control, 103–104
Deming prize
basics, 7
Florida Power & Light Company, 173
information systems, 43
ISO 9000, 359
Department of Defense, 107, 349
Design, process control and capability, 110, 405–410
Design for Manufacturability, 405
Design for Quality and Reliability, 407
Development, 86–87, 94n3, 362n12
Differentiation, 14, 67, 70, 150–151
Digital Equipment Corporation (DEC)
benchmarking, 152

objectives, 70
organizational structure, 170
quality function deployment, 108
Dimensions of quality, 125, 473–474
Direct labor, 188
Disney University, 153
Dispersion analysis cause-and-effect diagrams, 265
Dispersion errors, 246–247
Displays of data, 237–241
Distribution process, 413
Doctor's Hospital (Detroit), 16
Documentation, ISO 9000, 356–357
Domino's Pizza, 16, 65, 143
Donovan, M., 176
Drawing process, 411–412
Drucker, P.
Appley approach, 29
communication, 28
control, 35–36
human resource dimension, 114
information systems, 46
objectives, 70
organizational structure, 165
process quality management, 100
quality and productivity, 195
white-collar productivity, 192
DuPont company
culture beliefs, 31
ISO 9000, 347, 349, 356, 382
Dynatech company, 84

E

Eastman Kodak company, 134, 142, 349
Eaton company, 109
Economic Control of Quality of Manufactured Product, 99
Effectiveness, 66, 196, 323
Eisenhower, D. (President), 26
Electronic Data Interchange, 407
Emerson Electric company, 67
Employee input, 128–129
Engineering Information Architecture, 407
Engineers, ISO 9000, 351
Environment, strategic quality planning, 65
Environmental analysis, 45
Environmental Auditing Standards (ISO), 376–377
Environmental Labeling standards, 377
Environmental Management System (ISO), 376
Environmental Performance Evaluation, 377
Ernst & Young firm, 122, 144
Errors, data, 246–247
Europe
competition, 18n3
ISO 9000 adoption, 350, 361n3
quality and productivity, 187
European Community, 351, 361n3
European Quality Award, 137
European Union, 13, 347–348
Examples, see Case studies

F

Failure costs, 208
Failure rate, 6, 435–437
Failure-terminated tests, 435
Fat, 195
Federal Aviation Administration, 349
Federal Express company
Baldrige Award, 48, 171, 367
human resources, 83
information systems, 43, 48
organizational structure, 171
Federal Productivity Measurement System (FPMS), 200
Feedback, 36, 229
Feigenbaum, A.V.
approach, 10, 12
cost of quality, 206–207, 211
culture beliefs, 31
gross national product, 18n2
leadership, 23
Final Design Review, 406
"Find it, fix it" approach, 460–462
First National Bank of Chicago, 51, 105
Fishbone diagram, 33, see also Cause-and-Effect diagrams
Fitness for use, 9
Florida Power & Light Company, 173, see also Qualtec
Flowcharts, 256
Focus, customer needs, 125–127
Food and Drug Administration, 349
Ford Motor Company
Baldrige Award, 368
benchmarking, 143, 149
consistency, 64
consumer preferences, 69
culture beliefs, 5, 31, 59
customers, 51–52, 134
human resources, 82, 85

information systems, 51–52
morale, 131
QS-9000, 371–373
quality function deployment, 108
FPMS, see Federal Productivity Measurement System (FPMS)
Fraction non-conforming control charts, 295–301
Fram, E., 131
Freightliner Mack Trucks, 371
Front-line supervision, 171
Functional process, 228
Functional standards, 355
Future action, QI stories, 324

G

Gain sharing, 90
Galvin, B., 4, 445
Gap analysis, 154
Garvin, D., 15, 68
Gault, S., 82
Gemba no QC Shuho, 251
General Dynamics company, 20n23, 52–53
General Electric company
cost of quality, 213
culture beliefs, 31
customers, 66, 127
human resources, 82
Six Sigma, 445, 448
General Motors company
human resources, 84
organizational structure, 169
Pareto charts, 260
QS-9000, 371–373
quality and productivity, 186
quality function deployment, 331
Globe Metallurgical company
Baldrige Award, 164, 173, 368
human resources, 85
organizational structure, 164, 173
Godfrey, A. B., 26
Godiva (chocolate), 67
Goodyear Tire & Rubber Company, 63–64, 82
Gore, A. (Vice President), 17, 200
Grace Specialty Polymers, 381–382, 386
Graphical displays of data, 237–238
Graphs, 117n25, 256–259
Grayson, C.J., 141
Grey Poupon, 72
Growth, 187–188, 195
Guide to Quality Control, 263
Gulf Coast Health Systems, 136
Gumpert, D., 172

H

Harley-Davidson company, 153
Harleysville Insurance Company, 173
Harry, M., 403
Hawthorne Works, 6
Hay Group, 89
Healthcare services
attitudes, 471
basics, xix, 451, 474–475
behavior, 469–471
benchmarking, 149, 152
benefits, 454–455
components of quality, 463–469
conformance quality, 462–463
customers, 122, 124, 452, 456–458
dimensions of quality, 473–474
"find it, fix it" approach, 460–462
hospital implications, 454
input from patients, 138n9, 138n18
marketing research, 472–473
patient profiling, 459–460
process control, 105
profiles, 458–469
reliability, 431–432
service quality, 451–455, 460–462, 464–469
surveys, 471
TQM applicability, 13
TQSM, 471–474
Heat treating process, 412
Hewlett-Packard company
Baldrige Award, 368
cost of quality, 205, 210
customers, 127
just-in-time implementation, 112
quality function deployment, 110
Histograms
basics, 252, 258, 260
usage in Japan, 117n25
Hochschild, A.R., 465
Hockman, K., 347
Honda company, 186
Honeywell company, 448
Hospital industries, 152, 431–432, 454, see also Healthcare services
House of Quality, 108
Hughes Aircraft, 86
Human resource management

basics, 81–82, 91–92
benchmarking, 145
case studies, 93–94
compensation systems, 89–91
development, 86–87
involvement, 83–86
organizational structure, 84–86
performance appraisals, 88–89
process quality management, 114
selection, 87–88
training, 86–87

I

IBM company
Baldrige Award, 367–369
benchmarking, 142, 146
ISO 9000, 349
leadership, 24
process control, 106
Identification, QI stories, 321
Illustrative case studies, see Case studies
IMPACT, 49, 56n19
Implementation, QI stories, 322–323
Importance rating, QFD, 338
Improvement, 339, see also Productivity
Indicators, customer needs, 16–17, 127–129, see also Customers
Individual compensation, 90
Infant-mortality phase, 434
Inflation, 188, 202n14, see also Costs
Information and analysis
accounting systems, 45–47
advanced processes/systems, 49–50
basics, 41
case study, 54–55
costs, 213–214
customers, 43, 51–52, 55n8
data, 231–237
linkages, 47–49
organizational implications, 41–43
specialists, 52–53
strategic information systems, 44–45
system design, 53–54
Information industries, 13
Information technology manufacturers, 351, 362n11
In Search of Excellence, 16
Inspection, 100–102, 427
Instability, 288
Institute of Industrial Engineers, 183
Internal quality, 169–170, 353
Internal Revenue Service, 184
Internal services, 106
International Organization for Standardizaton (ISO), 348, 375, see also specific ISO program
International Quality Study, 87
Inverted organizational chart, 169
Involvement, 83–86, 172–173
Ishikawa, K., 251, 262, 331
Ishikawa (fishbone) diagram, 33
ISO 9000
accredited registration services, 358
Asia/Pacific region, 362n10
Baldrige Award comparison, 359–360
basics, 347, 351–355, 360–361, 381
benefits of certification, 355–356
brevity, 362n9
components, 352–353
cost of certification, 359, 363n23
documentation, 356–357, 385
European Community, 361n3
final assessment, 386–387
functional standards, 355
Grace Specialty Polymers, 381, 386
historical perspectives, 352–353
management responsibility, 353–354
post-certification, 357
pre-audit, 384–385
recommendations, 387–388
registration, 381–383
service areas and sector, 358–359
system implementation, 360
third-party audits, 356
training, 362n12
training program, 386
United Kingdom adoption, 348, 361n4
United States adoption, 349–351
world adoption, 347–349, 361n3
ISO 14000
14001, 377–378
basics, 375
benefits, 378–379
components, 375–377
registration, 378
Isuzu company, 110
Ivory soap, 67

J

Japan, see also Union of Japanese Scientists and Engineers (JUSE)
benchmarking, 141, 143

car manufacturers, 4
competition, 13, 18n3
cost of quality, 206–207
customer profiles, 458
Deming admiration, 12
information systems, 43
ISO 9000, 348, 350
kaizen, 81–82
kanban, 113
leadership, 23
quality and productivity, 16, 185–187
quality function deployment, 108, 331
statistical process control, 116n8, 117n25
statistical quality control, 103
Xerox competitors, 3
Jaworski, J., 23
JC Penney, 33
Job classifications, 168, 179n14
Johnson, J.A., 465, 472
Johnson and Johnson company, 33, 152
Johnson Sausage company, 153
Joint Commission on the Accreditation of Health Care Organizations (JCAHO), 462–464
Juran, J.M.
approach, 9–10, 12
benchmarking, 141
control, 74
cost of quality, 207
culture beliefs, 31
customers, 123
leadership, 24
Pareto charts, 260
process quality management, 100
quality and productivity, 184
Six Sigma, 447
statistical quality control, 103
Juran Institute, 10, 26, 169
JUSE, see Union of Japanese Scientists and Engineers (JUSE)
Just-in-case (JIC), 112–113
Just-in-time (JIT)
basics, 42, 50
plant maintenance, 118n34
processes, 110, 112–114
worker involvement, 84

K

Kaizen, 81
Kanban, 113
Kano, N., 458
Kaplan, R., 46
Kawasaki Steel Works, 262
Kearney, A.T., 187
Kearns, D., 25, 33
Kelsey Hayes company, 108, 331
Key success factors
benchmarking, 146–147, 149–151
information systems, 45
management systems, 34–35
organizational structure, 165
process quality management, 101
King, C., 464–465
Kobe Shipyard, 331
Koska, M., 471

L

La Quinta Motels, 67
Lawler, E., 91
Leadership
basics, 23–26
case studies, 37–38
communication, 27–30
control, 35–36
culture, 30–33
management systems, 33–35
top management, 26–27
Leebov studies, 469
Leverage effect, 185
Life cycle, 48, 216, 433–434
Life Cycle Assessment guidance, 377
Lifetime retention value, 133
Lincoln Electric company, 31
Line graphs, 256
L.L. Bean, 105, 142
Loews Hotels company, 186
Lorenz (Pareto charts), 261

M

Machining process, 412
Malcolm Baldrige National Quality Award
AT&T, 146
basics, 4, 183, 365–370
benchmarking, 141, 151–152
Building Products Operations Division, 115
customer focus and satisfaction, 121, 135
Federal Express company, 48, 171
federal support, 200
Globe Metallurgical company, 164, 173
human resources, 91

IBM, 24–25
information systems, 51, 53
ISO 9000, 349–350, 359–360
items and point values, 366–367, 370n3
Milliken & Company, 137
Motorola company, 4, 445
organizational structure, 170
process quality management, 97
quality function deployment, 107
quality management comparison, 19n12
Ritz-Carlton Hotel Company, 124, 128
service quality, 75
training, 86
Westinghouse Electric Company, 24–25, 60, 148
winners, 369
Xerox, 3
Maldistribution, 260
Management-by-drive, 195
Management systems
behavior, 6
inattention, 187
ISO 9000, 353–354
leadership, 33–35
process, 228
productivity, 185–187, 197–198
Managerial Breakthrough, 141
Managing Quality, 68
Manufacturer-based definitions, 69
Manufacturing, TQM applicability, 13
Marketing and sales, 130
Marketing research, 472–473
Market scope, 65–66
Market segmentation quality, 70
Martin Marietta company, 83, 85, 173
MBNA America, 106, 127, 133
McDonald's company
culture beliefs, 31
customers, 125–126
organizational structure, 165
policies, 72
McDonnell Douglas company, 20n23, 35
MCMC University, 152
Mean life, 435–437
Mean times between/to failures, 436
Measurements
costs, 211–213
customers, 129–130
productivity, 189–192
reliability, 435–437
Mercedes-Benz company, 67
Mercy Hospital (San Diego), 150
Merrill Lynch company, 165
Metropolitan Life Insurance Company, 170
Mid-Columbia Medical Center, 152
Middle managers, 171
Midstate University, 156
Milliken & Company, 137, 368
Mirage Resorts, 38
Mission of organization, 63–65, 72
Mitsubishi Heavy Industries, Ltd., 331
Monsanto company, 50, 184
Moore, B., 471
Motivation, 172, 180n27
Motorola company
Baldrige Award, 368, 445
benchmarking, 142–143, 152
cost of quality, 205
customers, 127, 130, 134
information systems, 49
ISO 9000, 349
Malcolm Baldrige National Quality Award, 4
organizational structure, 171
quality function deployment, 107
Six Sigma, 64, 179n20, 403, 443–445, 448
top management role, 171, 179n20
training, 86
Multiproduct lines costs, 216–217

N

NASA, 184, 313, 349
National Institute of Standards and Technology (NIST), 368
Navistar International Transportation, 54–55, 371
NCR company, 63, 146
Niche quality, 70
Noncompliance reports, 384
Non-conformance, 295–296, see also Six Sigma
Non-conformities control charts, 295, 304–309
Non-quality costs, see Costs
Northern Telecom Canada, Ltd., 30
Northwest Tool & Die Company, 153
Norton Manufacturing Company, 93
np charts, see Number non-conforming control charts
Number non-conforming control charts, 295, 304–309
Numerical description of data, 238–241

O

Objective data, 234
Objectives, strategic quality planning, 70–72
Observation, QI stories, 321
Omachonu, V., xxi
Operational analysis, 148
Operations Data System, 408
Oregon State University, 13
Organizational structure
  basics, 159–160
  case studies, 177–178
  cross-functional teams, 175–176
  customers, 167
  human resource management, 84–86
  internal quality, 169–170
  inverted organizational chart, 169
  involvement, 172–173
  quality circles, 174–175
  quality implementation, 165–168
  roles in transitions, 171–172
  systems approach, 160–165
  teams, 174–176
  transition changes, 168–172
Osada, H., 131
Out-of-industry companies, 152

P

PACCAR, 371
Pareto charts
  basics, 252, 258, 260–261
  usage n Japan, 117n25
Paring down, 195–196
Parkland Associates study, 469
Partial factor, 190
Parts per million method, 208, 211, 444–445
Pass Through Partnerships (PTP), 405, 423
Patient profiling, 459–460
p charts, see Fraction non-conforming control charts
PDSA cycle, see Deming cycle
Pentagon, 107
Pepsi-Cola company, 166
Performance appraisals, 88–89
Performance improvement, 145
Peters, T.
  cost of quality, 206
  culture beliefs, 31
  organizational structure, 159
  quality deterioration, 16
  strategic quality planning, 60
Peterson, D., 82
PIMS, see Profit Impact of Market Strategy (PIMS)
Pioneers of TQM, 7–9, 19n12, see also specific individual
PIQE, see Productivity Improvement and Quality Enhancement (PIQE) program, NASA
Pizza Hut, 126
Plan-Do-Study-Act (PDSA) cycle, see Deming cycle
Planning
  process control, 109–110
  QI stories, 322–323
  quality function deployment, 339
  strategic quality planning, 60–62
Policies, 33, 72–73, 353
Poor quality, see Costs
Porter, M.
  information systems, 45
  organizational structure, 161, 163, 166
  strategic quality planning, 61, 67
Porter Paint, 67
Post-certification, ISO 9000, 357
Preferred Customer Certification Program, 425
Preliminary Design Review, 405
Presentation of data, 237–241
President's Award, 200
President's Council for Management Improvement, 183
Preventative action, 322–323, 354
Prevention costs, 208
Price, see Costs
Price of Nonconformance, 208
Principles of Scientific Management, 99
Process capability, see also Colony Fasteners, Inc. (CFI)
  attribute data, 394
  basics, 393–394
  indexes, 399–403
  tolerance and specifications, 396–399
  variable data, 395–396
Process comparison, 123–124
Processes and quality tools, see also Quality control tools
  attributes, control charts, 295–312
  automation investment, 56n18
  basics, 225
  benchmarking, 149–151
  data, 231–248

examples, 226–228
feedback loop, 229
improvement stories, 319–329
process concept, 225–229
quality function deployment, 331–342
total system, 228–229
types, 228
usage, 313–317
variables, control charts, 275–292
Process quality management
basics, 97–98
case studies, 115–116
customers, 105–106
historical developments, 98–100
human resources impact, 114
inspection vs. process control, 100–102
internal services, 106
just-in-case, 112–113
just-in-time, 110, 112–113
quality function deployment, 107–110
service industries, 105–106
specifications vs. variation reduction, 104
statistical quality control, 102–104
Product-based definitions, 68
Production, process capability, 410–416
Production process classification cause-and-effect, 265
Production validation testing, 406
Productivity
activity analysis, 198–200
basics, 183–184
capital equipment, 197–198
case study, 200–201
costs, 150
improvement, 193–197
increase, 42, 55n3, 184, 201n4
leverage effect, 185
management systems, 185–187, 197–198
measurement, 189–192, 202n16
partial factor, 190
ratio of output to input, 189–192
service activity, 192–193
short-term financial gain comparison, 187, 202n11
slow growth, 187–188
technology, 185–187
total factor, 189
TPM model, 190–192
United States, 187–188
wheel, 194
white-collar workers, 192–193
Productivity Improvement and Quality Enhancement (PIQE) program, NASA, 184
Product life cycle curve, see Life cycle
Product quality, 15–17
Product scope, 65–66
Profiles, 52, 458–469
Profitability, 132–133
Profit Impact of Market Strategy (PIMS), see also Databases
basics, 20n27
benchmarking, 157n8
company characteristics, 45, 56n13
Profit sharing, 90
Program criteria
Baldrige Award, 365–370
ISO 9000, 347–363, 381–389
ISO 14000, 375–379
QS-9000, 371–373
PTP, see Pass Through Partnerships (PTP)
Public utilities, 13, 351, 362n11
Publishers Press, 173
Publix supermarkets, 33

Q

QA, see Quality assurance (QA)
QI stories, see Quality improvement (QI) stories
QS-9000, 371–373
Quality
acceptance and use, 12–13, 20n18
basics, 1–2, 5–6
business performance relationship, 13–15
costs, 5, 10, 14–15, 21n30, 101, 184, 206–207, 219n8, 219n18
historical developments, 3–12
measuring, 202n16
pioneers of TQM, 7–9
productivity comparison, 201n4
service vs. product, 15–17
Quality approaches, 68–73
Quality assurance (QA), 172, 463
Quality characteristics, 278, see also Variables control charts
Quality circles, 85, 174–175
Quality Control Handbook, 260–261
Quality control tools
basics, 251

cause-and-effect diagrams, 262–263, 265–269
check sheets, 252–255
control charts, 269–271
flowcharts, 256
graphs, 256–259
histograms, 260
Pareto charts, 260–261
scatter diagrams, 269
Quality function deployment (QFD)
basics, 42, 49, 50, 331–334, 342
benefits, 341–342
competitive technical assessment, 340–341
customers, 336
historical developments, 331
process quality management, 107–110
relationship strength, 337–338
technical requirements, 337
vertical entries, 338–340
voice of customer, 333, 335–336
Quality implementation, 165–168
Quality improvement (QI) stories
analysis, 321–322
basics, 319–320
data collection, 328–329
effectiveness check, 323
future action, 324
identification, 321
observation, 321
planning and implementation, 322–323
requirements, 324–325
standardization, 323–324
story line, 326–328
summary, 324
time frame, 324
Quality is Free, 10
Quality tools, see Processes and quality tools
Qualtec, 157n6
La Quinta Motels, 67

R

Random sampling, 242
Rate of improvement, QFD, 339
Ratio of output to input, 189–192
Receiving material check, 410–411
Re-engineering, 42
Reimann, C. W.
benchmarking, 154
biographical information, 370n1
information systems, 41, 53
leadership, 24
Relational subgroups, 278–279
Relationship strength, 337–338
Relevance Lost, 46
Reliability
availability, 435–437
basics, xix, 431–434
causes of unreliability, 434
data, 235
failure rate, 435–437
mean life, 435–437
measures, 435–437
product life cycle curve, 433–434
Requirements, QI stories, 324–325
Research and development, 188
Research Testing Laboratories Inc., 87
Retention of customers, 131–133, see also Customers; Defection of customers
Review process, 413
Revlon company, 166
Revson, C., 166
Ricoh company, 142
Ritz-Carlton Hotel Company
benchmarking, 153
customers, 124, 128
TQM adoption, 370n8
Rogers Corporation, 100
Rosander studies, 466, 469
Ross, J., 121

S

Sales process, 131, 138n22
Sampling, 241–247, 278–280
Samsung Electronics company, 448
Sara Lee Hosiery, 49
Savin Copiers, 97, 142
Scandinavian Airlines, 466
Scatter diagrams, 117n25, 252, 269
Sculley, J., 41
Selected sampling, 245
Selection, human resource management, 87–88
Self-analysis, benchmarking, 154
Self-managing work teams, 85, 175
Sequential tests, 435
Service activity, 192–193
Service America, 453
Service industry
ISO 9000, 358–359

process quality management, 105–106
TQM applicability, 13
Service quality
customer needs, 131–132
healthcare customer service, 451–455, 460–462, 464–469
importance, 16, 21n36
product quality comparison, 15–17, 21n41
strategic quality planning, 74–75
Service tracking report, 47
Shanghai-Foxboro Company, 361n3
Shawnee Mission Medical Center, 90
Shewhart, W., 99, 103, 275
Shingo, S., 113
Shipping process, 413
Short-term gains, 187
Siemens company, 348
Sigma Project, 43
Signaling, 33
Simple random sampling, 242–244
Simplon Devices, 37
Six Sigma
basics, xix, 208, 443–445
benefits, 447–448
features, 447
historical developments, 445
Motorola, 64, 179n20
phases, 446–447
Sloughing off, 195–196
Slow growth, productivity, 187–188
Smith, F., 43
Software vendors, 351
Southeast Freight Systems, 76
Southwest Airlines, 31, 64
Specialists, information systems, 52–53
Specifications, process capability, 396–399
Specifications vs. variation reduction, 104
Standardization, QI stories, 323–324
Statistical Process Control (SPC)
basics, 42, 49, 50
processes, 102, 106, 114
United States usage, 117n10
usage in Japan, 116n8, 117n25
Statistical Quality Control (SQC), 49, 50, 99, 102–104
Steiber, S.R., 471
Stewart, P. (Supreme Court Justice), 125
Stock ownership, 90
Story line, 326–328
Strategic analysis, 146–147
Strategic database, see Profit Impact of Market Strategy (PIMS)
Strategic information systems, 44–45
Strategic Planning Institute, 13–14
Strategic quality planning
accounting systems, 78n25, 221n29
activity-based costing, 217
basics, 59–60, 75
case studies, 76–77
control, 73–74
costs, 217
differentiation, 67
environment, 45, 65
market scope, 65–66
market segmentation quality, 70
mission, 63–65
niche quality, 70
objectives, 70–72
planning process, 60–62
policies, 72–73
product scope, 65–66
quality approaches, 68–73
service quality, 74–75
Stratification, 117n25, 235–237
Stratified random sampling, 244
Subjective data, 234
Summary, QI stories, 324
Supplier Corrective Action Request, 425
Supplier performance management, 423–424
Supplier Rating System, 425
Supply-side economics, 185
Support services, 44, 416–423
Surveying company, QFD, 338
Surveys
Chief Information Officers, 53, 57n27
cost of quality, 208, 211
employee behavior, 469
healthcare services, 452, 471
human resources utilization, 4, 19n6
measurement of quality, 202n15
quality importance, 4
shortcomings, 127
U.S.-made products, 4
SWOT analysis, 61–62
Systematic sampling, 245
System availability, 438–440
System design, 53–54
System implementation, 360
Systems approach, 160–165

T

Tabular displays of data, 237
Taco, Inc., 93
Taco Bell, 106, 133
Taguchi methods, 42, 117n12
Tally check sheets, 253–254
Target stores, 90
Task teams, 85, 175
Taylor, F., 6, 99
Teams, 90, 174–176, 325–326
Technical Assistance Research Program (TARP), 452–454
Technical Data System, 407
Technical requirements, QFD, 337
Technology, productivity, 185–187
Telecommunications, 13
Telecommunications Products Division, Corning Inc., 93
Texas Instruments, 370n8
Themes, customer needs, 127
Third-party audits, 356
Thor, C.G., 189
Threading process, 412
3M company
  Baldrige Award, 368
  consistency, 64
  cost of quality, 205
  culture beliefs, 31
  design process, 118n29
  leadership, 26
  strategic quality planning, 60
Thriving on Chaos, 206
Time frame, QI stories, 324
Timeliness, 150
Time-terminated tests, 435
Tisch, J.M., 186
Tolerance, process capability, 396–399
Tools, see Processes and quality tools
Top management leadership, 26–27, 171
Total factor, productivity, 189
Total Productivity Measurement (TPM) model, 190–192
Total quality management (TQM)
  applicability to all functions, 13
  basics, 5–6
  benchmarking, 141–158
  costs, 205–221
  customers, 121–139
  human resource development, 81–94
  implementation, 6
  information and analysis, 41–57
  leadership, 23–40
  organizational structure, 159–180
  processes and systems, 49–50, 97–119
  productivity, 183–202
  quality assurance comparison, 463
  quality revival, 3–22
  strategic quality planning, 59–79
Total Service Quality Management (TSQM), 471–474
Toyota company, 113, 331
TPM, see Total Productivity Measurement (TPM) model
Training, 86–87, 94n3, 362n12
Transition changes, 168–172
Transportation industry, 105
Travelers Insurance Company, 205
Tsuda, Y., 183

U

U charts, see Number non-conforming control charts
Union of Japanese Scientists and Engineers (JUSE), 9, 131, 331
United Kingdom, 349, 351, 355, 361n4
United States
  benchmarking, 141–142
  ISO 9000, 348–351
  productivity, 187–188
  quality deterioration, 16, 116
Universities, 13, 152
University of Japan, 262
Unreliability, see Reliability
Upsetting process, 411–412
User-based definitions, 68–69

V

Value, 66, 166, 178n10
Valued-based definitions, 69
Variables control charts
  applications, 277–280
  basics, 275–277, 290
  examples, 280–290
  interpretation, 287–290
  process capability, 395–396
  usage, 275–276
Variation reduction vs. specifications, 104
Vertical entries, 338–340
Voice of the customer, 333
Volvo company, 166, 371

W

Wal-Mart, 65, 143, 163, 165
Wawa Food Markets, 76
Wear-out phase, 434
Welch, J., 31, 82, 445
West Babylon School District, 18
Western Electric plant, 6, 99
Westinghouse Electric company
  Baldrige Award, 368
  benchmarking, 148
  just-in-time implementation, 112
  leadership, 24–25
  strategic quality planning, 60
White-collar workers
  information systems, 49
  productivity, 192–193
  quality, 15, 21n33
Whitney, Eli, 99
Willimatic Division, Rogers Corporation, 100
Winnebago Industries, 164
Working smarter, 195

X

Xerox
  Baldrige Award, 368–369
  benchmarking, 142–144, 149, 152
  culture beliefs, 31, 33
  customers, 134
  information systems, 54
  Ishikawa diagram, 33
  ISO 9000, 349
  leadership, 25
  Malcolm Baldrige National Quality Award, 3
  morale, 131
  quality program, 219n15
  quality revival, 3
  training, 86

Y

Yokogawa Hewlett-Packard company, 43

Z

Zemke, R., 453, 455, 466, 471
Zero defects concept
  basics, 11–12, 208
  Federal Express, 48
  process quality management, 104, 117n12